A Practical Guide to Using Glycomics Databases

Kiyoko F. Aoki-Kinoshita
Editor

A Practical Guide to Using Glycomics Databases

Springer

Editor
Kiyoko F. Aoki-Kinoshita
Faculty of Science and Engineering
Soka University
Hachioji
Tokyo, Japan

ISBN 978-4-431-56452-2 ISBN 978-4-431-56454-6 (eBook)
DOI 10.1007/978-4-431-56454-6

Library of Congress Control Number: 2016960320

Printed on acid-free paper

This Springer imprint is published by Springer Nature
The registered company is Springer Japan KK
The registered company address is: Chiyoda First Bldg. East, 3-8-1 Nishi-Kanda, Chiyoda-ku, Tokyo 101–0065, Japan

Preface

The idea for this book originated in a workshop held during the Annual Meeting of the Japanese Society for Carbohydrate Research in 2014. Subsequently, I attempted to contact the major database providers of glycoscience-related data and gratefully received very positive replies, resulting in this practical guide. I must note that I originally focused on databases and not Web-based tools when I first contacted authors for the book. Thus several useful Web tools have not been included. Readers are recommended to check the Glycoinformatics Consortium (GLIC) Web site at http://glic.glycoinfo.org/ to access the latest list of glycoinformatics databases and software tools. One major database that has not been included in this book is GlycomeDB. After discussion with the developer of this database, it was decided that GlyTouCan currently serves a similar role to that of GlycomeDB in that unique IDs are assigned and links to the same structures in other databases are provided. Moreover, all of the data in GlycomeDB has now been imported into GlyTouCan.

Although it took much longer than expected to complete this book, it was with the hope of aiding in advancing the field of glycoscience that readers may find this book useful for their research.

Kiyoko F. Aoki-Kinoshita　　　　Hachioji, Japan

Contents

Part I
Introduction

Chapter 1
Introduction

Nicolle H. Packer

The exciting emerging research area of glycoscience is focusing on how disease affects the sugar molecular (glycan) makeup of the cell membrane and how cells "talk" to each other. In particular, the sugar molecules that decorate the surface of every cell are being targeted for the development of new drugs, new imaging techniques and new antibiotics.

More than 50 years after the double helix was described, genomics and the reference DNA sequence of *Homo sapiens* are now available for downloading, and individuals can have their DNA sequenced for a few hundred dollars. More than 20 years ago, the word proteomics was coined to describe the products of the genome. However we now know there are only about 20,000 genes that are available to code for millions of different forms of the proteins that vary from cell to cell in time and space. Thus, genes and proteins are not enough to explain the complexity of biology. The cell adds complex sugar structures onto proteins and fats on its surface, modifications not directly coded for by the genome. Compared to genomics and proteomics, researchers in the field of glycoscience are restricted by a lack of experimental and informatics tools to explore the changes that occur to these sugars that adorn the outside of all cells. As was the case in genomics and proteomics, knowledge of the structure and function of these previously unexplored sugar structures (glycomics) holds huge promise to provide new drug targets and diagnostics for human, animal and plant health.

Therefore, glycomics examines the way in which the cell can modify the gene product proteins in a combinatorial way. However, since the synthesised glycan structures cannot be predicted from the genome, glycoscience relies on the analysis of the structure and function of the oligosaccharides (chains of sugars) attached

N.H. Packer (✉)
Biomolecular Frontiers Research Centre, Macquarie University, North Ryde, NSW 2109, Australia
e-mail: nicki.packer@mq.edu.au

K.F. Aoki-Kinoshita (ed.), *A Practical Guide to Using Glycomics Databases*, DOI 10.1007/978-4-431-56454-6_1

to biomolecules such as proteins and lipids. This emerging area of research is gaining momentum and will allow us to account for the millions of protein variants produced. *By enabling the acquisition, storage, consolidation and linking of data in the field and by collaboration with like-minded researchers on each continent, foundations are being built for the experimental and informatics advances required for the development of the glycoscience field.*

However, the difference between the proteomics knowledge base and glycomics knowledge base is that there is no template to predict sugar sequences from the genome – the only way knowledge on glycomics is obtained is by experiment, peer-reviewed publications and integration. Information in the public domain on the analytical properties (e.g. LC and CE retention times, MS/MS spectra, NMR spectrum), attached glycoconjugates (e.g. protein, lipid) and function (e.g. interaction partner, disease association, drug target) needs to be linked to each described glycan structure.

The synergy developed between those "at the coalface" who process the biological samples and obtain the experimental data and those who develop the bioinformatics tools to facilitate the data analysis is an essential relationship to guarantee successful research outcomes. As a consequence, experimental and informatics resources which are freely accessible to the international research community are being built, some of which are described in detail in this book. These resources will enable further biological discoveries to be made around the world.

Specifically, the aims of glycobioinformatics are to enable:

(a) Glycomics data to be tightly integrated with proteomics and genomics data as well as with biological information
(b) Glycobiologists engaged in systems biology or translational medicine research to be able to interpret the large, and rapidly expanding, quantities of molecular data that can be generated by modern analytical technologies
(c) The field of glycomics to become more cohesive by connecting the diverse, currently incompatible, glycobiology databases and by developing software-assisted data interpretation and mining tools

This book addresses the task of facilitating glycomics research by describing the infrastructure of some of the glyco-specific bioinformatics databases and tools that are currently in development and publically accessible. From assisting the analysis of targeted experimental data to curating collections of previously published data in the literature, the core connectivity of glycoknowledge is starting to be established.

Specifically, monosaccharide building blocks (MonosaccharideDB), the unique identifiers being ascribed to all experimentally determined glycans (GlyTouCan), and the collection of bacterial, plant and fungal carbohydrates that have been recorded in the literature (CSDB), are described in Part II. Part III presents databases that contain glyco-related genomic information (GGDB) as well as the 3D Protein Data Bank structures of known glycan-binding proteins (lectins) (Glyco3D Portal) and a catalogue of the over 250 enzymes responsible for synthesising glycans (CAZy). Glycomics data in the form of a collection of curated glycan structures reported in the literature, with links to glycomics experimental data (UniCarbKB), also connects the glycans on specific protein sites, where known, to the proteomics database UniProt, whilst GlycoProtDB also provides experimentally determined protein glycosylation site information (Part IV). The specific glycan epitopes that have been found to be recognised in protein interactions are found in GlycoEpitope in Part V, as are the reported glycan structures to which pathogens bind (SugarBind) and the diseases that result from this pathogen adherence (PACOnto). Finally, new and past glyco-specific informatics tools (RINGS, GlycoSciences.de, TogoTable and the Semantic Web) currently being developed are described in Part VI.

The overall aim of the these resources, as well as other glycoinformatic tools now in development, to make experimental research in glycomics and glycoproteomics faster, more reliable, simpler and thoroughly integrated with other biomolecular research data.

Chapter 2
Development of Carbohydrate Nomenclature and Representation

Serge Perez and Kiyoko F. Aoki-Kinoshita

Abstract This chapter offers a general background for researchers embarking in glycoscience to grasp the evolution and present status of the nomenclature(s) and representation(s) of glycans and complex carbohydrates. The availability of high-performance computing and the application of data mining are opening new paths to discovery. The field of structural glycobiology has benefited from such advances with the development of tools and databases for the structural and functional analysis of carbohydrates. There is a need to conform to the recommendations of nomenclatures of carbohydrates while the constraints are required by the developing field of glycobiology in terms of visualization and encoding. The present chapter describes the nomenclatures, symbols, and presentations that form part of the "language" used to communicate more effectively and used in different databases. Besides, some issues related to the interoperability of glycan databases throughout glycan databases are also addressed. The semantic web approach promotes further the description and integration of structural and experimental metadata throughout the development of ontologies for domain knowledge representation.

Keywords Glycans • Nomenclature • Graphical representations • Three dimensions • Encoding • Databases

2.1 Introduction

Monosaccharides are the chemical units from which all members of the major family of natural products, the carbohydrates (or sugars), are built. They are the individual carbohydrate building blocks. Most recent results suggest that the generation of numerous monosaccharides, including ribose, may be possible from photochemical and thermal treatment of cosmic ices in the late stages of the

S. Perez (✉)
Department of Molecular Pharmacochemistry, CNRS, University Grenoble Alpes, Grenoble, France
e-mail: spsergeperez@gmail.com

K.F. Aoki-Kinoshita
Department of Bioinformatics, Graduate School of Engineering, Soka University, Tokyo, Japan

K.F. Aoki-Kinoshita (ed.), *A Practical Guide to Using Glycomics Databases*,
DOI 10.1007/978-4-431-56454-6_2

solar nebula (Meinert et al. 2016). On earth, monosaccharides are constituents of more complex macromolecular architectures. They will be referred to as glycans, an assembly of sugars either in free forms or attached to another molecule or macromolecule. Glycans occur as:

(i) Oligosaccharides (comprising 2–10 monosaccharides linked together either linearly or branched)
(ii) Polysaccharides (for glycan chains built up from more than 10 monosaccharides, but the distinction with oligosaccharides is not strictly drawn)
(iii) Glycoconjugates (when the glycan chains are covalently linked to proteins (glycoproteins), lipids (glycolipids), or naturally occurring aglycones (e.g., as in antibiotics, saponins, alkaloids))

Glycoscience is the study of structure, chemistry, biosynthesis, and biological functions of glycans and their derivatives.

As chemical compounds, monosaccharides and glycans may be represented following different levels of description. Starting from *chemical formulas*, which have a limited number of symbols and limited descriptive power, they are ideally better represented using *structural formulae*, i.e., a graphic representation of the molecular structure, displaying the three-dimensional arrangement of the atoms and subsequent explicit or implicit arrangement in the molecule. Several systematic chemical-naming formats, as in chemical databases, are used that are equivalent to geometric structures. These systematic chemical names can be converted to structural formulas and vice versa. As components of bio(macro)molecules, the shapes of monosaccharides along with their sequential arrangements generate several hierarchic levels of structures. They can range from the atomic and molecular to the cellular, tissue organ, organismic, population, and ecosystem level.

2.2 Monosaccharides: From Chemical Formula to Structural Formula

2.2.1 Chemical Evidence

Emil Fischer elucidated the structure of glucose and its isomers using ingenious chemical and polarimetric methods (Fischer 1890), the work being recognized as one of the outstanding achievements of early structural work (Lichtenthaler 2002). Monosaccharides with an aldehydic carbonyl (or potential aldehydic) group are called aldoses; with a ketonic carbonyl (or potential ketonic carbonyl) group are called ketoses. Glyceraldehyde ("glycerose" in carbohydrate terms) is the simplest aldose (a triose containing an aldehydic group) having one asymmetric center and therefore two stereoisomers (enantiomers); there are 4 aldotetroses, 8 aldopentoses, and 16 aldohexoses. Fischer projection chemical formulas of the D-enantiomers of the common aldotriose, aldotetroses, aldopentoses, and aldohexoses are presented in

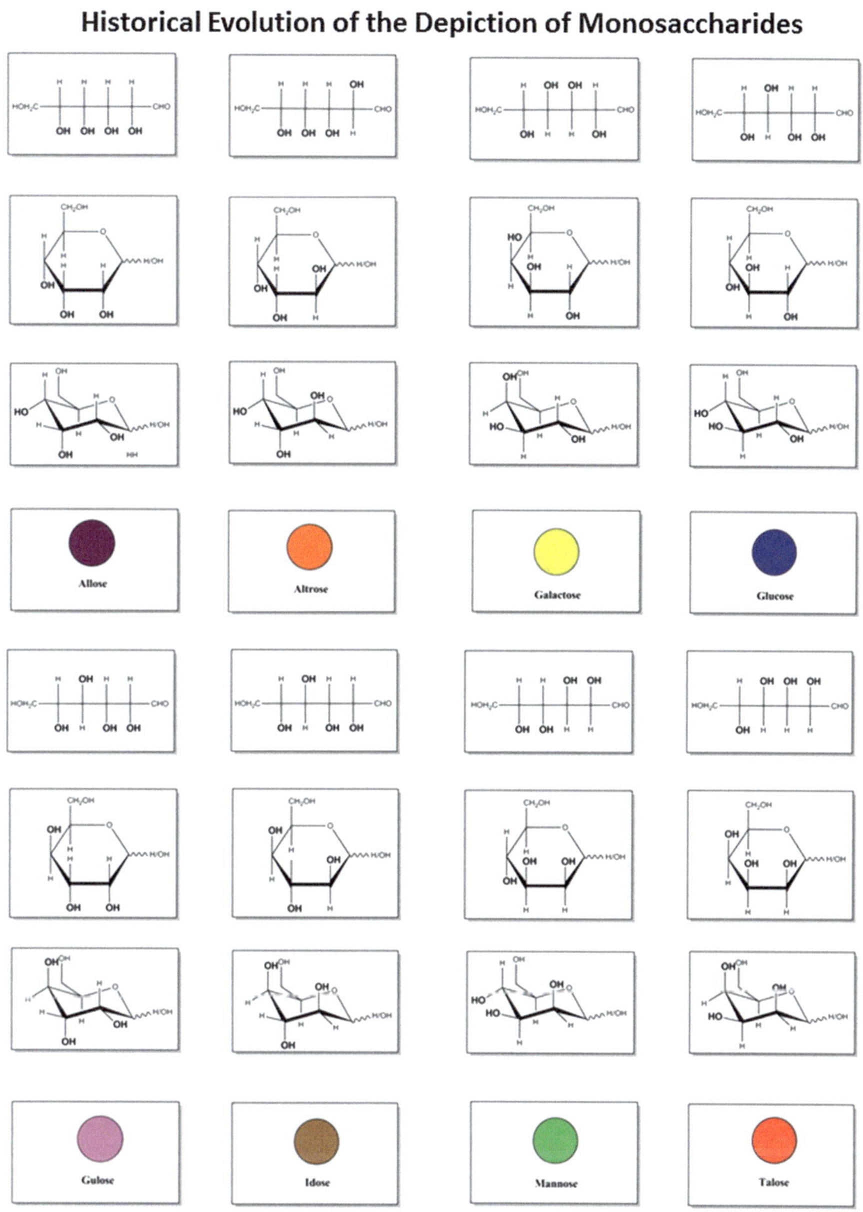

Fig. 2.1 Historical evolution of the representation of monosaccharides, from Emile Fisher depiction, Haworth depiction to Symbol Nomenclature for Glycans

Fig. 2.1, including their trivial names, and their abbreviations when defined. Fischer assigned to the dextrorotatory glucose (via glucaric acid) the projection with the OH group at C-5 pointing to the right. But the absolute configuration was not established until 1951 by Bijvoet et al. (1951) proving the correctness of the assignment by Fisher.

2.2.2 Cyclization of Monosaccharides

In aqueous medium, monosaccharides with a suitable carbon-chain length, having both hydroxyl and carbonyl functions, undergo intramolecular (cyclic) hemiacetal formations. The equilibrium, of which the formation is accelerated under weak acidic or alkaline conditions, favors cyclic forms, and "open-chain" forms occur only in trace amounts. Stable five (4 C and 1 O atom)- and six (5 C and 1 O atom)-membered ring forms are the result. The drawing of the cyclic forms of the Fischer projection formulas does not provide a realistic representation. More realistic drawings of the cyclic forms were introduced by Haworth in the 1920s and are referred to as Haworth representations (Haworth 1929). A perspective drawing of the ring offers a simplified model. The ring is oriented almost perpendicular to the plane of the paper, but viewed from slightly above so that the edge closer to the viewer is drawn below the more distant edge, with the intra-cyclic oxygen behind and the anomeric carbon at the right-hand end. To define the perspective, the ring bonds closer to the viewer are often thickened.

A schematic representation of a pyranose ring closure in D-glucose that shows the reorientation at C5 necessary to allow ring formation is given in Fig. 2.2, along with the conventional labeling of atoms. In the case of D-glucose, the hydroxyl group at C5 reacts intramolecularly with the aldehyde group at C1. As the carbonyl carbon atom C1 of the open-chain form becomes an additional asymmetric carbon

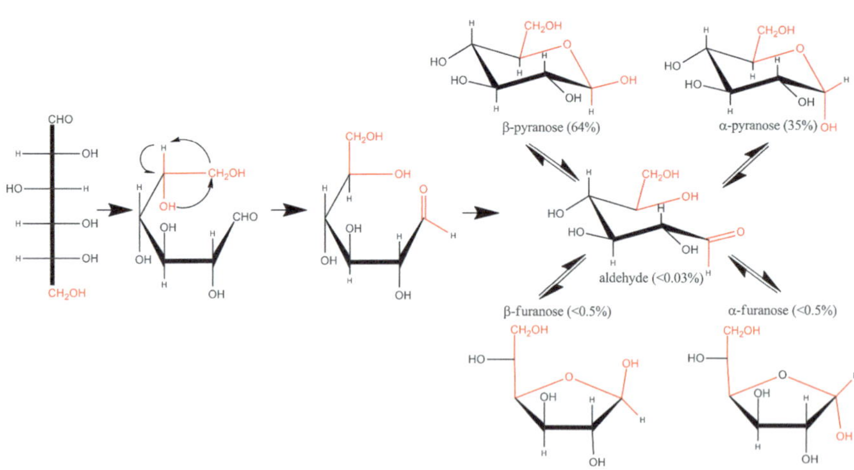

Fig. 2.2 From linear representation to cyclic structure and the equilibrium resulting from mutarotation (example of glucose)

in the hemiacetal formation, two pyranose rings are formed. To describe the stereochemistry around C1 (denoted by the anomeric carbon atom), the terms α and β have been chosen. In the α-anomer, the exocyclic oxygen atom at the anomeric center is formally *cis* in the Fisher projection, to the oxygen attached to the anomeric reference atom; in the β anomer these oxygen atoms are formally *trans*.

In solution, most simple sugars and many of their derivatives occur in a monosaccharide-specific equilibrium of α-pyranose, β-pyranose, α-furanose, β-furanose, and acyclic (open-chain) forms. This process is called mutarotation as it refers to changes in optical rotation to an equilibrium value when pure anomeric forms of monosaccharides are dissolved in water. In general the acyclic forms are only present in trace amounts (e.g., <0.03 % in the case of D-glucose). The presence of a mixture of two anomers of the same ring size may be indicated in the name by the notation α, β, e.g., α, β-D-glucopyranose. In formulae, the same situation may be expressed by use of a wavy line.

2.2.3 *Naming Monosaccharides and Derivatives*

The IUPAC-IUBMB document entitled "Nomenclature of Carbohydrates" (McNaught 1997) provides recommendations in giving systematic names to monosaccharides and their derivatives. (Kamerling 2007) Nevertheless, the use of trivial names still persists. The hydroxyl groups of monosaccharides can undergo a series of chemical modifications: methylation, esterification (phosphate, acyl esters, sulfate esters, etc.), and deoxygenation to form deoxy-sugars. Many monosaccharides have *N*-acetamido groups, such as GlcNAc, GalNAc, and Neu5Ac. In rare cases, the *N*-acetamido group is de-N-acetylated to form amino groups. These are found in heparan sulfate, glycosylphosphatidylinositol (GPI) anchors, and many bacterial glycan structures. Amino groups can be modified with sulfates, similar to hydroxyl groups, as found in heparan sulfate. Of special biochemical importance are the esters of monosaccharides with phosphoric and sulfuric acid. They are generally termed as phosphates and sulfates (regardless of the state of ionization or the counter ions). In abbreviations, an italic capital P (*P*) is used to indicate a –PO3H2 group or 'PO2H- group. An italic capital S (*S*) is used in the case of sulfuric acid preceded by the appropriate locant after the carbohydrate name.

2.2.4 *Molecular Shapes of Monosaccharides*

Because structure and behavior are so intertwined, a large part of carbohydrate chemistry has been devoted to the determination of geometries that characterize monosaccharides (Fig. 2.3). The shapes of acyclic aldehyde and keto carbohydrates are described in terms of linear (planar zigzag) or sickle shapes. The most common

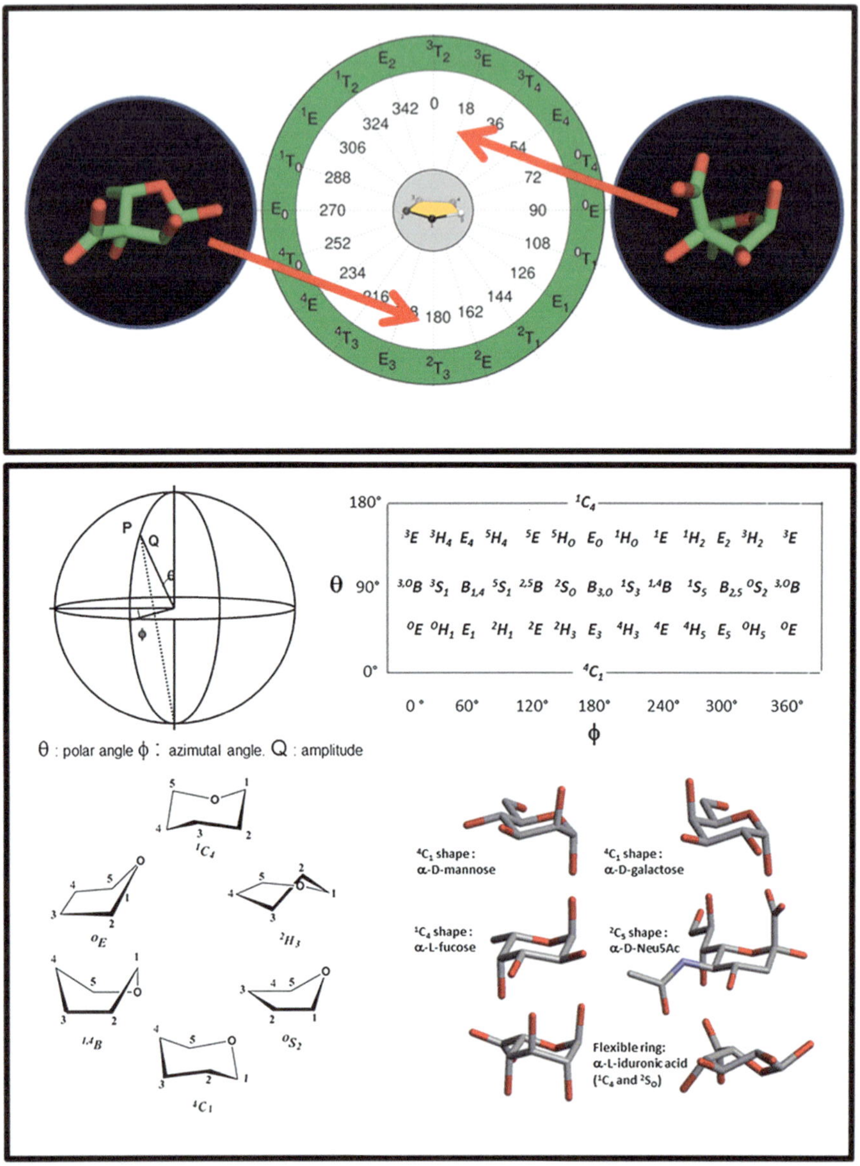

Fig. 2.3 The conformational descriptors. (**a**) Schematic representation of the five-membered ring structures, following the pseudo-rotational wheel of five-membered ring that encompasses 20 twist and envelope shapes. The molecular drawings of the low-energy conformers north, and south, are presented. (**b**) Representation of the puckering parameters describing the conformations of six-membered ring: (θ, ϕ, Q), along with five low-energy conformations O_E, $^{1,4}B$, $^{O}S_2$, $^{2}H_3$, and ‘C_1. The low-energy shapes taken by common monosaccharides are shown

rings are furanoid or pyranoid, with five and six members, respectively. Furanose ring structures occur in envelope (E) and twist (T) conformations which can be represented on a pseudo-rotational wheel. As the difference in energy between the different conformations on the wheel is generally low, two regions having low-energy conformations occurring in the north and south can be identified. They are referred to as N (north) and S (south) forms. Because furanoses can adopt several low-energy conformations, the Haworth projection still appears to be the simplest means to avoid the complexity of structural representation. Six-membered ring structures can occur in two chair (C), six boat (B), six skew (S), and twelve half-chair (H) conformations. These variants are defined by the locants of the ring atoms that lie outside a reference plane. In practice, the two chair conformations have the lowest energy, and strongly dominate. The preference for these low-energy conformations is dictated by the relative orientations of the hydroxyl groups. In the case of D-glucopyranoses, only the $^4C_1$ conformation is of importance, whereas the $^1C_4$ conformation dominates in α-D-idopyranose. Cases occur as in β-D-arabinopyranose where both chair conformations are in equilibrium. In monosaccharides with an exocyclic hydroxymethyl group, three staggered situations about the CH-CH_2OH bond (the C5-C6 bond in aldohexoses) can be considered. As the CH-CH_2OH bond is prochiral, the two hydrogen atoms need to be differentiated based on the R/S system. The description of the three rotamers orientations of O6 with respect to O5 and C4 is GG (*gauche-gauche*), GT (*gauche-trans*), and TG (*trans-gauche*) depending upon the choice of atom of reference (Marchessault and Perez 1979).

2.2.5 *Monosaccharides as Components of Oligosaccharides, Polysaccharides, and Complex Glycans and Glycoconjugates*

Carbohydrates have a potential information content that is several orders of magnitude higher than any other biological macromolecule. The diversity of carbohydrate structures results from the broad range of monosaccharides (>100) of which they are composed and the different ways in which these monomers are joined. Thus, even a small number of monosaccharide units can provide a large number of different oligosaccharides, including branched structures, a unique feature among biomolecules. For example, the number of all possible linear and branched isomers of a hexasaccharide exceeds 10^{12} (Laine 1994). At the present time, the number of glycans that comprise the human genome is still unknown, and it may be acknowledged that sequencing the human "glycome" may still be unrealistic given the current technology for glycoproteomics, glycosaminoglycanomics, or glycolipidomics. The number of glycan determinants likely to be important in their interactions with proteins is estimated to be about 7000.

2.3 Representing and Encoding Carbohydrate Structures

2.3.1 Representation

The variety in nomenclature and structural representations of glycans makes it complex to decide the best approach for illustrating these complex structures. The choice of notation is frequently based on whether the study is focused on the chemistry or has a more biological approach. Representation in text of the primary structure, or sequence, of complex carbohydrates was first described following the IUPAC-IUBMB (International Union for Pure and Applied Chemistry and International Union for Biochemistry and Molecular Biology) terminology, in its extended and condensed forms (McNaught 1997). These forms are used within the carbohydrate community and are adequate for describing complex sugar sequences, including monosaccharide stereochemistry, anomeric configuration, and linkage information as well as for naming derivatized oligosaccharides. Recommendations also apply to the description of polysaccharides, glycoproteins, and glycolipids.

Whereas these rules govern the naming of all carbohydrates and carbohydrate containing molecules, their complexity remains a major hurdle to their use in common practices and exchange within and outside the glycoscience community. One of the major difficulties originates from the nonlinear nature of glycans, a unique feature among occurring biological (macro)-molecules, which requires a tree-like representation of the structures. Several developments have been proposed to deal with such an inherent complexity essentially throughout the linearization of the structures. Such examples are (i) the LINUCS format (Bohne-Lang et al. 2001) and (ii) the Bacterial Carbohydrate Structure DataBase sequence format (Toukach 2011) or the linear code adopted by the Consortium of Functional Genomics (Banin et al. 2002). These formats provide rules to extract the structure of the branches and create a unique sequence for the glycan. The limitations of the linear encoding may be bypassed throughout the use of connection tables exemplified by KCF (KEGG Chemical Function (Aoki et al. 2004)) and GlycoCT (Herget et al. 2008) formats, at the expense of becoming practically incomprehensible by human beings. With the same aim of exchanging glycan structure information, other XML formats such as CabosML (Kikuchi et al. 2005) and GLYDE (Sahoo et al. 2005) have been developed.

2.3.2 Graphical Representations

In addition to the sequence representation of structures, several graphical representations have been developed in the field of glycobiology, favoring cartoon representations that facilitate the visualization of the monosaccharides present. Given the complexity of many glycan structures, it became difficult to represent such molecules in publication figures, and various investigators began to use symbols to represent monosaccharides in glycans. In 1978, Kornfeld and colleagues put forward an elegant and simple system for representation of vertebrate glycans (Kornfeld et al. 1978), and it entered into popular use over the next two decades. This system was eventually adopted and standardized by the editors of the first edition of the textbook Essentials of Glycobiology (Varki et al. 1999). The symbolic representation dealing essentially with mammalian glycans gained wide acceptance and is being used to describe glycan composition and biosynthesis for N- and O-linked glycans. The symbol notation has been extended to describe the structure of pathogen polysaccharides (Berger et al. 2008) with the objective of providing a quick and easy way to visually distinguish between polysaccharides. Those glycans that are encountered in pathogens are more diverse and contain greater structural variability than glycans found in mammals. There is a need to identify monosaccharides (by shape and color) and with the connectivity and substituent differences between polysaccharides. An extra level of complexity is found, occurring from the fact that pathogens utilize a greater diversity of monosaccharides with multiple modifications. An extension of the graphical representation of glycans, called SNFG (Symbol Nomenclature for Glycans) (Varki et al. 2015), has been proposed as the result of a concerted international agreement with the hope that it will cope better with the rapidly growing information on the structure and functions of glycans in chemical and biological systems (Fig. 2.4). This newly adopted extension of picture representation of monosaccharides is compatible with graphical representation of glycans such as the Essential cartoon representation (Varki et al. 1999) and the Oxford cartoon representation (Harvey et al. 2009). In the first one, linkage information along the glycosidic bonds connecting the sugar symbols is indicated; α and β are used to represent the two stereochemical types of glycosidic bonds and numbers to denote the ring position of the carbon on the sugar on which the glycosidic linkage originates. In the original Oxford system, some of the scheme's monosaccharide symbols differ from those in the essential scheme. Linkages between sugars are encoded as dashed lines for α stereochemistry and solid lines for β. The angles of the lines denote the ring position where the bond originates. Figure 2.5, illustrates these two types of representation for a glycan, along with the IUPAC-like representation using textual names and a chemical representation favored by carbohydrate chemists.

It is fully recognized that no symbol system will ever convey a full appreciation of the conformational structures of monosaccharides as components of complex structures, essential information for understanding their three-dimensional structures, and the interactions they take part in. It is nevertheless possible to concatenate the symbolic representation of monosaccharides with a limited number of structural descriptors to achieve a fairly exhaustive nomenclature which can be further used in constructing three-dimensional structures of glycans. Such an extension, originally presented in Glycopedia (Perez 2014), requires a limited set of rules (as illustrated on Fig. 2.6). While maintaining the spirit of using the symbolic representation for monosaccharides (and toward glycans), this set of rules provides the necessary extension to the construction of three-dimensional structures, allowing encoding for computational manipulation while maintaining IUPAC nomenclature.

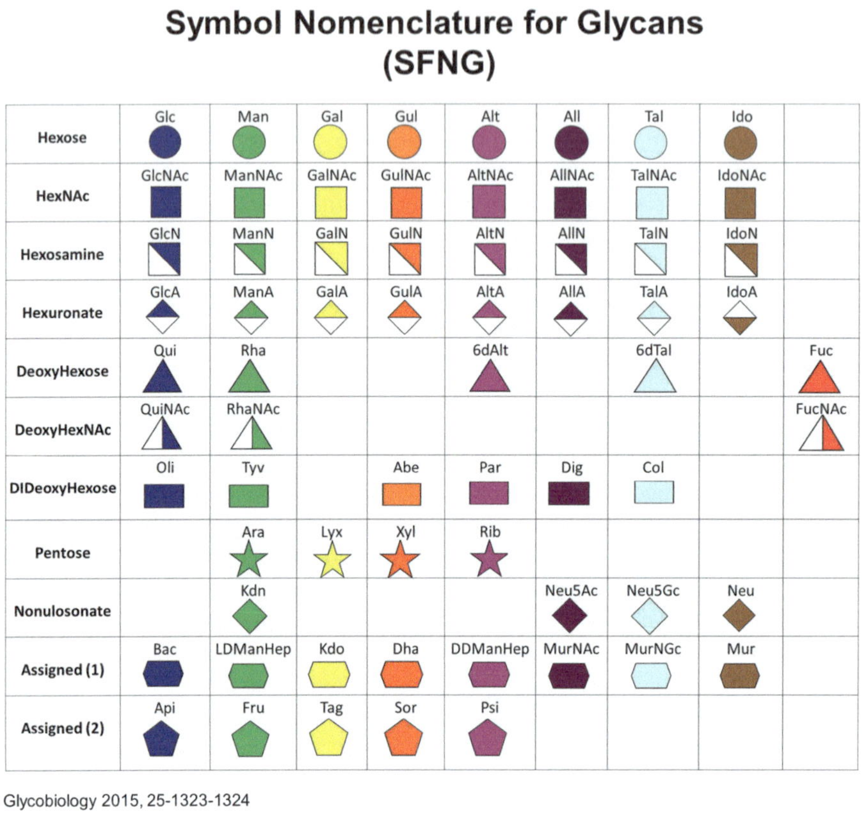

Fig. 2.4 Symbol representation of monosaccharides (SNFG) (Taken from Varki et al. 2015)

2.4 From Graphical Input to Three-Dimensional Structures

Graphical input tools have been developed to provide representation of chemical formulas by using the cartoon representation. The two most widely used tools for graphical input are GlycanBuilder (Ceroni et al. 2007) and DrawRINGS (Akune et al. 2010). GlycanBuilder allows the input of glycan structures in all cartoon notations, whereas DrawRINGS is a Java applet that searches for already known glycan structures that are similar to the drawn glycan structure.

Translating such chemical formulas into three-dimensional structures is far from being a trivial exercise, and only a limited set of applications based on computer simulation is available. These procedures can convert sequence information into reliable 3D models prior any optimization throughout molecular mechanics and molecular dynamics methods. All these constructions are based on the linking of preconstructed 3D molecular templates of monosaccharides (Engelsen et al. 2013; Lutteke et al. 2006; Woods 2014). Another source of information comes

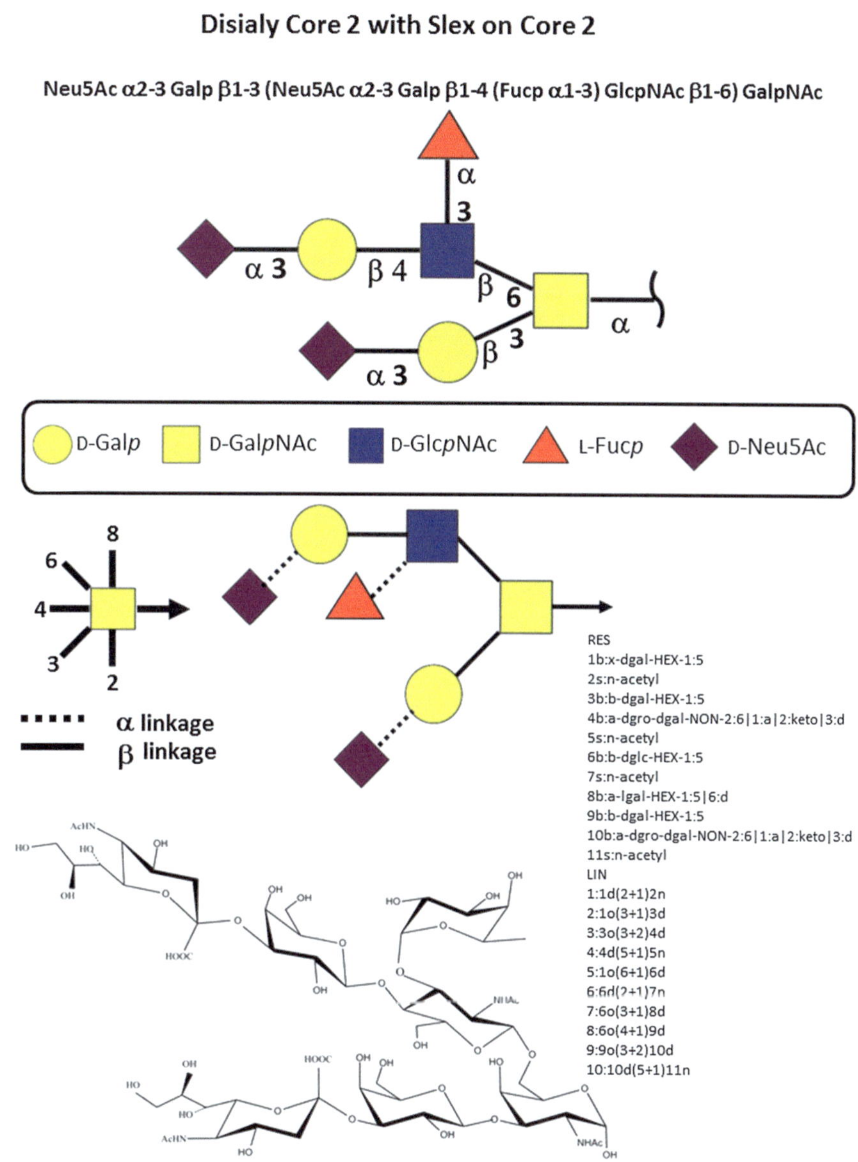

Fig. 2.5 Graphical representation of glycan (disialyl core 2 with Slex on core 2) according to the scheme used in Essential of Glycobiology (Varki et al. 1999) and the Oxford System (Harvey et al. 2009) using the newly adopted Symbol Nomenclature for Glycans (Varki et al. 2015)

from the increasing number of crystal structures that are reported for glycoproteins and protein-carbohydrate complexes. In both cases the 3D structures/architectures are defined by a set of final atomic coordinates, from which molecular structural

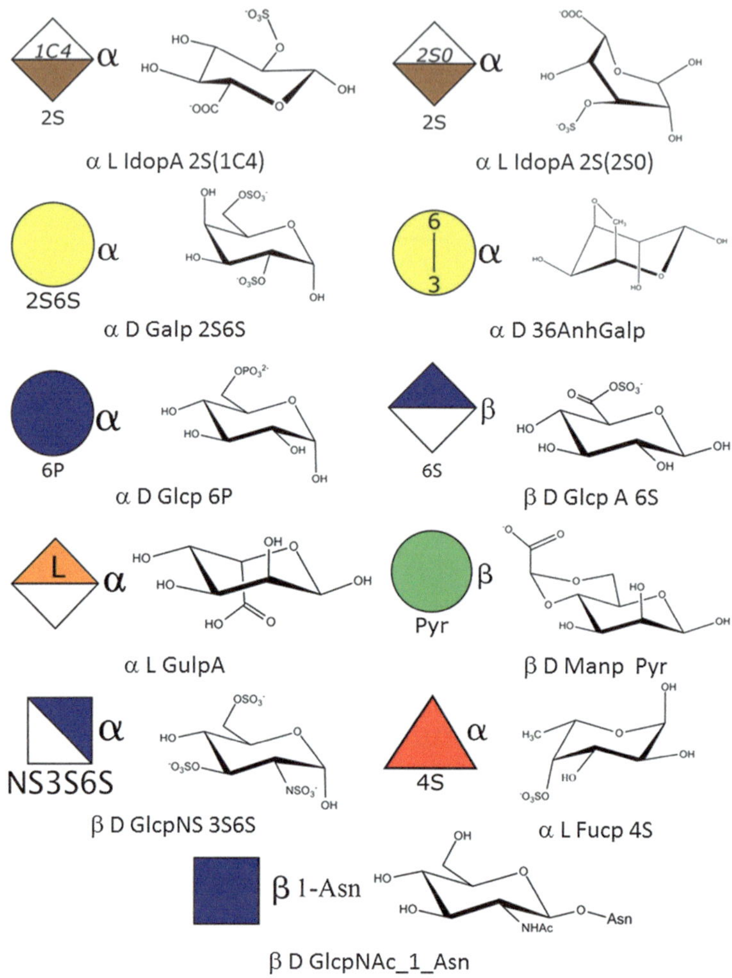

Fig. 2.6 Proposed extension of symbol representation following the set of rules indicated below
Residue letter name: Rib, Ara, Xyl, Lyx, All, Alt, Glc, Man, Gul, Ido, Gal, Tal; and *abbreviated trivial name*:
[O-ester and ethers] (when present) are shown attached to the symbol with a number, e.g., 6Ac for 6-*O*-acetyl group; 3S for 3-*O*-sulfate group, 6P for 6-*O*-phosphate group, 6Me for 6-*O*-methyl group, 36Anh for 3,6-anhydro, and Pyr for pyruvate group
Absolute configuration: The D configuration for monosaccharide and the L configuration for fucose and idose are implicit and do not appear in the symbol. Otherwise the L configuration is indicated in the symbol, as in the case of arabinose or L-galactose. For those occurring in the furanose form, a letter *N* or *S* is inserted in the symbol, indicating the northern (*N*) or southern (*S*) conformation of the five-membered ring.
Anomeric configuration: The nature of the glycosidic configuration (α or β) is explicitly set within the symbol.
Ring conformation: All pyranoses in the D configuration are assumed to have $^{4}C_{1}$ chair conformation; those in the L configuration are assumed to have $^{1}C_{4}$ chair conformation. Otherwise, the ring conformation is indicated in the symbol, as $^{2}S_{0}$ in the case of α-L-idopyranose. *N* or *S* indicates the conformation of the five-membered rings on the conformational wheel

features (i.e., bond lengths, bond angles, torsion angles, hydrogen bonds, intermolecular distances, etc.) can be directly computed. Several distinct repositories hold 3D structural information which follow the Protein Data Bank (Berman et al. 2003), i.e., pdb file format which is a textual file format describing the three-dimensional structures of molecules. The rapid extension of these sets of structural information requires the development of appropriate tools capable of representing and visualizing the range of carbohydrate structures from the simplest oligosaccharide structures to the most complex macromolecular structures and assemblies. Molecular graphics software can generate PDB files from other formats. Widely used examples include Chimera (Petersen, et al. 2004), Jmol (Hanson 2010), Python Molecular Viewer (Sanner 1999), PyMol (DeLano 2002), and Visual Molecular Dynamics (Humphrey et al. 1996). Whereas these standard molecular visualizations can be used, the complexity of glycan structures is best depicted throughout ad hoc visualization tools (Kuttel et al. 2006, 2011; Cross et al. 2009; Perez 2014). They provide a continuous linkage between the most popular ways to depict the primary structure of glycan to their 3D structures while giving the users many options to select the most appropriate modes of representations with respect to their ongoing scientific endeavor (Fig. 2.7).

2.5 In Search of Standards for Representing and Encoding Glycan-Related Information

From the standpoint of bioinformatics, it is impractical to encode glycans (composed of more than 100 monosaccharide units) into distinct graphical symbols. Moreover, to establish effective interoperability of databases, a simple representation in a common/standard format is essential. This would facilitate computational processing and ensure that the data content is nonredundant.

The first approach to encode a carbohydrate molecule is to connect atom sets through chemical bonds. This approach, commonly followed in chemo-informatics and chemical file formats like InChi (McNaught 1997) and SMILES (Weininger 1988) have been developed to aid storing of molecule information in chemical databases like PubChem (Wang et al. 2010) or ChEBI (Degtyarenko et al. 2008). IUPAC (extended), InChi, and SMILES encoding are computed from the chemical drawing (ring structure), and thus, autogeneration of these encodings is possible. Yet, there are severe limitations that do not make this kind of encoding the favored choice. The second approach is based on the connection of building blocks (monosaccharides) through glycosidic linkages. Like nucleic acids and proteins, it is far more efficient to encode carbohydrates using a residue-based approach (Frank and Schloissnig 2010). However, as compared to nucleic acids or proteins, there are a far greater numbers of building blocks (monosaccharides); they arise due to the frequent modifications occurring on the parent monosaccharides. Also, since carbohydrates are frequently found to have branched structures, most of them

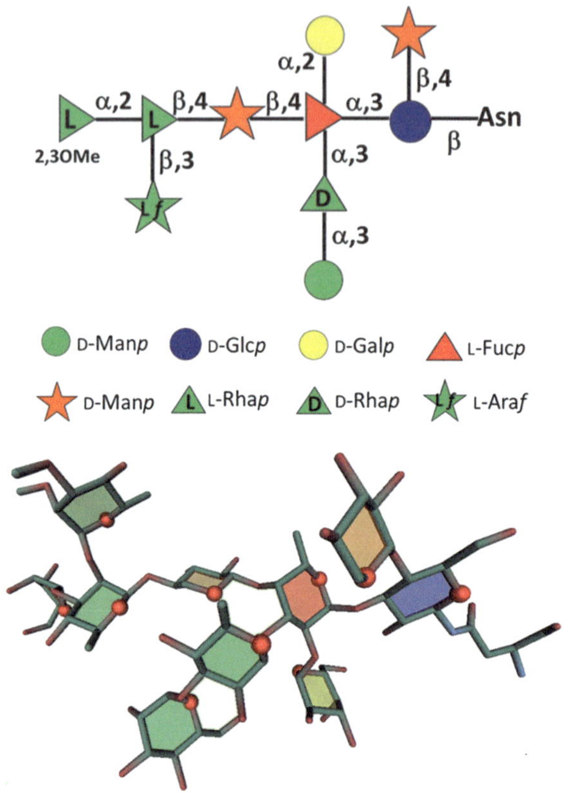

Fig. 2.7 Schematic and three-dimensional representations of a new type of N-glycan core common to all chloroviruses (De Castro et al. 2016). The core element is a pentasaccharide with a β-glucose linked to an asparagine residue which is not located in the typical sequon N-X-T/S. The glucose is linked to a terminal xylose unit and a hyperbranched fucose, which is in turn substituted with a terminal galactose and a second xylose residue. The third position of the fucose unit is always linked to a rhamnose, which is a semiconserved element because its absolute configuration is virus dependent. Additional decorations occur on this core N-glycan and represent a molecular signature for each chlorovirus. The schematic representation uses the Symbol Nomenclature for Glycans and its proposed extension presented in Fig. 2.6 (Perez 2014). Depiction of a tentative three-dimensional model of the N-glycan using the option of the SweetUnityMol software (Perez et al. 2015) where each monosaccharide ring is filled with the Symbol Nomenclature for Glycans color code

are tree-like molecules, unlike nucleic acids and proteins. The prerequisite for a residue-based encoding format is a controlled vocabulary of its residue names. For practical reasons, it makes sense to restrict the number of residues to as low a number as possible. Yet, the lack of clear rules to subscribe atoms of a molecule to one particular monosaccharide, and not to a substituent, poses the main hurdle in encoding monosaccharide names.

2.6 Databases

Due to the development of glycan databases at approximately about the same time in various geographical locations on the globe, but essentially independent of each other, several formats for representing glycan structures have been developed. The major formats for representing glycans that have been developed over these years are described in Table 2.1.

Table 2.1 Major glycan structure formats and their characteristics

Representation format	Pros	Cons
IUPAC condensed/IUPAC extended	Fairly human readable; standardized by the IUPAC commission	Somewhat computer readable, but variations exist between databases
LINUCS	Somewhat human readable	Can only represent completely defined structures (no ambiguity allowed)
CarbBank 2D notation	Human readable	Difficult for computer to process
BCSDB linear code	Can represent rare and complex oligosaccharides including those found in bacteria and plants	Difficult for humans to read
KCF	Uses graph notation; can represent ambiguous structures	Monosaccharide representation is ambiguous, making integration with other databases difficult
Linear code ®	Compact representation	Not human readable
GlycoCT	Uses graph notation; has strict rules to represent monosaccharides	Uses a library to represent substituents, which makes it difficult to update and integrate with other databases
GLYDE-II	Uses XML notation, making database integration and exchange easier	Not human readable
WURCS	Can uniquely represent any sequence, including those containing rare monosaccharides and ambiguous linkages	Not human readable

2.7 Integration

The integration of biological data in the life sciences has been a major challenge in recent years (Katayama et al. 2014). Web services using technologies such as SOAP and REST were attempted to ease such integration, but difficulties arose when data were modified; the maintenance of web services became a bottleneck in the ever-changing field of the life sciences. Thus, a more flexible system was required that could adapt to changes in database content while not requiring other dependent databases to update their systems simultaneously (either by downloading the updated data or reprogramming the web services). Semantic Web technologies thus gained focus in allowing database providers to keep their databases maintained, while other dependent web resources dynamically adapted to the changes. Moreover, the TogoTable service, introduced in Sect. 2.6, provides a way for users to provide their own data easily using Semantic Web technologies.

In short, data on the Semantic Web uses the Resource Description Framework (RDF) to format their data in such a way that the meaning behind the data and their relationships to other data can be stored in a computer-readable format. Resource Description Framework (RDF) and property graph databases are emerging technologies that are used for storing graph-structured data. They are particularly well suited to encode the structure of glycans into a direct acyclic graph where each node represents a building block and each edge serves as a chemical linkage between two building blocks (Fig. 2.8). In this context, graph databases are possible software solutions for storing glycan structures and graph query languages. When RDF data are stored in a triplestore, they can be queried using the SPARQL language from any computer on the web. Moreover, the SPARQL language allows users to perform federated queries by which multiple triplestores at different locations can be queried simultaneously. Thus, databases on the Semantic Web can be "integrated" while still being independent, as long as their RDF data are compatible with others. Compatibility is obtained by the usage of a common ontology such that the data having the same meaning are defined as the same class of data. GlycoRDF (Aoki-Kinoshita et al. 2013; Ranzinger et al. 2008) is the first standardized ontology for representing glycan structures and their pertinent metadata. Many of the databases introduced in this book, including CSDB, MonosaccharideDB, GlycoEpitope, UniCarbKB, and GlyTouCan, use GlycoRDF and are thus interoperable databases. Thus, even though independent databases may use their own specialized representation format for storing their glycan structures, they can potentially be easily integrated with other databases by conforming to GlycoRDF and the Semantic Web.

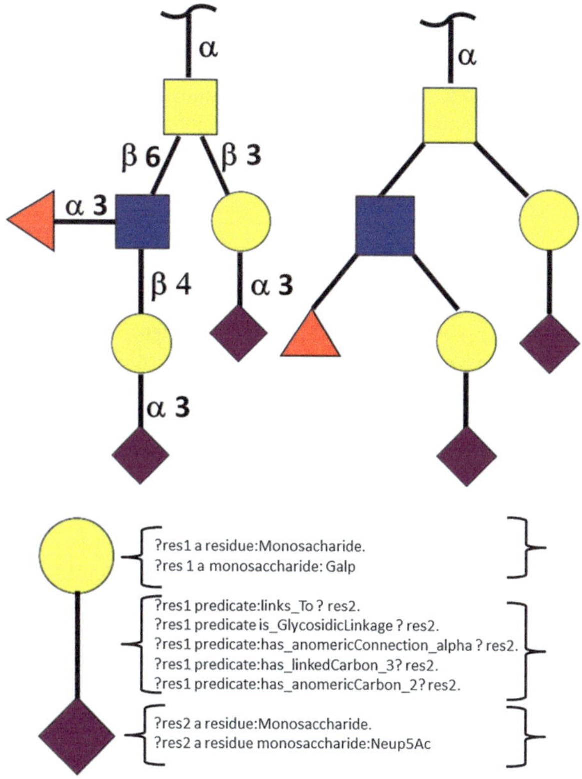

Fig. 2.8 Encoding the structure of glycan (disialy core 2 with Slex on core 2) on the left-hand side into a graph (right-hand side) using the Symbol Nomenclature for Glycans. Each monosaccharide or substituent becomes a node, and each glycosidic bond becomes an edge in the graph. Without any loss of information, all the properties of each monosaccharide or substituent are converted in node properties, whereas glycosidic bond properties are translated in edge properties. Graph is an ordered pair $G = (V, E)$; V set of nodes and E set of edges. Example of use of the RDF model to build a SPARQL query from a glycan substructure focusing on the translation process (Adapted from Alocci et al. 2015)

References

Akune Y, Hosoda M, Kaiya S, Shinmachi D, Aoki-Kinoshita KF (2010) The RINGS resource for glycome informatics analysis and data mining on the web. Omics 14(4):475–486

Alocci D, Mariethoz J, Horlacher O, Lisacek F (2015) Property graph vs RDF triple store: a comparison on glycan substructure search. PLoS One 10:e0144578

Aoki-Kinoshita KF, Yamaguchi A, Ueda N, Akutsu T, Mamitsuka H, Goto S, Kanehisa M (2004) KCaM (KEGG Carbohydrate Matcher): a software tool for analyzing the structures of carbohydrate sugar chains. Nucleic Acids Res 32:W267–W272

Aoki-Kinoshita KF, Bolleman J, Campbell MP, Kawano S, Kim JD, Lütteke T, Matsubara M, Okuda S, Ranzinger R, Sawaki H, Shikanai T, Shinmachi T, Suzuki Y, Toukach P, Yamada Y, Packer YH, Narimatsu H (2013) Introducing glycomics data into the Semantic Web. J Biomed Semant 2013(4):39

Banin E, Neuberger Y, Altshuler Y, … Dukler A (2002) A novel linear code nomenclature for complex carbohydrates. Trends Glycosci Glycotechnol 14: 127–137

Berger O, McBride R, Razi N, Paulson J (2008) Symbol notation extension for pathogen polysaccharides. The Scripps Research Institute, Consortium for Functional Glycomics

Berman H, Henrick K, Nakamura H (2003) Announcing the worldwide protein data bank. Nat Struct Biol 10:980

Bijvoet JM, Peerdeman AF, van Bommel AJ (1951) Determination of the absolute configuration of optically active compounds by means of X-rays. Nature 168:271–272

Bohne-Lang A, Lang E, Forster T, von der Lieth C (2001) Linucs: linear notation for unique description of carbohydrate sequences. Carbohydr Res 336:1–11

Ceroni A, Dell A, Haslam SM (2007) The GlycanBuilder: a fast, intuitive and flexible software tool for building and displaying glycan structures. Source Code Biol Med 2:3–10

Cross S, Kuttel M, Stone JE, Gain J (2009) Visualisation of cyclic and multi-branched molecules with VMD. J Mol Graph Modell 28:131–139

DeLano WL (2002) ThePyMolmoleculargraphicssoftware. DeLanoScientific. http://www.pymol.org

De Castro C, Speciale I, Duncan G, Dunigan DD, Agarkova I, Lanzetta R, Sturiale L, Palmigiano A, Garozzo D, Molinaro A, Tonetti M, Van Etten JL (2016) N−linked glycans of Chloroviruses sharing a core architecture without precedent. Angew Chem 128:664–668

Degtyarenko K, de Matos P, Ennis M, … Ashburner M (2008) Chebi: a database and ontology for chemical entities of biological interest. Nucleic Acids Res 36:D344–D350

Engelsen SB, Hansen P, Perez S (2013) POLYS: an open source software package for building three-dimensional structures of polysaccharides. Biopolymers 101:733–743

Fischer E (1890) Synthesen in der Zukergruppe. Ber Dtsch Chem Ges 23:2114–2141

Frank M, Schloissnig S (2010) Bioinformatics and molecular modeling in glycobiology. Cell Mol Life Sci 67:2749–2772

Hanson RM (2010) Jmol – a paradigm shift in crystallographic visualization. J Appl Crystallogr 43:1250–1260

Harvey DJ, Merry AH, Royle L, Campbell MP, Dwek RA, Rudd PM (2009) Proposal for a standard system for drawing structural diagrams of N- and O-linked carbohydrates and related compounds. Proteomics 9:3796–3801

Haworth WN (1929) The constitution of sugars. Edward Arnold & Company, London (Longmans, Green & Co, New York City)

Herget S, Ranzinger R, Maass K, von der Lieth C (2008) Glycoct – a unifying sequence format for carbohydrates. Carbohydr Res 343:2162–2171

Humphrey W, Dalke A, Schulten K (1996) VMD-visual molecular dynamics. J Mol Graph 14:33–58

Kamerling JP (2007) Basic concepts and nomenclature recommendations in carbohydrate chemistry. In: Kamerling JP (ed) Comprehensive glycosciences, vol 1. Elsevier, Oxford, pp 1–37

Katayama T, Wilkinson MD, Aoki-Kinoshita KF, Kawashima S, Yamamoto Y, Yamaguchi A, Okamoto S, Kawano S, Kim JD, Wang Y, Wu H, Kano Y, Ono H, Bono H, Kocbek S, Aerts J, Akune Y, Antezana E, Arakawa K, Aranda B, Baran J, Bolleman J, Bonnal RJ, Buttigieg PL, Campbell MP, Chen YA, Chiba H, Cock PJ, Cohen KB, Constantin A, Duck G, Dumontier M, Fujisawa T, Fujiwara T, Goto N, Hoehndorf R, Igarashi Y, Itaya H, Ito M, Iwasaki W, Kala M, Katoda T, Kim T, Kokubu A, Komiyama Y, Kotera M, Laibe C, Lapp H, Lütteke T, Marshall MS, Mori T, Mori H, Morita M, Murakami K, Nakao M, Narimatsu H, Nishide H, Nishimura Y, Nystrom-Persson J, Ogishima S, Okamura Y, Okuda S, Oshita K, Packer NH, Prins P, Ranzinger R, Rocca-Serra P, Sansone S, Sawaki H, Shin SH, Splendiani A, Strozzi F, Tadaka S, Toukach P, Uchiyama I, Umezaki M, Vos R, Whetzel PL, Yamada I, Yamasaki C, Yamashita R, York WS, Zmasek CM, Kawamoto S, Takagi T (2014) Penetration of ontology and linked data in life science domains. BioHackathon series in 2011 and 2012. J Biomed Semant 5:5

Kikuchi N, Kameyama A, Nakaya S, ... Narimatsu H (2005) The carbohydrate sequence markup language (cabosml): an XML description of carbohydrate structures. Bioinformatics 21:1717–1718
Kornfeld S, Li E, Tabas I (1978) The synthesis of complex-type oligosaccharides. II. Characterization of the processing intermediates in the synthesis of the complex oligosaccharide units of the vesicular stomatitis virus G protein. J Biol Chem 253:7771–7778
Kuttel M, Gain J, Burger A, Eborn E (2006) Techniques for visualization of carbohydrate molecules. J Mol Graph Modell 25:380–388
Kuttel M, Mao Y, Widmalm G, Lundborg M (2011) CarbBuilder: an adjustable tool for building 3D molecular structures of carbohydrates for molecular simulation. In: Seventh IEEE international conference on eScience, pp 395–402
Laine RA (1994) Invited commentary: a calculation of all possible oligosaccharide isomers both branched and linear yields 1.05×10^{12} structures for a reducing hexasaccharide: the isomer barrier to development of single-method saccharide sequencing or synthesis systems. Glycobiology 4:759–767
Lichtenthaler FW (2002) Emil Fischer, his personality, his achievements, and his scientific progeny. Eur J Org Chem 24:4095–4122
Lutteke T, Bohne-Lang A, Loss A, ... von der Lieth C (2006) Glycosciences.de: an internet portal to support glycomics and glycobiology research. Glycobiology, 16:71R–81R
Marchessault RH, Perez S (1979) Conformations of the hydroxymethyl group in crystalline aldohexopyranoses. Biopolymers 18:2369–2374
McNaught AD (1997) International union of pure and applied chemistry and international union of biochemistry and molecular biology. Joint commission on biochemical nomenclature. Nomenclature of carbohydrates. Carbohydr Res 297:1–92
Meinert C, Myrgorodska J, de Marcellus P, Buhs T, Nahon L, Hoffmann SV, Le Sergeant d'Hendecourt L, Meierhenrich UJ (2016) Ribose and related sugars from ultraviolet irradiation of interstellar ice analogs. Science 352:208–212
Sanner MF (1999) Python: a programming language for software integration and development. J Mol Graph Model 17:57–61
Perez S (2014) The symbolic representation of monosaccharides in the age of glycobiology. http://glycopedia.eu/e-chapters/the-symbolic-representation-of/Abstract
Perez S, Tubiana T, Imberty A, Baaden M (2015) Three-dimensional representations of complex carbohydrates and polysaccharides. SweetUnityMol: a video game based computer graphic software. Glycobiology 25:483–491
Petersen EF et al (2004) UCSH chimera – a visualization system for exploratory research and analysis. J Comp Chem 25:1605–1612
Ranzinger R, Herget S, Wetter T, von der Lieth CW (2008) Glycomedb – integration of open-access carbohydrate structure databases. BMC Bioinf 9:384
Sahoo SS, Thomas C, Sheth A, Henson C, York WS (2005) GLYDE-an expressive XML standard for the representation of glycan structure. Carbohydr Res 340:2802–2807
Toukach PV (2011) Bacterial carbohydrate structure database 3: principles and realization. J Chem Inf Model 51:159–170
Varki A, Cummings RD, Esko JD, Freeze H, Hart GW, Marth JD (1999) Essentials of glycobiology. Cold Spring Harbor Laboratory Press, Plainview
Varki A, Cummings RD, Aebi M, Packer NH, Seeberger PH, Esko JD, Stanley P, Hart G, Darvill A, Kinoshita T, Prestegard JJ, Schnaar RL, Freeze HH, Marth JD, Bertozzi CR, Etzler ME, Frank M, Vliegenthart FG, Lütteke T, Perez S, Bolton E, Rudd P, Paulson J, Kanehisa M, Toukach P, Aoki-Kinoshita KF, Dell A, Narimatsu H, York W, Taniguchi N, Kornfeld S (2015) Symbol nomenclature for graphical representations of glycans. Glycobiology 25:1323–1324
Wang Y, Bolton E, Dracheva S, ... Bryant SH (2010) An overview of the PubChem bioassay resource. Nucleic Acids Res, 38(suppl 1):D255–D266
Weininger D (1988) Smiles, a chemical language and information system. 1. Introduction to methodology and encoding rules. J Chem Inf Comput Sci 28:31–36
Woods R (2014) GLYCAM web. Complex Carbohydrate Research Center, University of Georgia, Athens. (http://www.glycam.com)

Part II
Monosaccharide and Glycan Structures

Chapter 3
Translation and Validation of Carbohydrate Residue Names with MonosaccharideDB Routines

Thomas Lütteke

Abstract Various glycoinformatics resources developed their individual notations for encoding of glycan structure information. Therefore, translation of glycan structures is required for an efficient use of multiple resources and for data exchange. A major problem for translations is residue notation, because individual notations use different names to encode the same residues, and the number of different monosaccharides is too large to quickly build a translation table manually. MonosaccharideDB offers various means to perform translation of carbohydrate residue names. This chapter illustrates the usage of the MonosaccharideDB web interface for both manual and automated conversion and validation of glycan residues.

Keywords Carbohydrate notation • Nomenclature • Monosaccharide • Glycoinformatics • Sequence format conversion • Web interface • Database

3.1 Introduction

Carbohydrate residues are multifaceted. Hundreds of different monosaccharide residues are present in carbohydrate databases and often only differ in the stereochemistry of individual atoms (Werz et al. 2007; Herget et al. 2008b). Many monosaccharide residues are generated by modifications of a variety of parent monosaccharides. These modifications can be divided into two subclasses: *core modifications* such as deoxygenation, double bonds, or carboxyl groups affect the atoms that are involved in the main $[CH_2O]_n$ scaffold of a monosaccharide and often have an effect on the stereochemistry of a monosaccharide, whereas *substitutions* are formed by chemical groups that are linked to an oxygen atom of a hydroxyl group or directly to the carbon backbone by replacing a hydrogen atom or an entire hydroxyl

T. Lütteke (✉)
Institute of Veterinary Physiology and Biochemistry, Justus-Liebig-University Giessen, Frankfurter Str. 100, 35392 Giessen, Germany
e-mail: thomas.luetteke@vetmed.uni-giessen.de

K.F. Aoki-Kinoshita (ed.), *A Practical Guide to Using Glycomics Databases*,
DOI 10.1007/978-4-431-56454-6_3

group (Herget et al. 2009). These modifications have to be represented in carbohydrate nomenclature, which can result in rather complex residue names, especially when stereochemistry is affected by core modifications. IUPAC recommendations are available for carbohydrate nomenclature (McNaught 1997). However, these are not always followed, and they allow multiple names in some instances. For example, trivial names are available for some frequently occurring modified residues, such as "fucose" for "6-deoxy-galactose" or "KDO" ("keto-deoxy-octulosonic acid") for "D-3-deoxy-manno-octulosonic acid," and both trivial names and systematic names can be used.

Another feature of carbohydrates that contributes to their complexity is found in different ways to link individual residues to each other, including the ability to form branched structures. In IUPAC extended nomenclature, which is frequently used to represent glycan structures in the literature, linkages are denoted as a 2D graph using arrows to indicate linkages (McNaught 1997). Alternatively, cartoon representations of glycans can be used, where residues are encoded by symbols of different shape and color (Campbell et al. 2014). These formats are easy to survey for humans but hard to handle by computers. Therefore, they are often used within I/O interfaces of glycoinformatics applications, but internally the tools use different representations with multifaceted residue names (see Table 3.1). Complex Carbohydrate Structure Database (CCSD) (Doubet et al. 1989), better known by the name of its querying software CarbBank (Doubet and Albersheim 1992), used textual 2D graphs to store glycan structures. Residue names in CarbBank notation resemble IUPAC names, in which the Greek characters ("α/β") to indicate the anomeric state of a residue are replaced by Latin letters ("a/b"). The LINUCS notation (Bohne-Lang et al. 2001) that is used in GLYCOSCIENCES.de (Lütteke et al. 2006) converts CarbBank 2D graphs to a linear format with clear rules for sorting of branches; the residue names that are used in LINUCS are basically the same as in CarbBank notation. The glycan sequence formats of the Russian Carbohydrate Structure Database (CSDB), formerly Bacterial Carbohydrate Structure Database (BCSDB) (Toukach 2011), as

Table 3.1 Representations of 4-o-sulfono-β-D-GalpNAc in various notations

Notation	Base monosaccharide	Separate substituents
BCSDB	bDGalpN	(2-1) Ac
		(4-1) S
CarbBank/LINUCS	b-D-GalpNAc	(4-1) sulfate
CarbBank/LINUCS*	b-D-GalpNAc4SO3	–
CFG (LinearCode®)	AN[4S]b	–
GlycoCT	b-dgal-HEX-1:5	(2d-1) n-acetyl
		(4o-1) sulfate
MonosaccharideDB	b-dgal-HEX-1:5||(2d:1)n-acetyl|(4o:1)sulfate	–
PDB	ASG	–
PDB*	NGL	–

Please note that in some notations, multiple representations are in use. Residue names that are marked by an asterisk (*) are non-preferred secondary alias names well as the KEGG Chemical Function (KCF) format (Aoki et al. 2004) used in KEGG Glycan (Hashimoto et al. 2006) and RINGS (Akune et al. 2010) also employ residue names that are close to IUPAC rules, whereas GlycoMinds LinearCode® (Banin et al. 2002) defines a one-letter code to represent monosaccharide residue types. The LinearCode® approach requires a dictionary table where residue names are explicitly defined. Therefore, only glycan structures that consist of residues that are listed in the dictionary can be encoded with LinearCode®, and the user needs to know the definitions to be able to interpret glycan structures encoded in this format because the codes do not resemble IUPAC nomenclature. On the other hand, this clearly defined dictionary produces nonambiguous names and is significantly less error-prone than formats such as CarbBank notation or LINUCS that allow free-text residue definitions. The GlycoCT format (Herget et al. 2008a) attempts to combine the advantages of both systems by defining strict rules to encode monosaccharides in a form that still shows some resemblance to IUPAC nomenclature but is easily computer-parsable and thus easier to interpret and to validate automatically than "free-text" formats. GLYDE-II (Sahoo et al. 2005) follows a similar concept as GlycoCT but allows to include aglycone parts. The recently introduced WURCS notation (Tanaka et al. 2014) defines glycans in a more chemistry-based fashion.

Notations do not only differ in the description of a residue but also in the definition of what is to be considered a residue. GlycoCT, for example, treats each substituent as a separate residue, whereas in CarbBank notation or LINUCS, most substituents are considered to be part of the monosaccharide and thus are included in the residue names. In some cases, substituents can even be fragmented. For example, CSDB includes amino groups in the monosaccharide names, whereas acetyl groups are handled as separate residues. Consequently, N-acetyl modifications are split into an amino modification of the parent monosaccharide and an acetyl substituent in this format.

This variety of glycan notation formats imposes several problems on the developers of glycoinformatics databases and tools as well as on scientists who want to make use of these resources. Apparently, cross-linking and exchange of data between individual applications that make use of different notations is difficult, because translation from one format to another one includes conversion of residue names. Moreover, inconsistent residue nomenclature can cause problems also within individual resources. If multiple names (such as trivial names and systematic names) are used for a single residue, then several distinct database entries can describe chemically identical structures, and users might miss some of these redundant entries and the data stored therein when querying a database.

These problems can be addressed by usage of a standard dictionary that defines a primary alias name for each glycan, secondary alias names (if applicable) that are mapped to the primary alias, and translations to or from other notations. Such an approach is used by GlycomeDB (Ranzinger et al. 2011) for conversion of glycan structures from various resources to GlycoCT, the format internally used in GlycomeDB. Such a dictionary can be very useful to handle the residues that are known to be present in the databases but needs to be curated each time a new residue is found or a new alias name is used for an existing residue. MonosaccharideDB attempts to provide a dynamic dictionary by implementing name parsing and encoding routines that can handle residue names that are not yet in the dictionary, provided that the individual elements of a residue names (such as monosaccharide stem type name, modification descriptors, or substituent names) can be parsed or encoded correctly.

MonosaccharideDB also provides the namespace of GLYDE-II (Sahoo et al. 2005) and supplies monosaccharide information to the GlycoRDF project (Aoki-Kinoshita et al. 2013; Ranzinger et al. 2015).

This chapter describes some use cases that employ MonosaccharideDB functionality to explore properties of specific monosaccharides, to translate residue names from one notation to another one, and to create unique residue names within one notation. Translation of complete glycan structures beyond the residue level has been described recently in (Lütteke 2015).

3.2 Use Cases

Many monosaccharide-related questions can be answered by accessing individual MonosaccharideDB entries. Therefore, the following sections will first introduce some query options of the MonosaccharideDB web interface (http://www.monosaccharidedb.org) and then describe some concrete questions that can be answered by performing one of the queries. Individual pages are referred to in the following text in the form →*A*→*B*→*C*, which means that you need to select menu item *A*, then submenu item *B*, and then pick *C*. (In this context *B* or *C* can also refer to links on the retrieved page instead of items of the submenu.)

3.2.1 Query Options

3.2.1.1 Query by Residue Name

Notation-based queries will be the most important ones for the specific tasks below. The corresponding search form is found at → *Database*→*Query Monosaccharide*. In the section *Query MonosaccharideDB by name*, you can enter a residue name and specify the notation scheme that is used to encode the residue (Fig. 3.1). For example, select *CFG* notation to define a residue in LinearCode®, enter the monosaccharide name *GNb* (case sensitive), and press the submit button. This will direct you to the entry page of β-D-GlcpNAc (Fig. 3.2), where you can find information on this residue including chemical properties (such as molecular weight, atomic composition, absolute configuration, ring type, etc.); a list of the

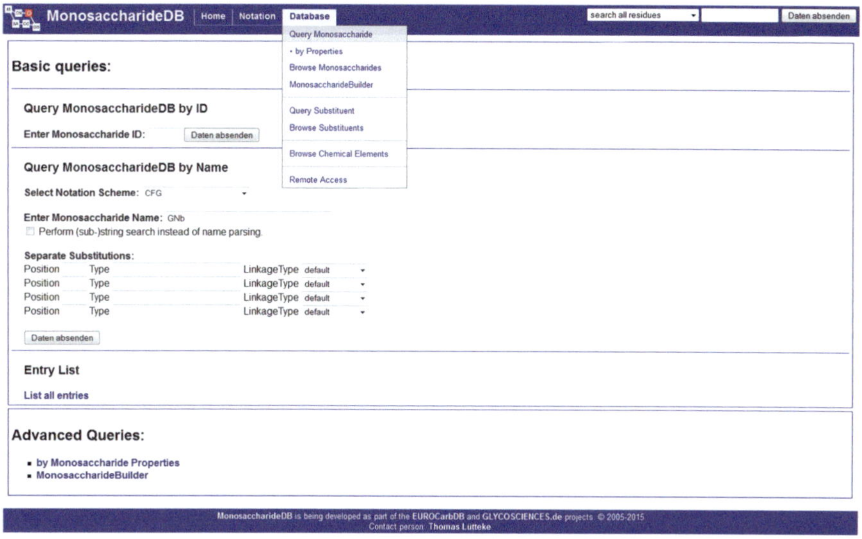

Fig. 3.1 MonosaccharideDB query interface for notation-based queries

linkage positions, via which this residue can be linked to other monosaccharides (which can be used, e.g., for validation of glycan structures, i.e., to test if a specified glycosidic linkage is possible); and the atoms that constitute the monosaccharide including their names and connections, graphical representations of the monosaccharide, and a list of alias names in various notations. In this list you can also find your query string "GNb" as primary alias in CFG notation. Go back to the query form, select *CarbBank* notation, enter *4-deoxy-b-D-Glcp* (not case sensitive) to search for a glucose that has a deoxy-modification at position 4, and submit this query. This time you will reach the entry of *b-D-4-deoxy-xylHexp*. Your query string is not present in the alias name list, but the name could be successfully parsed and interpreted correctly Many residues can be encoded in various ways, especially if they include modifications that affect stereochemistry. In this case, for example, you also could have entered, e.g., *4-deoxy-b-D-Galp* (Glc and Gal differ in the stereochemistry of C4, i.e., the orientation of the OH group at C4, and thus their 4-deoxy variants are identical) or *b-4-deoxy-D-xylHexp* (the order of the elements *4-deoxy-*, *b-*, and *D-* is not relevant here) as alias names to reach this entry.

The query form also contains fields to enter separate substituents for those notations that handle (some) substituents as distinct residues. For example, select the notation scheme "BCSDB," enter the residue name *bDGlcpN* (case sensitive) and a separate substitution with position *2*, type *Ac*, and linkage type *default*. This query will also direct you to the β-D-GlcpNAc entry.

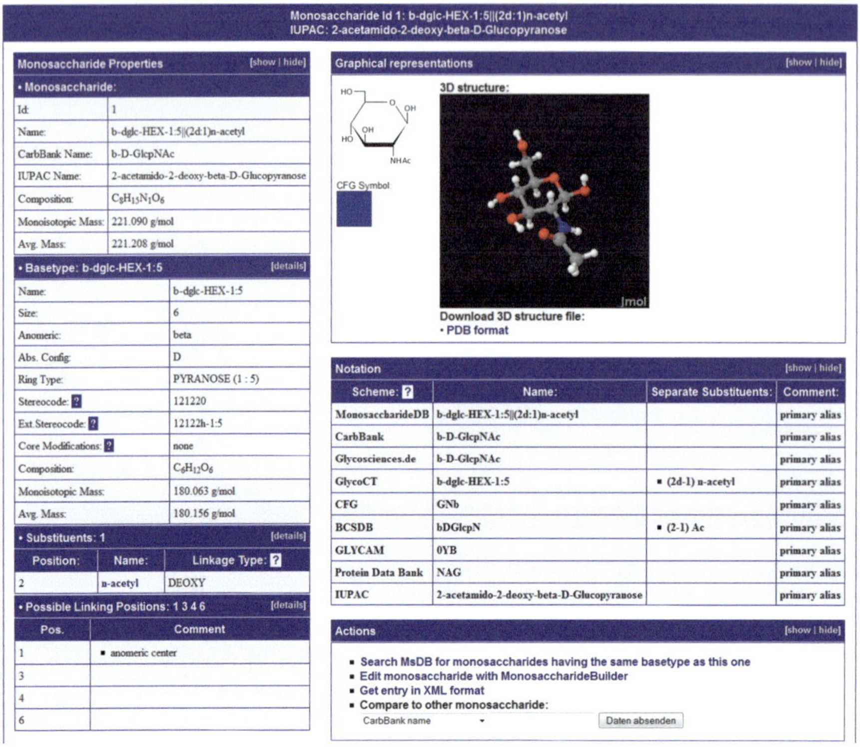

Fig. 3.2 MonosaccharideDB entry page of β-D-GlcpNAc. A list of the atoms that constitute the monosaccharide is omitted from this screenshot for brevity

3.2.1.2 Substituent Query

Individual substituents can be queried in a similar way. Go to →*Database*→*Query Substituent*, select the notation scheme *GlycoCT*, enter the substituent name *methyl*, and submit the query. The retrieved page summarizes the data that are stored about this substituent, in particular the primary and secondary alias names of the substituent in the covered notation schemes. The alias names also include a *linkage type*. This is required because a methyl group can be linked to the oxygen atom of a hydroxyl group, replacing the hydroxyl hydrogen (commonly referred to as "O-methyl," linkage type "H_AT_OH"), but it can also replace the hydrogen that is directly linked to a backbone carbon ("C-methyl," linkage type "H_LOSS"). In most notations the linkage type is implied in the alias name (e.g., *O-Me* vs. *C-Me*). Therefore, you will find multiple primary alias names for individual notations here, but only one for a combination of notation scheme and linkage type. A description of the individual linkage types can be found at →*Notation*→*Substituents*.

3.2.1.3 Quick Access

To quickly access a residue, you can type in its name into the query field in the upper right corner of each MonosaccharideDB page.

3.2.2 Residue Name Translation

As mentioned above various notations that are used in glycoinformatics use different residue namespaces. Therefore, translation from one notation to another requires translation of residue names as well. Such translations can be performed manually using the residue name queries described above. If the residue can be encoded in the target notation of the translation, the monosaccharide entry page that is retrieved by the query lists the primary alias name that is to be used (Fig. 3.2).

However, manual translations are error-prone and hardly applicable to large data sets. Thus, translations are rarely performed manually. Instead, computer programs are employed to automate this process. Looking up the names in HTML web pages is not useful in that case. Therefore, MonosaccharideDB also provides XML output of its entries. To see an example of this output, you can access any monosaccharide entry and click the *Get entry in xml format* link in the *Actions* section of the entry. This can also be achieved by adding a parameter "output=xml" to the entry URL. For example, the entry of b-D-GlcpNAc (CarbBank notation) in XML format can be retrieved via the following URL: http://www.monosaccharidedb.org/display_monosaccharide.action?scheme=carbbank&name=b-d-glcpnac&output=xml

There is also a specific HTTP service available for translations. It accepts the residue name, source and target notation schemes, and potential external substituents as parameters and returns an XML file that contains the primary alias name of the residue in the requested target notation, including (if applicable) a list of substituents that are handled as separate residues in the target notation, and some further properties of the monosaccharide. For example, translation of the residue of Table 3.1 from CarbBank to BCSDB format is achieved with the following parameters: *sourceScheme*="carbbank," *name*="b-D-GalpNAc4SO3," and *targetScheme* "bcsdb." The resulting URL to call the translator is as follows: http://www.monosaccharidedb.org/convert_residue.action?name=b-D-GalpNAc4SO3&sourceScheme=carbbank&targetScheme=bcsdb. From the resulting XML file (Fig. 3.3), you can see that the residue name in BCSDB

```
<?xml version="1.0"?>
<monosaccharide_exchange_object
    ms_name="b-dgal-HEX-1:5||(2d:1)n-acetyl|(4o:1)sulfate" >
     <namescheme>BCSDB</namescheme>
     <monosaccharide_name>bDGalpN</monosaccharide_name>
     <residue_type>monosaccharide</residue_type>
     <external_substituents count="2">
          <substituent>
               <name>Ac</name>
               <original_name>nac</original_name>
               <linkage n="1">
                    <linkage_position_ms>2</linkage_position_ms>
                    <linkage_type>H_AT_OH</linkage_type>
                    <linkage_position_subst>1</linkage_position_subst>
                    <original_linkage_type></original_linkage_type>
               </linkage>
          </substituent>
          <substituent>
               <name>S</name>
               <original_name>so3</original_name>
               <linkage n="1">
                    <linkage_position_ms>4</linkage_position_ms>
                    <linkage_type>H_AT_OH</linkage_type>
                    <linkage_position_subst>1</linkage_position_subst>
                    <original_linkage_type></original_linkage_type>
               </linkage>
          </substituent>
     </external_substituents>
     <orientationChanged>false</orientationChanged>
</monosaccharide_exchange_object>
```

Fig. 3.3 XML output of the HTTP interface to residue translation: result of translation of "b-D-GalpNAc4SO3" from CarbBank to BCSDB notation

notation is "bDGalpN" with two separate substituents: Ac linked to position 2 and S linked to position 4 of the monosaccharide. See http://www.monosaccharidedb.org/remote_access.action for further details of the service.

3.2.3 Creating Unique Residue Names

The usage of multiple alias names for a single residue within one resource causes inconsistencies within that resource and imposes challenges on the users. The residue translation service of MonosaccharideDB described above can also be used to obtain the primary alias of a residue name within one notation scheme. You simply need to use the same notation as source and target notation of the translation, e.g., you translate a residue name from CarbBank notation to CarbBank notation. If your input name already is the primary alias, then the result will be identical with your input, and you do not need to perform any changes, whereas in case of a secondary alias as input name, also the primary alias name will be returned, which means that you have to change your residue name and use the primary alias to create a glycan structure with consistent residue nomenclature.

3.2.4 *MonosaccharideBuilder*

The routines described so far require a residue name as query input. Sometimes, however, you might not have such a name available, e.g., if you see a graphical representation of a monosaccharide and need to find the correct name for it or if you want to answer a question such as *what residue do I get if I change the stereochemistry of C3 of glucose?*, or in other words *what is the 3-epimer of glucose?* In such cases you can use the MonosaccharideBuilder (→*Database*→*MonosaccharideBuilder*) to quickly find the answer. In the first step, you need to select the size of the carbon backbone, and then you can build your monosaccharide by specifying the local properties of each carbon atom, i.e., the stereochemistry (by selecting *HO-C-H [laevus pos.]* or *H-C-OH [dexter pos.]*), or modifications such as deoxy (*H-C-H [deoxy position]*) or a terminal acid group (*COOH [carboxyl group]*). You also need to indicate ring properties. When changing from open chain to ring form and vice versa, not only the ring positions but also the local properties of the anomeric carbon need to be adjusted. This can be achieved conveniently by using the *quick changes* menu, e.g., by selecting *set ring: open chain* and clicking the *Set* button.

Once you have entered all properties, you can press the *preview* button to create or adjust a graphical representation (Haworth formula for ring structures, or Fischer formula in case of open-chain residues) and see the name of the residue. If necessary, you can adjust the properties until you get the residue that you want to represent. The *finish* button directs you to the MonosaccharideDB entry of the monosaccharide that you have built.

This manual specification of each position can be useful if you need to build a monosaccharide but do not have any idea of the correct name of this residue (or of a similar one). However, it can be tedious to find the correct settings. Therefore, you can preset the fields by entering a valid residue name in the start form of MonosaccharideBuilder or alternatively find a MonosaccharideDB entry and click the *Edit monosaccharide with MonosaccharideBuilder* link in the *Actions* section of the entry. For example, to answer the question on the 3-epimer of glucose, you can start with a glucose entry of MonosaccharideDB or by, e.g., entering *b-D-Glcp* in *CarbBank* notation on MonosaccharideBuilder's start page (Fig. 3.4a). In the result page (Fig. 3.4b), you only need to change the property of carbon 3 from *HO-C-H [laevus pos.]* to *H-C-OH [dexter pos.]* (Fig. 3.4c) and click *preview* to see the modified residue within MonosaccharideBuilder (Fig. 3.4d) or *finish* to go to the corresponding entry. This way you will find that the 3-epimer of *b-D-Glcp* is *b-D-Allp*; that means that the 3-epimer of glucose is allose.

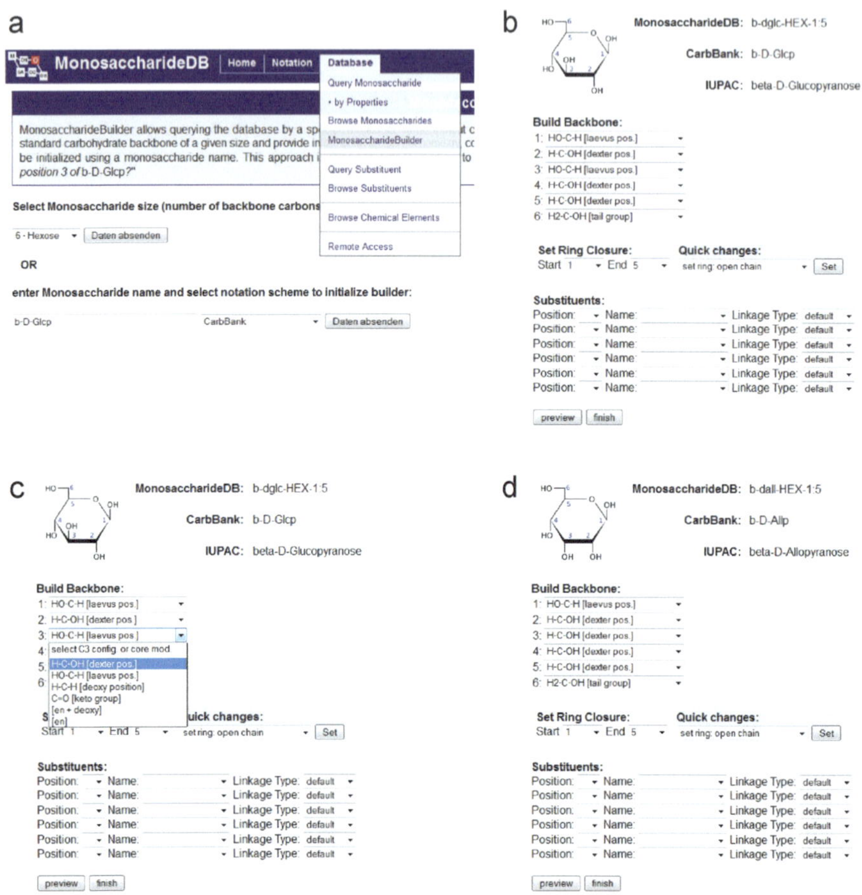

Fig. 3.4 MonosaccharideBuilder: Steps to find 3-epimer of glucose. (**a**) Start by selecting residue size or by entering a valid monosaccharide name. (**b**) Interface for definition of monosaccharide properties. Backbone carbon positions are indicated by *blue* numbers in the Haworth representation of the monosaccharide. (**c**) Change stereochemistry of backbone position 3 and click "preview." (**d**) Result of the alteration. Now you can further modify the residue or click "finish" to go the corresponding MonosaccharideDB entry

3.3 Conclusions

The MonosaccharideDB web interfaces offer various means for accessing monosaccharide properties and for translating carbohydrate residue names from one notation format to another. Currently, its role in translation as well as creation of unique names within individual notations is the predominant function of MonosaccharideDB. In the long run, access to monosaccharide properties will gain importance, because it offers possibilities for the development of algorithms that group carbohydrates by residue similarity as well as for enhanced query options of existing databases.

References

Akune Y, Hosoda M, Kaiya S, Shinmachi D, Aoki-Kinoshita KF (2010) The RINGS resource for glycome informatics analysis and data mining on the Web. Omics 14(4):475–486. doi:10.1089/omi.2009.0129

Aoki KF, Yamaguchi A, Ueda N, Akutsu T, Mamitsuka H, Goto S, Kanehisa M (2004) KCaM (KEGG Carbohydrate Matcher): a software tool for analyzing the structures of carbohydrate sugar chains. Nucleic Acids Res 32 (Web Server issue):W267–W272.

Aoki-Kinoshita KF, Bolleman J, Campbell MP, Kawano S, Kim JD, Lütteke T, Matsubara M, Okuda S, Ranzinger R, Sawaki H, Shikanai T, Shinmachi D, Suzuki Y, Toukach P, Yamada I, Packer NH, Narimatsu H (2013) Introducing glycomics data into the Semantic Web. J Biomed Semant 4(1):39. doi:10.1186/2041-1480-4-39

Banin E, Neuberger Y, Altshuler Y, Halevi A, Inbar O, Dotan N, Dukler A (2002) A novel linearcode(R) nomenclature for complex carbohydrates. Trends Glycosci Glycotechnol 14(77):127–137

Bohne-Lang A, Lang E, Forster T, von der Lieth CW (2001) LINUCS: linear notation for unique description of carbohydrate sequences. Carbohydr Res 336(1):1–11

Campbell M, Ranzinger R, Lütteke T, Mariethoz J, Hayes C, Zhang J, Akune Y, Aoki-Kinoshita K, Damerell D, Carta G, York W, Haslam S, Narimatsu H, Rudd P, Karlsson N, Packer N, Lisacek F (2014) Toolboxes for a standardised and systematic study of glycans. BMC Bioinf 15(Suppl 1):S9

Doubet S, Albersheim P (1992) CarbBank. Glycobiology 2(6):505

Doubet S, Bock K, Smith D, Darvill A, Albersheim P (1989) The complex carbohydrate structure database. Trends Biochem Sci 14(12):475–477

Hashimoto K, Goto S, Kawano S, Aoki-Kinoshita KF, Ueda N, Hamajima M, Kawasaki T, Kanehisa M (2006) KEGG as a glycome informatics resource. Glycobiology 16(5):63R–70R

Herget S, Ranzinger R, Maass K, von der Lieth CW (2008a) GlycoCT-a unifying sequence format for carbohydrates. Carbohydr Res 343(12):2162–2171

Herget S, Toukach P, Ranzinger R, Hull W, Knirel Y, von der Lieth CW (2008b) Statistical analysis of the Bacterial Carbohydrate Structure Data Base (BCSDB): characteristics and diversity of bacterial carbohydrates in comparison with mammalian glycans. BMC Struct Biol 8(1):35

Herget S, Ranzinger R, Thomson R, Frank M, von der Lieth CW (2009) Introduction to carbohydrate structure and diversity. In: von der Lieth CW, Lütteke T, Frank M (eds) Bioinformatics for glycobiology and glycomics – an introduction. Wiley, Chichester, pp 23–47. doi:10.1002/9780470029619.ch2

Lütteke T (2015) Handling and conversion of carbohydrate sequence formats and monosaccharide notation. Methods Mol Biol 1273:43–54. doi:10.1007/978-1-4939-2343-4_4

Lütteke T, Bohne-Lang A, Loss A, Goetz T, Frank M, von der Lieth CW (2006) GLYCOSCIENCES.de: an internet portal to support glycomics and glycobiology research. Glycobiology 16(5):71R–81R

McNaught AD (1997) Nomenclature of carbohydrates (recommendations 1996). Adv Carbohydr Chem Biochem 52:43–177

Ranzinger R, Herget S, von der Lieth CW, Frank M (2011) GlycomeDB–a unified database for carbohydrate structures. Nucleic Acids Res 39 (Database issue):D373–376. doi:10.1093/nar/gkq1014

Ranzinger R, Aoki-Kinoshita KF, Campbell MP, Kawano S, Lütteke T, Okuda S, Shinmachi D, Shikanai T, Sawaki H, Toukach P, Matsubara M, Yamada I, Narimatsu H (2015) GlycoRDF: an ontology to standardize glycomics data in RDF. Bioinformatics 31(6):919–925. doi:10.1093/bioinformatics/btu732

Sahoo SS, Thomas C, Sheth A, Henson C, York WS (2005) GLYDE-an expressive XML standard for the representation of glycan structure. Carbohydr Res 340(18):2802–2807

Tanaka K, Aoki-Kinoshita KF, Kotera M, Sawaki H, Tsuchiya S, Fujita N, Shikanai T, Kato M, Kawano S, Yamada I, Narimatsu H (2014) WURCS: the Web3 unique representation of carbohydrate structures. J Chem Inf Model 54(6):1558–1566. doi:10.1021/ci400571e

Toukach PV (2011) Bacterial carbohydrate structure database 3: principles and realization. J Chem Inf Model 51(1):159–170. doi:10.1021/ci100150d

Werz DB, Ranzinger R, Herget S, Adibekian A, von der Lieth CW, Seeberger PH (2007) Exploring the structural diversity of mammalian carbohydrates ("glycospace") by statistical databank analysis. ACS Chem Biol 2(10):685–691

Chapter 4
Using GlyTouCan Version 1.0: The First International Glycan Structure Repository

Daisuke Shinmachi, Issaku Yamada, Nobuyuki P. Aoki, Masaaki Matsubara, Kiyoko F. Aoki-Kinoshita, and Hisashi Narimatsu

Abstract Glycans are known as the third major class of biopolymers next to DNA and proteins and have many biological roles by structural properties. The structure of glycans differs greatly from DNA and proteins in that they are branched structures of monosaccharides, as opposed to linear sequences of amino acids or nucleotides. Therefore, the assignment of glycan structure information has been a difficult problem. In order to solve this problem, an international team of glyco-scientists has collaborated to develop this repository, called GlyTouCan, to provide a centralized resource to deposit glycan structures and obtain unique accession numbers. GlyTouCan can accept glycan structures in any form, including ambiguous structures consisting of compositions and topologies. Users can register new glycan structures and additionally search for glycan structures that have been registered into this repository. All of these tools are freely available at https://glytoucan.org/. This will enable glycomics researchers to easily identify glycan structures by accession number. This chapter describes the procedures for the registration and search methods of glycan structures and provides an overview of the entry pages. Furthermore, troubleshooting tips and cautionary notes for using GlyTouCan are also included.

Keywords Glycan structure • Accession number • Structure repository • Database • Structure search • Glycan sequence • WURCS • GlycoCT

D. Shinmachi • N.P. Aoki
Department of Bioinformatics, Graduate School of Engineering, Soka University, Tokyo, Japan

I. Yamada • M. Matsubara
The Noguchi Institute, Tokyo, Japan

K.F. Aoki-Kinoshita (✉)
Department of Bioinformatics, Graduate School of Engineering, Soka University, Tokyo, Japan

Research Center for Medical Glycoscience (RCMG), National Institute of Advanced Industrial Science and Technology (AIST), Tsukuba, Japan
e-mail: kkiyoko@soka.ac.jp

H. Narimatsu
Research Center for Medical Glycoscience (RCMG), National Institute of Advanced Industrial Science and Technology (AIST), Tsukuba, Japan

K.F. Aoki-Kinoshita (ed.), *A Practical Guide to Using Glycomics Databases*,
DOI 10.1007/978-4-431-56454-6_4

4.1 Introduction

As the third major molecule of biopolymers next to DNA and proteins, glycans have been found to be involved in various important biological functions. The structure of glycans, however, differs greatly from DNA and proteins in that they are branched, as opposed to linear sequences of amino acids or nucleotides. Therefore, the storage of glycan information in databases, let alone their curation, has been a difficult problem.

This has caused efforts in the integration of glycan data between different databases difficult, making an international repository for glycan structures, where unique accession numbers are assigned to every identified glycan structure, necessary. As such, an international team of developers and glycobiologists have collaborated to develop this repository, called GlyTouCan, which has been released in 2015 and is freely available at https://glytoucan.org/, to provide a centralized resource for depositing glycan structures, compositions, and topologies and to retrieve accession numbers for each of these registered entries (Aoki-Kinoshita et al. 2016). As a result, GlyTouCan enables researchers to reference glycan structures simply by accession number, as opposed to by chemical structure or text string, which has been a burden to integrate glycomics databases in the past. Thus, in the future, not only can GlyTouCan serve as a central registry, but it can serve as a portal to search for glycan-related publications as well as other biological information.

4.2 Signing In to GlyTouCan

When a user first clicks on the Sign In button, they are prompted with a request to allow access to their email address (Fig. 4.1). Once permissions are granted, GlyTouCan stores the user information and allows registration functionality. Details regarding the architecture and registration process are available on the GlyTouCan system architecture web page.

4.3 Registration

Registration of glycan structure(s) is the main functionality of GlyTouCan. Registered users can choose the registration method depending on the structure informa-

Fig. 4.1 Sign In (*blue square*) button is provided in the menu bar at the top of GlyTouCan. In order to register new glycan structures, users need to "Sign in" using their email address

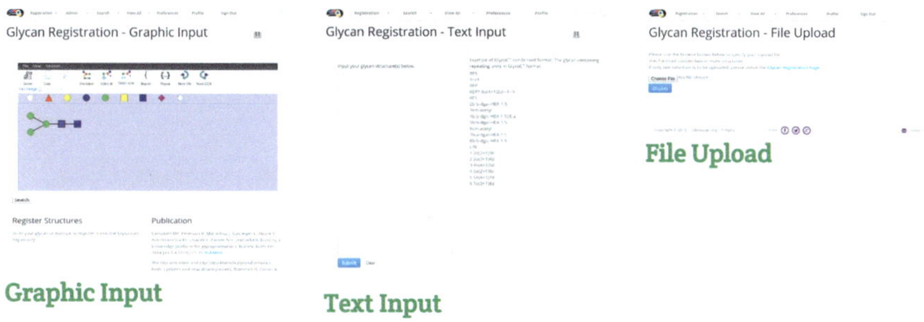

Fig. 4.2 Three interfaces for the glycan structure registration. GlyTouCan provides three methods of registration: (1) Graphic Input, (2) Text Input, and (3) File Upload

tion and the number of glycan structure(s) to register. The options are (1) Graphical Input, (2) Text Input, and (3) File Upload (Fig. 4.2). (1) Graphical Input allows users to enter glycan structures individually by drawing them on a canvas in the browser. (2) Text Input allows users to enter or copy and paste individual glycan structures specified using a text format, such as GlycoCT. (3) File Upload allows users to register groups of glycan structures specified in a text file. Each of these registration procedures is described in detail below. Once the registration is complete, users can obtain accession number(s) of the registered glycan structure(s), which can be downloaded into spreadsheet format for review.

4.3.1 Graphical Interface for Registration

Submitting a glycan structure through a graphical interface is based on GlycanBuilder (Damerell et al. 2012) (Fig. 4.3). It is possible to extend predefined glycan structures by clicking on the Structure->Add Structure menu option.

After pressing the submit button, the input structure is translated into a text format and sent to the GlyTouCan database to check if it was previously registered. If the structure is found in the database, the accession number and a link to the structure resource entry will be displayed; otherwise, a final confirmation screen will be shown (Fig. 4.4). In the latter case, if the submit button is entered to indicate confirmation, the newly registered accession number and graphical representation will be displayed. The glycan structure will be stored in GlycoCT{condensed} format (Herget et al. 2008), the sequence initial input will be displayed under "Original Structure," while the sequence converted into WURCS (Tanaka et al. 2014) will be displayed under "Structure." The image generated will be shown on the right. This screen is necessary to confirm that there are no issues in the conversion process.

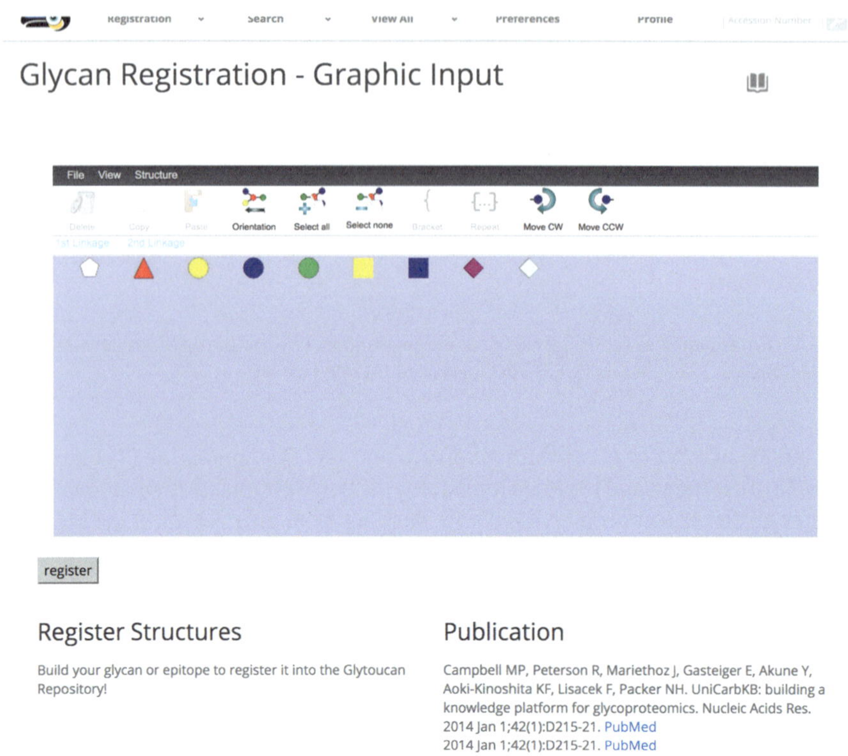

Fig. 4.3 The user interface for drawing a glycan structure using GlycanBuilder. The GlycanBuilder can be used to draw a glycan structure for registration

4.3.2 Text Input

In GlyTouCan, a user can register glycan structures specified in a text format (Fig. 4.5). Click the "Registration" pull-down menu in the menu bar at the top of the page and then the "Text Input" menu button. The "Text Input" page has a text box that can take structures in GlycoCT format (other formats are planned to be supported in the future). Once a valid sequence is entered into the text box, click the "Submit" button, and the "Text Input" page will display a confirmation page. This confirmation page is the same as that of the "Graphical Interface" registration.

4.3.3 File Upload

In order to register glycans using the "File Upload" system, a text file needs to be prepared as follows (Fig. 4.6):

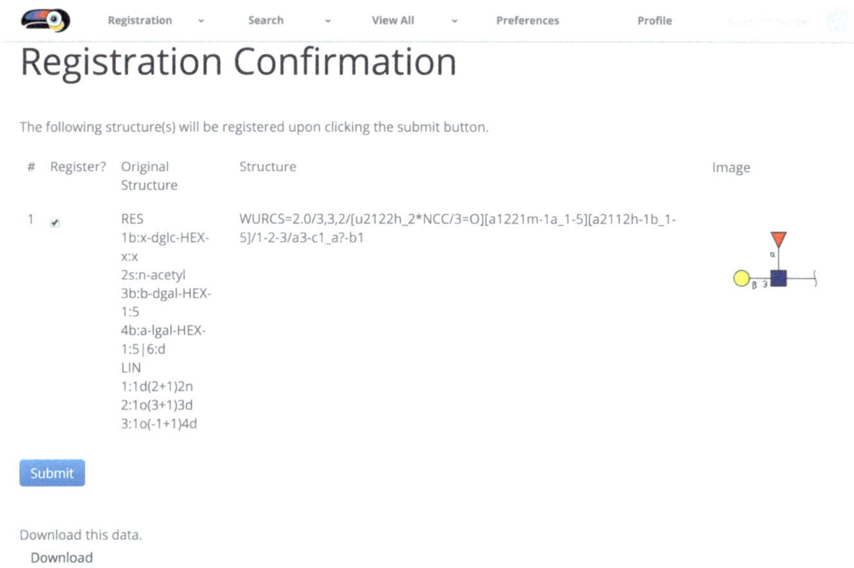

Fig. 4.4 A snapshot of the screen for registration confirmation. The registration confirmation page lists four items: a checkbox to select the glycan structure(s) to register, the "Original Structure" in the format originally provided, the "Structure" in WURCS format, and the "Image" of the glycan in graphical format

1. Glycan structures in GlycoCT sequence format.
2. Blank lines separating glycan structures.
3. One file upload will have a limitation; however, it is recommended to initially upload at most 500 sequences.

In order to use file upload, click the "File Upload" button under the "Registration" menu bar. Click the "Choose File" button to select the glycan sequence file from the local computer. After confirming the file, click the "Submit" button. The same confirmation page as used in the "Graphical Interface" registration process will be displayed.

4.3.4 *Registration Results*

The "Graphical Input," "Text Input," and "File Upload" registration procedures all provide a confirmation page after structure input. Once confirmed, the list of glycan structures with their unique accession numbers is displayed (Fig. 4.7). For each glycan structure, the theoretical mass is computed along with the structure in WURCS format. Such automatically computed data can be confirmed on the glycan entry page containing the detailed information for each glycan structure.

Fig. 4.5 The user interface for Text Input. To register a glycan structure in text format such as GlycoCT, this text form can be used

4.4 Search

4.4.1 Graphic Input

In GlyTouCan, a user can use structure search by drawing glycan structures using GlycanBuilder (Damerell et al. 2012) to find glycans containing the drawn structure. Click the "Graphical Input" under the "Search" menu. The design of the "Graphical Input" interface is the same as the graphical input for registration. Please refer to "Drawing glycan structure by GlycanBuilder" for details on how to use

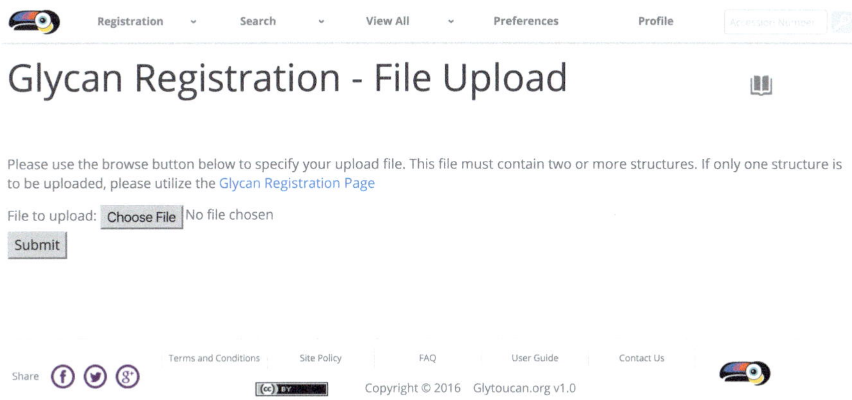

Fig. 4.6 The user interface for File Upload. A text file containing glycan sequences in GlycoCT format can be uploaded with this interface

Fig. 4.7 A snapshot of a successful registration. After the registration of glycan structure, the complete registration page displays the newly added accession number, which links to its entry page

GlycanBuilder. After drawing a glycan structure, click the “Search” button. As a result, all the glycan structures that have the query as a substructure are listed.

4.4.2 *Text Input*

The “Text Input” interface allows users to input a query with glycan text sequence formats. Click the “Text Input” option under the “Search” menu. Currently, GlycoCT, LinearCode™ (Banin et al. 2002), and KCF (Aoki et al. 2004) are supported. After entering a glycan sequence text in the text box, click the “Search” button. As a result, all of the targeted glycan structures that have the query as a substructure will be displayed. When the data is pasted in the text box, please make sure that there is no blank space in the first line.

4.4.3 *By Motif*

Currently, 61 glycan motifs are registered in GlyTouCan. A user can use any of the motifs as a query. Click the “By Motif” menu option under the “Search” menu (Fig. 4.8).

The existing motif structures are displayed in a table. This table has four items: “Name,” “Sequence,” “Reducing end,” and “Frequency.” “Reducing end” indicates whether or not the glycan motif is found at the reducing end. Frequency indicates the number of glycans in GlyTouCan that contain the motif. This motif table is sorted by frequency.

In order to search for glycan structures by motif, click a motif name. For example, by clicking “Sialyl Lewis X,” a list will be displayed in a “Glycan List” view (Fig. 4.9; in this case the hit count is 977 as of Feb 2016). Details regarding how to use the “Glycan List” view are described in the “Glycan List” subsection of the “View” section.

4.4.4 *By Accession Number*

If the user already wants to search for glycans using their accession numbers, a search option is available to perform an exact search by accession number. The search form is available in the top menu bar (Fig. 4.10).

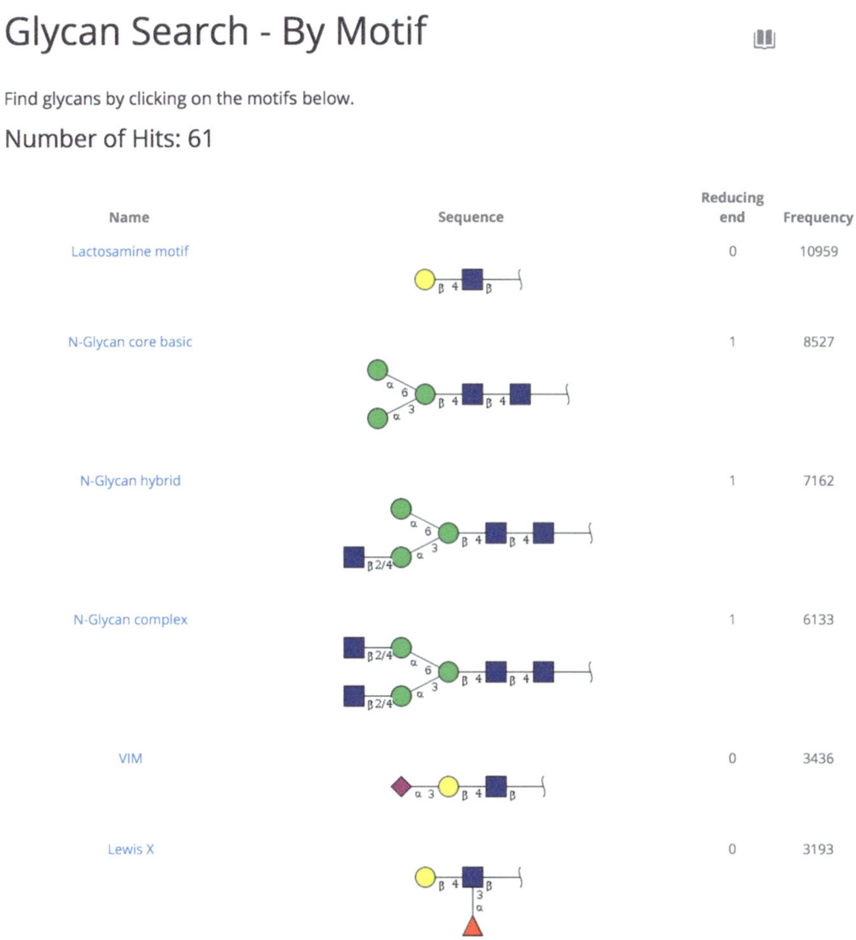

Fig. 4.8 Motif list view for glycan structure search. Currently, 61 motifs are displayed in the "Glycan Search - By Motif" page

4.5 Viewing GlyTouCan Data

4.5.1 Glycan Entry

GlyTouCan provides individual glycan entry pages for each glycan, which displays registration information and related information. Each entry page consists of three sections: "Overview," "Related Data," and "Linked DB." The Overview section (Fig. 4.11) is located at the top of the entry page and lists the registration information about the glycan structure. It displays the accession number, mass value (calculated monoisotopic mass), contributor, contribution time, and descriptor (GlycoCT, WURCS) of the selected glycan structure. The default notation for the

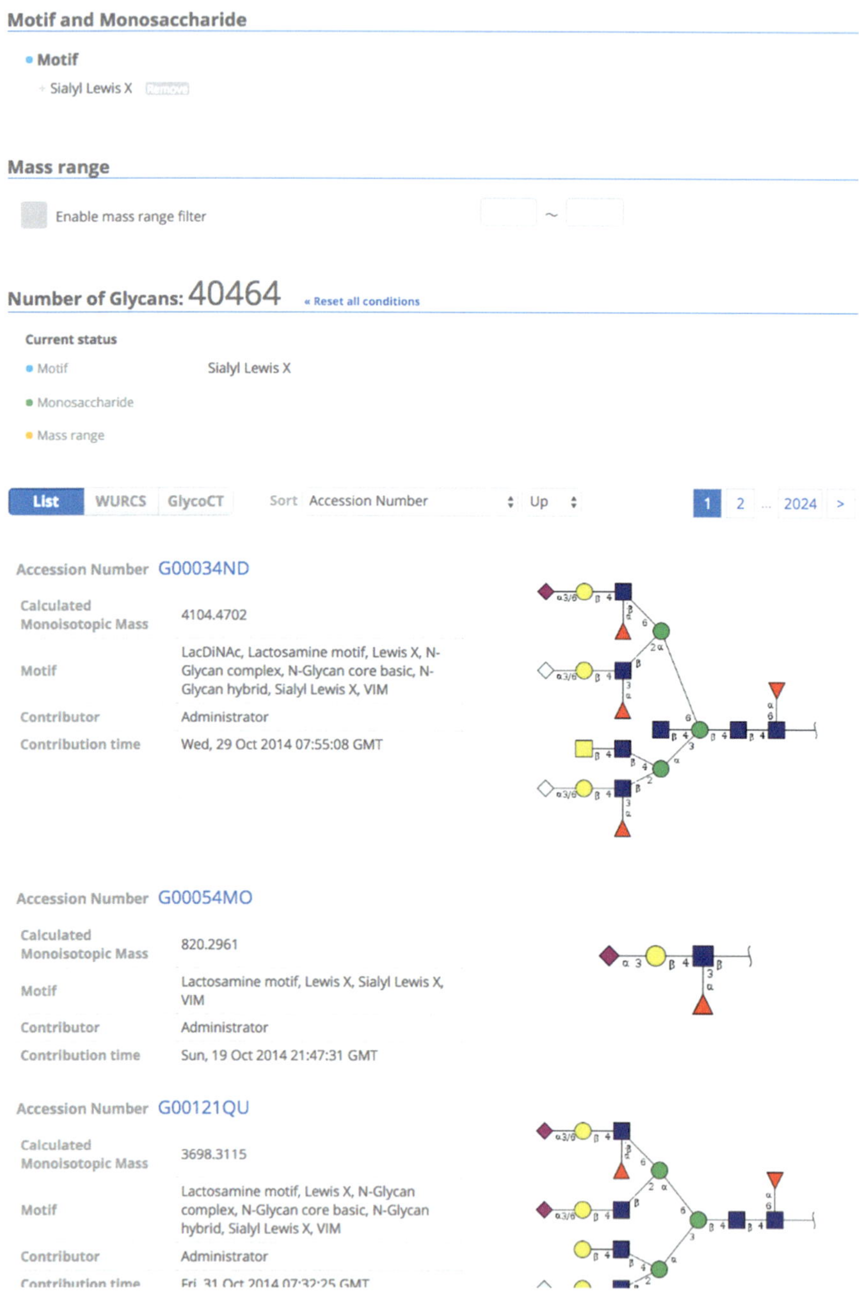

Fig. 4.9 "Glycan List" view of glycan structure search result. The result of a substructure search by motif structure. The user can add conditions to filter these results

Fig. 4.10 Entry form for a search by accession number. This form located in the GlyTouCan menu bar

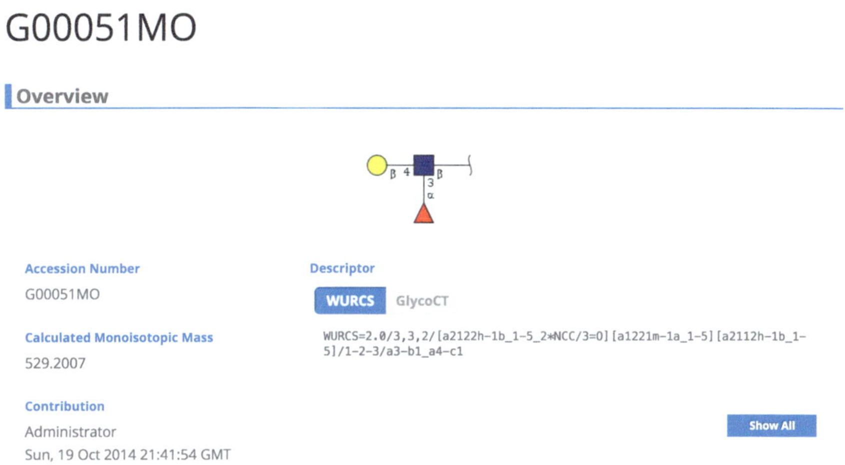

Fig. 4.11 Overview section of a glycan entry page (e.g., G00051MO)

glycan image uses the CFG symbol (Varki et al. 2015). This notation can be changed using the "Preferences" menu.

In the Related Data Section (Fig. 4.12), structural information such as "Motif" and "Monosaccharide Composition" are displayed. Click the "Motif" tab to find the motif structures included in the selected glycan structure. Click the "Monosaccharide Composition" tab to display the monosaccharide(s) and their frequency in the glycan structure.

The "Visualization Tool" displays a viewer that allows the user to browse through the structurally related glycan structures in GlyTouCan. First, click the Visualization Tool tab. The screen will display the relationships between the selected structure and other related structures registered in GlyTouCan, such as its motif structures (Fig. 4.13). When the user hovers the mouse pointer over a motif structure in the screen, the accession number and an enlarged view of the motif structure are displayed. When the user right-clicks on an image of a glycan structure, a drop-down menu will be displayed. This drop-down menu has the following four functions: (1) Copy this glycan ID. (2) Center on this glycan. (3) Open this glycan's entry page. (4) Show this glycan image. As the relationships between glycans in GlyTouCan are more clearly defined, other relationships will be viewable from this tool, such as substructures, superstructures, linkage isomers, and subsumed structures.

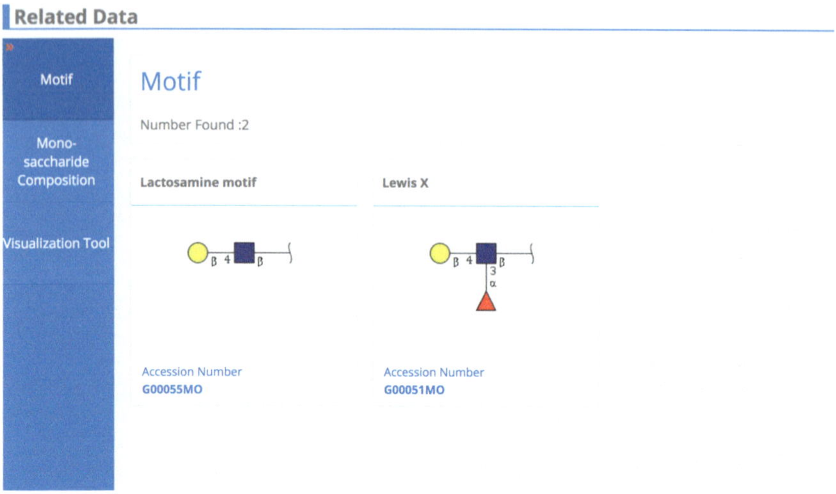

Fig. 4.12 Related Data section containing motif, monosaccharide composition, and visualization tool

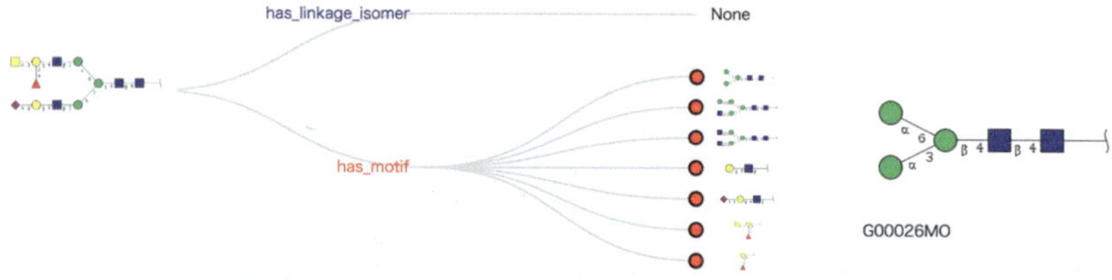

Fig. 4.13 Visualization tool of structurally related glycan structure. The "Visualization Tool" displays the structurally related glycan structures registered in GlyTouCan. In this example, the selected structure (*left*) contains six motifs

4.5.2 Linked Databases

The "Linked DB" section is located at the bottom of each entry page (Fig. 4.14). This section displays IDs and related information found in other glycan-related databases for the selected glycan structure. At the time of this writing, this section includes IDs from glycan-related databases such as BCSDB (Toukach et al. 2016), GlycomeDB (Ranzinger et al. 2009), GlycoEpitope (Kawasaki et al. 2006), and UniCarbDB (Campbell et al. 2014). BCSDB is a database of carbohydrates in plant, fungi, and bacteria. From GlyTouCan, the BCSDB entry displays the taxon and references for the glycan structure, as registered in BCSDB. GlycomeDB is a database of carbohydrate structures and links to other carbohydrate databases. The GlycomeDB entry in GlyTouCan displays IDs and their links in other related databases. GlycoEpitope is a database of carbohydrate epitopes and antibodies that

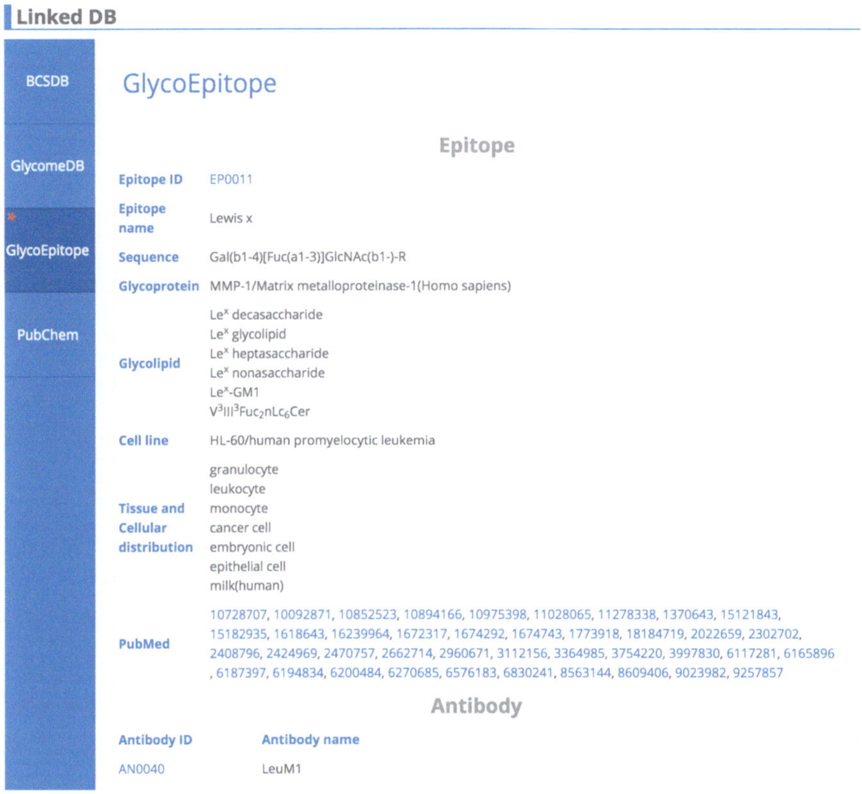

Fig. 4.14 Linked DB section to external databases. The Linked DB section provides meta-information that are stored in other curated databases. In this example, this structure is an epitope that is also registered in the GlycoEpitope database. The BCSDB tab and GlycomeDB tab indicate that this entry is also registered in these other databases

recognize glycans. From GlyTouCan, the GlycoEpitope entry displays the epitope name, sequence, related glycoproteins, related glycolipids, related antibodies, and publication references. UniCarbDB is a database of analytical data mainly focused on mass spectrometry. GlyTouCan provides links to UniCarbDB entries matching those in GlyTouCan.

4.5.3 Motif List

At the time of this writing, GlyTouCan contains 61 motif structures, which can be seen by clicking "Motif List" under the "View All" menu (Fig. 4.15). This list displays items such as "Name," "Sequence," "Reducing end," and "Frequency," which are the same as the list in Motif Search. Detailed information about a

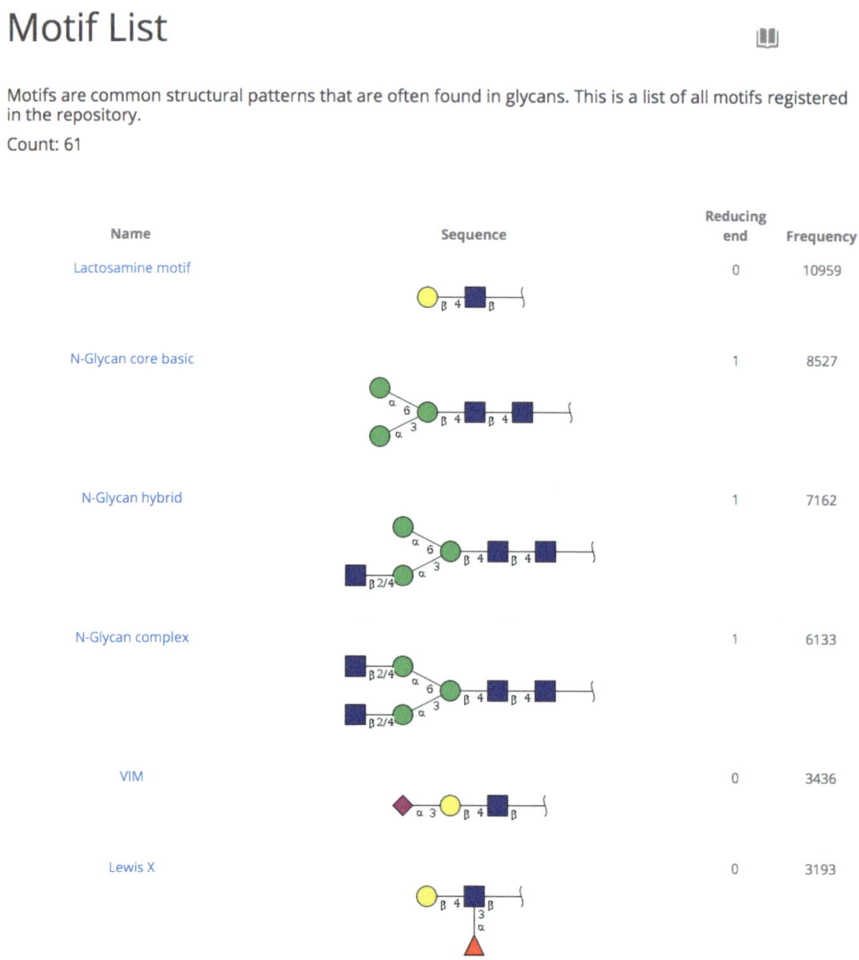

Fig. 4.15 List view of glycan motifs. There are currently 61 motifs with particular properties defined, such as its location found in glycans (reducing end, terminal end, or internal). The frequency of the motifs as found in GlyTouCan is also listed on this page

particular motif structure can be found by clicking the motif name, which will jump to its entry page.

4.5.4 Glycan List

In GlyTouCan, the entire list of registered glycan structures can be seen from the "Glycan List" option under the "View All" menu. From this view, glycans can be filtered to find a particular glycan structure matching specific conditions. Figure 4.16

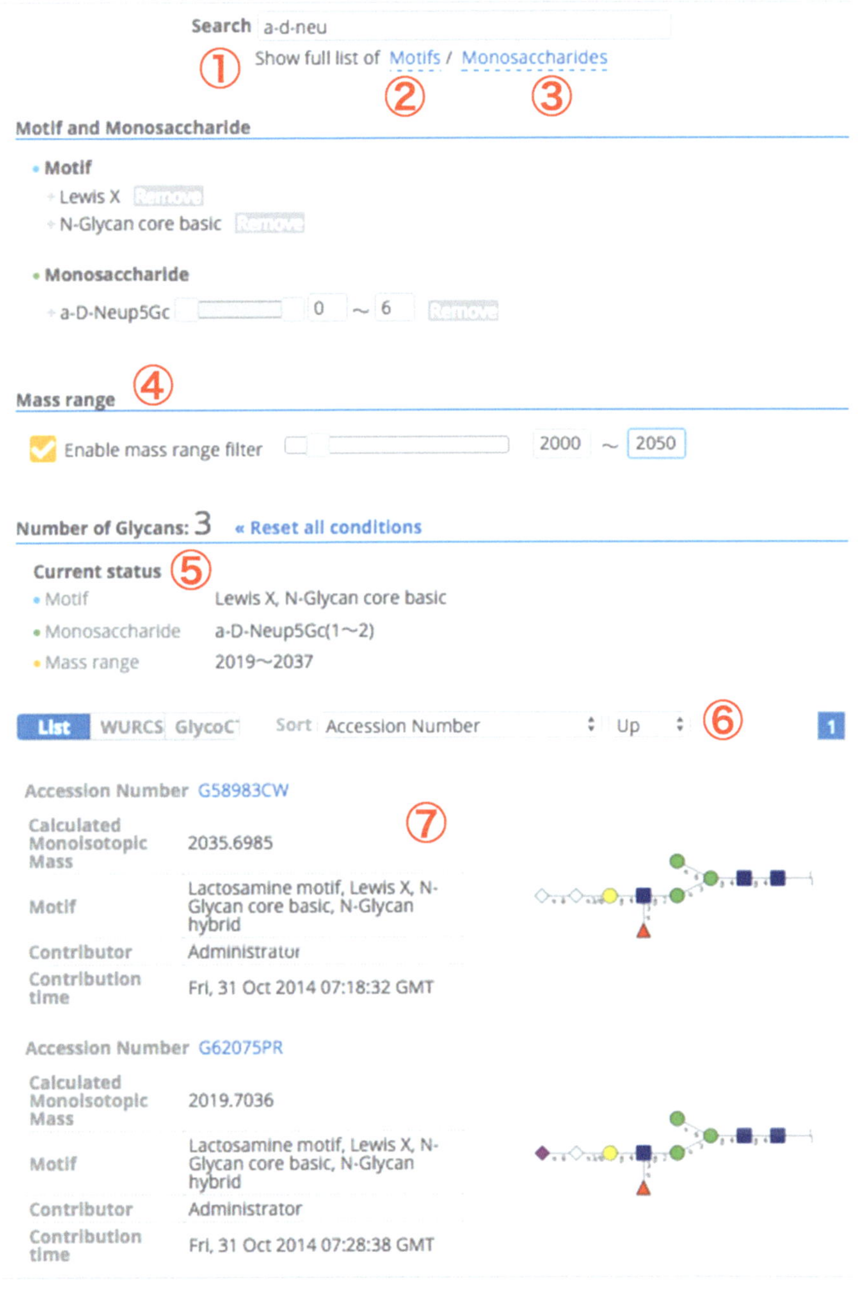

Fig. 4.16 Filtering of glycan structure in a Glycan List. The Glycan List displays the entire list of registered glycan structures. This screen can be used to filter by specific conditions such as selected motifs, monosaccharides, and mass

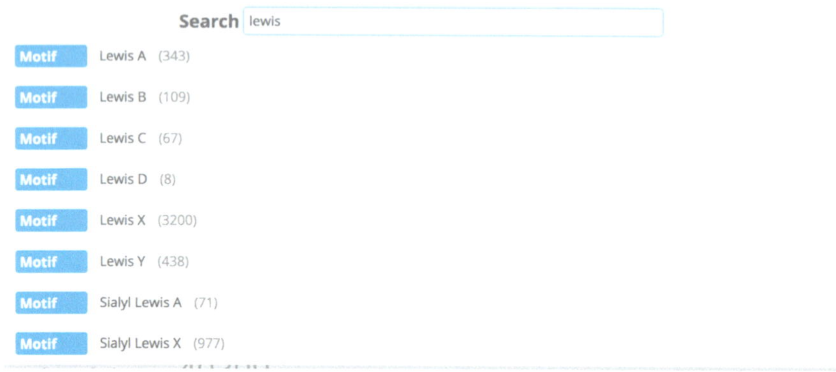

Fig. 4.17 Search process by motif name. When a motif name is typed in the form, a list of candidate names will pop up automatically when more than three characters are entered

is a snapshot of the Glycan List screen. The following describes each numbered section of Fig. 4.16.

1. Search

This search form allows users to search for particular glycans having specific motifs and/or monosaccharides. When a motif or monosaccharide name is typed in the form, a list of candidate names will pop up automatically when three or more characters are entered (see Fig. 4.17). If a candidate name is chosen, the Glycan List is filtered to display only those glycans containing the selected name or monosaccharide. Note that currently, this filter option is limited to only motifs and monosaccharides. Keyword search will be made available in the future.

2. Motifs

When the "Motifs" label is clicked, the motif list window is displayed (Fig. 4.18a). The number to the right of the motif name indicates the corresponding number of glycan structures containing the motif. When a motif name is clicked, the Glycan List is filtered to display only those glycans containing the selected motif.

3. Monosaccharides

When the "Monosaccharides" label is clicked, the monosaccharide list window is displayed (Fig. 4.18b).

The number to the right of the monosaccharide name indicates the number of glycan structures containing the monosaccharide. When a monosaccharide name is clicked, the Glycan List is filtered to list only those glycans containing the selected monosaccharide. In addition, the number (cardinality) of monosaccharides can be specified as a range (e.g., glycans containing between 3 and 5 galactoses).

Fig. 4.18 Windows for glycan structure filtering by motifs and monosaccharides: (**a**) motif list, (**b**) monosaccharide list

4. Mass range

Optionally, a range of values for the calculated mass of glycan structures can be used to filter the list of glycans. In order to use this option, the checkbox to "Enable mass range filter" must be checked.

Note that glycans containing repeating units of unspecified number are excluded when this option is enabled because their mass cannot be calculated.

5. Current status

The "Current status" section lists the filters that have been applied to the Glycan List. Motif names, monosaccharides and their ranges, and mass ranges are listed if they have been selected.

6. Sort

Sorting options to sort the list of glycans by accession number, mass, contributor, and date are available. Moreover, each option can be sorted in ascending or descending order.

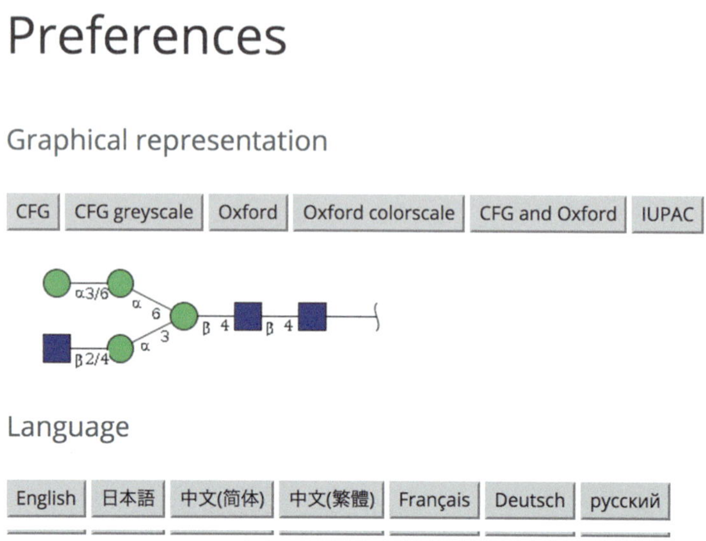

Fig. 4.19 User Preferences for changing the glycan structure representation and language throughout GlyTouCan during the duration of the user's logged-in session

7. Structure entry

Each glycan structure is displayed along with details such as accession number, calculated monoisotopic mass, motif, contributor, and contribution date/time.

4.5.5 User Preferences

The User Preferences screen (Fig. 4.19) is where it is possible to configure the graphical representation of structures and language used for all pages of GlyTouCan. These configurations will only last for the session during which the user is logged in.

4.5.6 Graphic Representation

The default glycan structure representation in GlyTouCan is the CFG nomenclature or the Essentials of Glycobiology nomenclature. The following graphical representation formats are available:

- Consortium for Functional Glycomics (CFG)
- CFG gray scale
- Oxford

- Oxford color scale
- CFG and Oxford
- IUPAC

Once a representation is selected, the image used to display glycan structures will change to the selected format. Once the browser is closed or if the user session times out, this selection will no longer be in effect, and it will have to be selected again.

4.5.7 Language

The language buttons allow the user to select a specific language to be used in GlyTouCan. At the time of this writing, the following are available:

- English
- Japanese
- Chinese
- Chinese (traditional)
- French
- German
- Russian

4.6 Drawing a Glycan Structure by GlycanBuilder

GlycanBuilder is a user-friendly interface for users to search, save, and generate images of glycans.

The following briefly describes how to use the GlycanBuilder interface.

4.6.1 File Menu

This menu allows users to input/output glycan structure data in a variety of formats, including text as well as images, and to restart GlycanBuilder (Fig. 4.20).

4.6.2 View Menu

The user can select a symbol notation to use to represent glycans. Options to display a symbol at the reducing end and mass values are also available (Fig. 4.21).

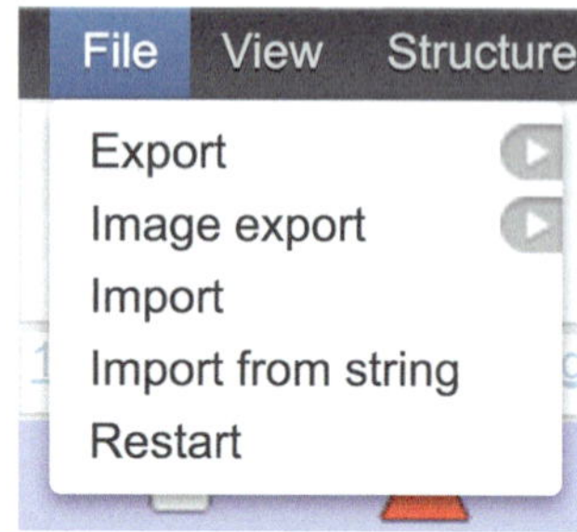

Fig. 4.20 File menu in GlycanBuilder. The file menu includes five options. For example, "Image export" can be used to export images of glycans drawn on GlycanBuilder

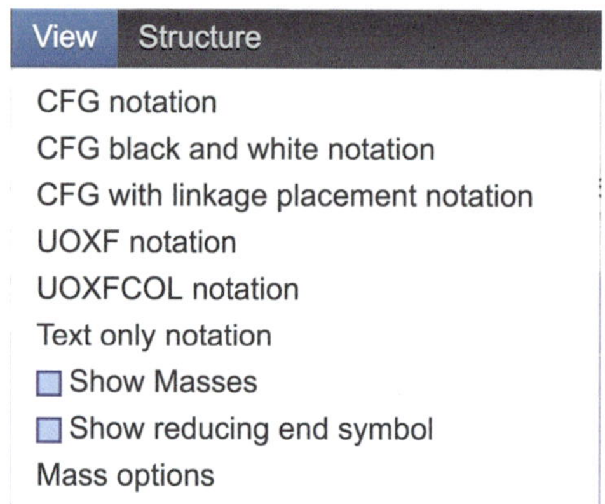

Fig. 4.21 View menu in GlycanBuilder. The view menu provides options to change the symbol notation

4.6.3 Structure Menu

Predefined glycan structures that are commonly used can be drawn on the canvas by the click of the mouse. Available structures are N-glycans, O-glycans, glycosphingolipids, glycosaminoglycans, and milk sugars (Fig. 4.22).

In the row under the menu button, rows are links to specify glycosidic bond configurations.

Clicking on the "1st Linkage" link, a window to specify the linkage configuration is displayed (Fig. 4.23).

The following parameters are available:

Anomeric state: "a" for alpha, "b" for beta, and "?" for unknown anomer.
Anomeric carbon: 1, 2, or 3 for the number of the anomeric carbon or "?" for unknown.
Linkage position: any combination of numbers 1–6 or "?" for unknown can be used.
Chirality: D, L, or ? for unknown.
Ring: "p" for pyranose, "f" for furanose, "o" for open, and "?" for unknown ring.

An example for drawing glycan structure (G45262QY in Fig. 4.24) will be described next. There are two ways to draw this structure in GlycanBuilder. The first is to draw each monosaccharide onto the canvas and to add all glycosidic linkages

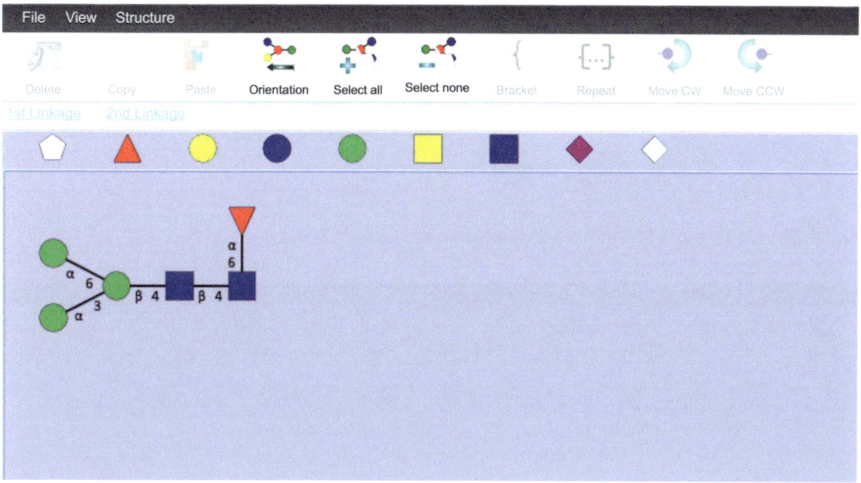

Fig. 4.22 Structure menu in GlycanBuilder. The structure menu includes templates of glycan structures and residues

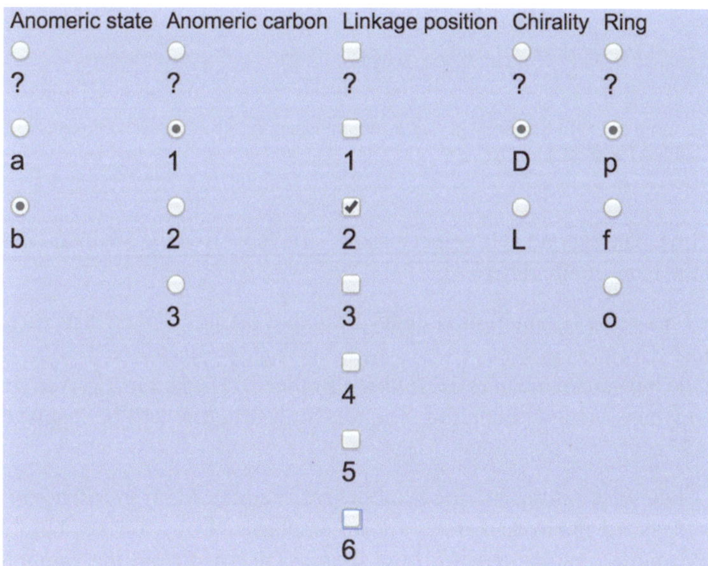

Fig. 4.23 Linkage window in GlycanBuilder. The linkage configuration displays the available parameters such as "Anomeric state," "Anomeric carbon," "Linkage position," "Chirality," and "Ring"

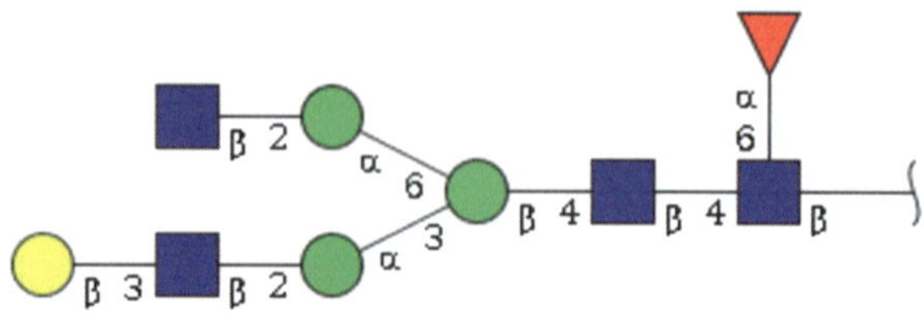

Fig. 4.24 An example for drawing glycan structure (G45262QY)

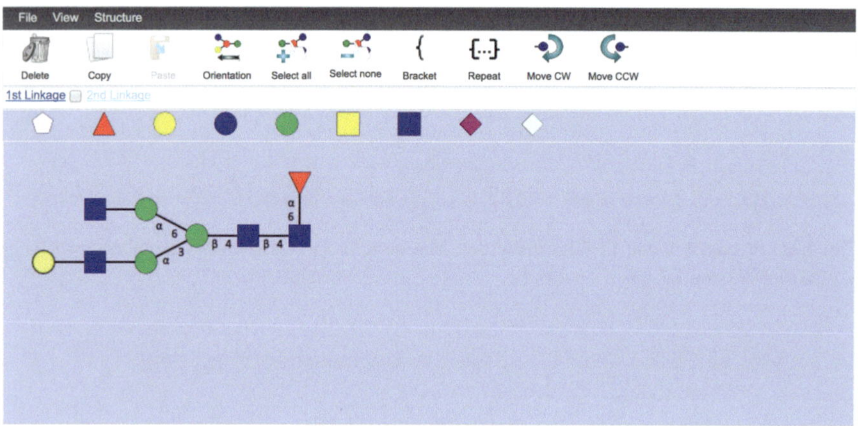

Fig. 4.25 Example of building an *N*-glycan structure using the structure menu option, with linkage positions defined for those known

one at a time. The second is to select a motif structure from the Structure menu. We describe the latter method here.

Click the "ncorefuc" menu option under Structure menu - -> Add structure - -> N-glycans.

Click on the pull-down menu in the following order: (1) Structure, (2) Add structure, (3) N-glycans, and (4) ncorefuc. The resulting structure will be displayed as in Fig. 4.22.

Next, click the monosaccharide symbol (green circle: Man) to extend on the left side. The selected monosaccharide will be emphasized by a thicker border. Then click the monosaccharide symbol (blue square: GlcNAc) from the symbols listed across the top to add onto the selected mannose. GlcNAc will then be attached to the mannose. In the same way, add the other monosaccharides by repeatedly selecting monosaccharides on the glycan and clicking the monosaccharide to add from the top row.

After the monosaccharides have all been attached, the structure should look like Fig. 4.25. Next, linkage information can be specified.

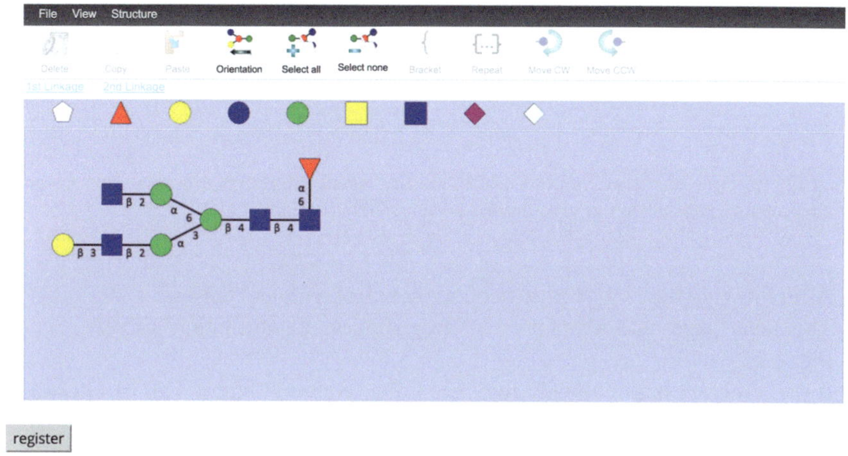

Fig. 4.26 The result of the *N*-glycan structure (G45262QY) drawn by GlycanBuilder

First, select one of the GlcNAcs on the nonreducing end and then click the "1st Linkage" link.

Select the following from the linkage configuration window (see Fig. 4.23):

Anomeric state: "b"
Anomeric carbon: "1"
Linkage position: "2"
Chirality: "D"
Ring: "p"

Repeat the procedure for each linkage by selecting the appropriate parameters in the linkage configuration window. The result of the glycan structure should look like Fig. 4.26.

4.7 Use Cases

No 1: How to assign an accession number to an N-glycan structure using GlycanBuilder

This use case describes the registration process of an N-glycan structure using GlycanBuilder.

1. Log in to GlyTouCan via the "Sign In" button.
2. Click the "Graphic input" under the "Registration" menu.
3. Draw the N-glycan structure on the canvas. Please refer to "Drawing glycan structure by GlycanBuilder" for how to use GlycanBuilder (Sect. 4.6).

#	Original Glycan Sequence	Sequence	Image	Accession number
1	RES 1b:b-dglc-HEX-1:5 2s:n-acetyl	WURCS=2.0/5,9,8/[a2122h-1b_1-5_2*NCC/3=O][a1122h-1b_1-5][a1122h-1a_1-5][a2112h-1b_1-5][a1221m-1a_1-5]/1-1-2-3-1-4-3-1-5/a4-b1_a6-i1_b4-c1_c3-d1_c6-g1_d2-e1_e3-f1_g2-h1		G45262QY

Fig. 4.27 Example of registration process for already registered glycan structure. This screen is the confirmation page when the structure has been registered already

4. After drawing the N-glycan structure, click the "register" button.
5. The next page will display the "Registration Confirmation" screen, as in Fig. 4.27.
6. If the structure is not already registered, the "Submit" button can be clicked to register the structure and obtain an accession number (Fig. 4.7).

No 2: How to assign an accession number to an N-glycan structure specified by text format

GlyTouCan provides a mechanism to register glycans in text format. Note that at the time of this writing, only glycans in GlycoCT can be registered. For this example, the GlycoCT shown in Fig. 4.28 will be used (G45262QY).

First, click the "Text Input" button under the "Registration" menu.
Input the GlycoCT text into the text form (Fig. 4.29) and click the "Submit" button.

The next page displays the Registration Confirmation screen, containing the original glycan sequence, the sequence in WURCS format, and the image of the sequence. If the sequence is already registered, the accession number will be displayed (Fig. 4.27). On the other hand, if the sequence is unregistered, the final confirmation screen will be shown (Fig. 4.4).

No 3: How to search for a registered N-glycan structure using GlycanBuilder

GlycanBuilder can be used to search for registered N-glycan structures. Details about how to use GlycanBuilder to draw structures are described in "Drawing glycan structure by GlycanBuilder" in Sect. 4.6.

Note that this search method uses substructure search such that all structures containing the query structure are returned in the results.

In this example, the N-glycan core structure will be used as a query. Click "ncore" under the structure menu (Structure → Add structure → N-glycans). The N-glycan core structure will be drawn on the canvas (Fig. 4.30).

The N-glycan core structure can be modified at this point. To use the drawn structure as a query, click the search button under the GlycanBuilder.

All registered glycan structure containing the query as a substructure will be retrieved by clicking the search button. The results will display the query structure and search results in a list (Fig. 4.31). Note that there is currently no scoring system for searches, so the order of the results is not necessarily by "similarity."

```
RES
1b:b-dglc-HEX-1:5
2s:n-acetyl
3b:b-dglc-HEX-1:5
4s:n-acetyl
5b:b-dman-HEX-1:5
6b:a-dman-HEX-1:5
7b:b-dglc-HEX-1:5
8s:n-acetyl
9b:b-dgal-HEX-1:5
10b:a-dman-HEX-1:5
11b:b-dglc-HEX-1:5
12s:n-acetyl
13b:a-lgal-HEX-1:5|6:d
LIN
1:1d(2+1)2n
2:1o(4+1)3d
3:3d(2+1)4n
4:3o(4+1)5d
5:5o(3+1)6d
6:6o(2+1)7d
7:7d(2+1)8n
8:7o(3+1)9d
```

Fig. 4.28 An example of a structure in GlycoCT format

No 4: How to search for glycan structure represented by text format

GlyTouCan provides a method to search for glycan structure by text format, using substructure search. The GlycoCT sequence of G08520NM is used as an example.

To obtain the GlycoCT sequence for an existing structure, enter the accession number (G08520NM in this example) in the search form in the menu bar. The screen displays the entry page of G08520NM and the GlycoCT can be obtained in the Overview section. Select "GlycoCT" and "Show" to display the entire GlycoCT sequence. Then the GlycoCT text can be copied as shown in Fig. 4.32.

To perform a search by text, click the "Text input" menu under the "Search" menu bar. Enter the text in the text form and click the search button (Fig. 4.33). All glycans having G08520NM as a substructure will be retrieved. Similar to Use Case No. 3, the query structure and search results of structures that contain G08520NM will be displayed as a list (Fig. 4.34).

No 5: How to search *N*-glycans that include the "Lewis X motif" and "a-D-Neup5Gc" and have a range of mass between 2000 and 2050, using the Glycan List

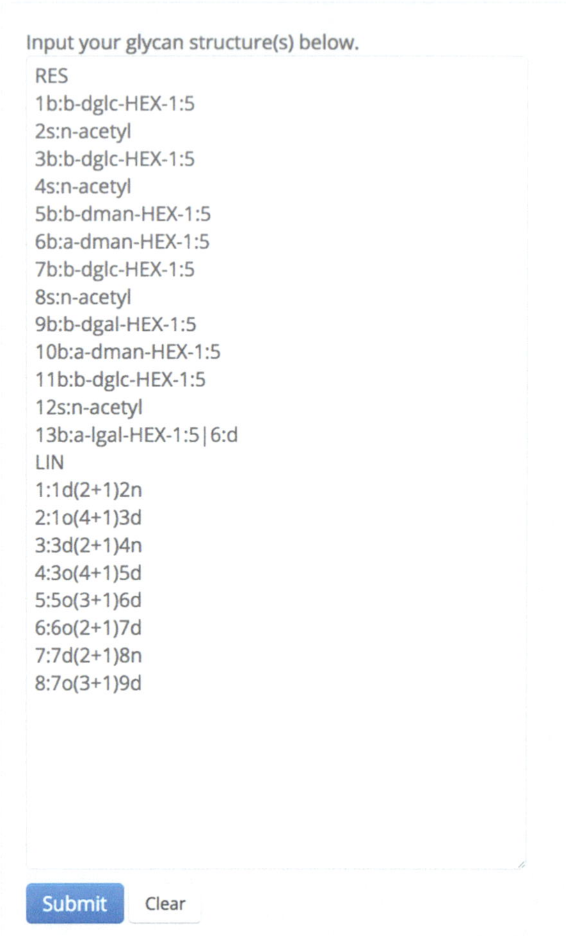
Input your glycan structure(s) below.

```
RES
1b:b-dglc-HEX-1:5
2s:n-acetyl
3b:b-dglc-HEX-1:5
4s:n-acetyl
5b:b-dman-HEX-1:5
6b:a-dman-HEX-1:5
7b:b-dglc-HEX-1:5
8s:n-acetyl
9b:b-dgal-HEX-1:5
10b:a-dman-HEX-1:5
11b:b-dglc-HEX-1:5
12s:n-acetyl
13b:a-lgal-HEX-1:5|6:d
LIN
1:1d(2+1)2n
2:1o(4+1)3d
3:3d(2+1)4n
4:3o(4+1)5d
5:5o(3+1)6d
6:6o(2+1)7d
7:7d(2+1)8n
8:7o(3+1)9d
```

Submit Clear

Fig. 4.29 A snapshot of the text form after GlycoCT text has been pasted into the text form

In this example, the following four conditions will be used to filter glycan structures in GlyTouCan:

1. Contains Lewis X
2. Is an *N*-glycan
3. Contains sialic acid "a-D-Neup5Gc"
4. Range of Mass: 2000–2050

First, click the "Glycan List" under the "View All" menu bar ("View All" --> "Glycan List").

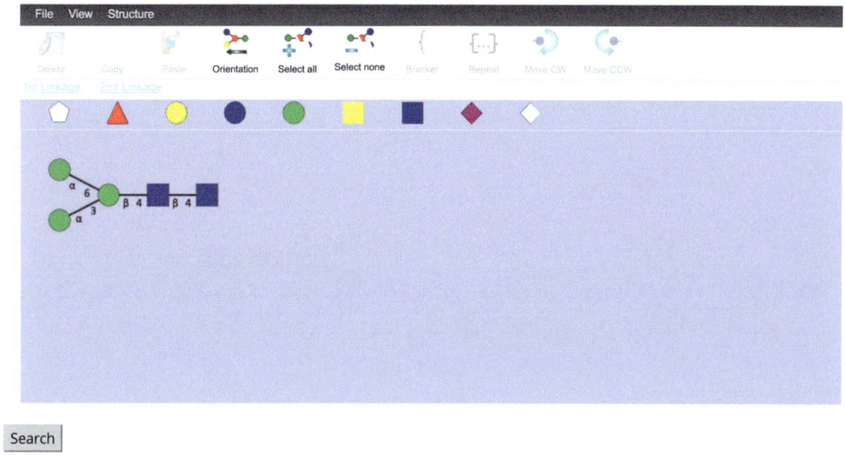

Fig. 4.30 Snapshot of GlycanBuilder with a glycan structure (*N*-glycan core) drawn

Click the "Motifs" button under the search form to display the list of motifs.
Click "Lewis X" and "N-Glycan core basic" in the motif list.
This will first retrieve glycan structures that are N-glycans and contain the Lewis X motif.
It will be displayed in the Glycan List.

Enter "a-D-Neup5Gc" in the search form. This will display some candidate names as a list.
Click "a-D-Neup5Gc" in the candidate list (Fig. 4.35).

Click the checkbox for "Enable mass range filter" and enter the mass range: 2000–2050.

As a result, the Glycan List will display N-glycan structures that include the Lewis X motif and a-D-Neup5Gc and have a mass in the range 2000–2050. The "Sort" option can be used to sort the structures in ascending or descending order of "Calculated Monoisotopic Mass."

No 6: How to obtain an image file of a structure drawn using GlycanBuilder.

GlycanBuilder provides a mechanism to export image files.

After drawing a structure on the canvas (see Sect. 4.6 for details), click "Image export" in the file menu (Fig. 4.36).

Supported formats for images are BMP, EPS, JPG, PDF, PNG, PS, and SVG. After selecting a format, the selected image data can be stored on the local drive.

Fig. 4.31 Example of a result of searching a structure containing the *N*-glycan core structure. The result page displays the retrieved structures containing the *N*-glycan core structure

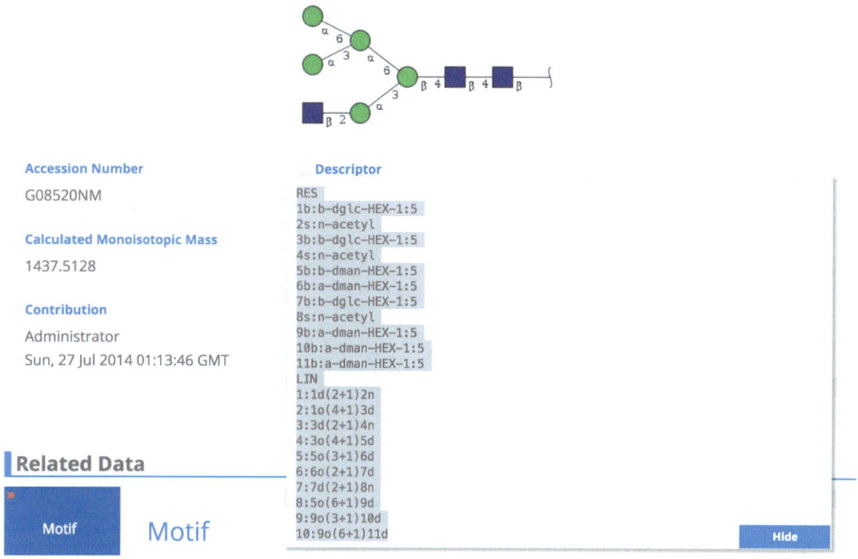

Fig. 4.32 Example to get a GlycoCT sequence from the glycan entry page. The GlycoCT sequence of G08520NM is available in the entry page

4.8 Troubleshooting Tips

Q: How to download a glycan structure image from the GlyTouCan entry page?
A: Right-click the glycan structure image. The web browser should provide an option to "Save as" a file onto the computer.

Q: I cannot seem to export an image file of a glycan drawn using GlycanBuilder.
A: Users may need to give permission to the web browser to allow files to be stored on the local computer.

Q: I am having problems logging in.
A: Ensure that your email address is valid. Otherwise, contact the support staff at support@glytoucan.org.

Q: I cannot see the structure image in the glycan entry view.
A: Some structures cannot be drawn due to limitations in the graphical output program. In most cases, this may be because the structure contains monosaccharides that are not assigned any graphical notation.

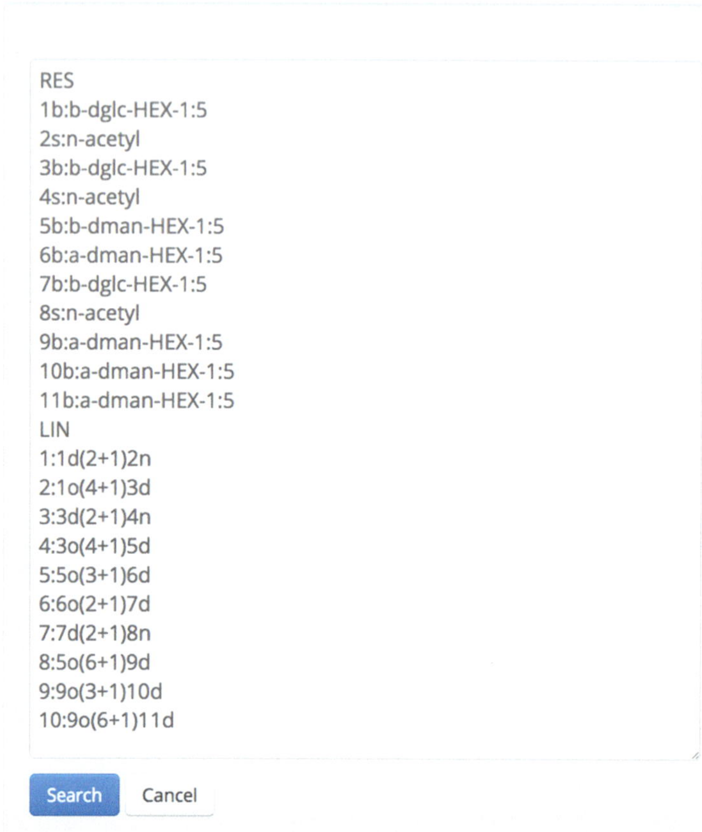

Fig. 4.33 Enter the GlycoCT sequence of G08520NM in the text form

Fig. 4.34 Structure search using the text format of the glycan structure (G08520NM). The result page displays the retrieved structures having G08520NM as a substructure

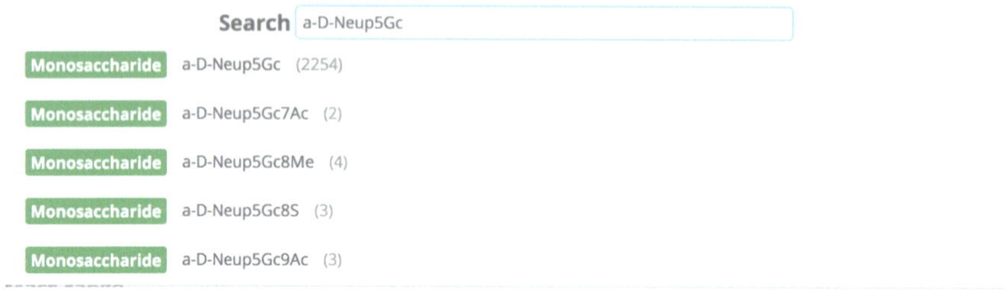

Fig. 4.35 Examples of the monosaccharide candidate list while inputting text into the search form. The candidate list of "a-D-Neup5Gc." The number 2254 indicates the number of registered glycans that contain "a-D-Neup5Gc" as a component

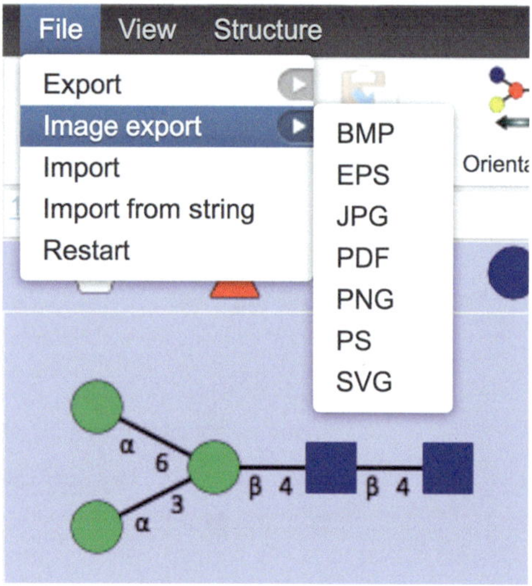

Fig. 4.36 Save an image file using GlycanBuilder. Formats such as BMP, EPS, JPG, PDF, PNG, PS, and SVG are supported by GlycanBuilder

References

Aoki K et al (2004) KCaM (KEGG Carbohydrate Matcher): a software tool for analyzing the structures of carbohydrate sugar chains. Nucleic Acids Res 32:W267–W272. doi:10.1093/nar/gkh473

Aoki-Kinoshita K et al (2016) GlyTouCan 1.0 – the international glycan structure repository. Nucleic Acids Res 44(Database issue):D1–D6. doi:10.1093/nar/gkv1041

Banin E et al (2002) A novel linear code(r) nomenclature for complex carbohydrates. Trends Glycosci Glycotechnol 14:127–137. doi:10.4052/tigg.14.127

Campbell MP et al (2014) Validation of the curation pipeline of UniCarb-DB: building a global glycan reference MS/MS repository. Biochim Biophys Acta 1844(1 Pt A):108–116

Damerell D et al (2012) The GlycanBuilder and GlycoWorkbench glycoinformatics tools: updates and new developments. Biol Chem 393(11):1357–1362. doi:10.1515/hsz-2012-0135

Herget S et al (2008) GlycoCT – a unifying sequence format for carbohydrates. Carbohydr Res 343(12):2162–2171. doi:10.1016/j.carres.2008.03.011

Kawasaki T et al (2006) GlycoEpitope: the integrated database of carbohydrate antigens and antibodies. Trends Glycosci Glycotechnol 18:267–272. doi:10.4052/tigg.18.267

Ranzinger R et al (2009) Glycome-DB.org: a portal for querying across the digital world of carbohydrate sequences. Glycobiology 19(12):1563–1567. doi:10.1093/glycob/cwp137

Tanaka K et al (2014) WURCS: the Web3 unique representation of carbohydrate structures. J Chem Inf Model 54(6):1558–1566. doi:10.1021/ci400571e

Toukach PV et al (2016) Carbohydrate structure database merged from bacterial, archaeal, plant and fungal parts. Nucleic Acids Res 44(D1):D1229–D1236. doi:10.1093/nar/gkv840

Varki A et al (2015) Symbol nomenclature for graphical representations of glycans. Glycobiology 25(12):1323–1324. doi:10.1093/glycob/cwv091

Chapter 5
Carbohydrate Structure Database (CSDB): Examples of Usage

Ksenia S. Egorova and Philip V. Toukach

Abstract The main goals of glycoscience are elucidation of carbohydrate features responsible for cellular processes, pathogenicity of microorganisms, and immunological properties of higher organisms, as well as application of glycans as diagnostic and therapeutic agents and classification of natural glycans and glycoconjugates. These goals are hardly achievable without freely available, regularly updated, and cross-linked databases, which provide data accumulated in glycoscience and allow tracking of their quality.

The Carbohydrate Structure Database is a curated data repository developed for provision of structural, bibliographic, taxonomic, NMR spectroscopic, and other related information on published carbohydrates and derivatives. Currently it covers ca. 90 % of published primary structures of bacterial and archaeal origin and ca. 30 % of published primary structures of plant and fungal origin. The data in bacterial part of CSDB are regularly updated. The expansion of plant and fungal coverage is expected in the future. The project aims at coverage close to complete in selected taxonomic domains and at high data quality achieved by manual literature analysis, annotation, verification, and data approval. CSDB is freely available on the Internet as a web service at http://csdb.glycoscience.ru.

This chapter presents a step-by-step guide to use CSDB for solving everyday tasks typical for carbohydrate research.

Keywords CSDB • Carbohydrate Structure Database • Bacterial • Plant • Fungal • Tutorial • Model problems • User manual • Search • Statistics • NMR simulation

---

K.S. Egorova (✉) • P.V. Toukach (✉)
N.D. Zelinsky Institute of Organic Chemistry, Russian Academy of Sciences, Leninskiy prospect 47, 119991 Moscow, Russia
e-mail: egorova-ks@ioc.ac.ru; netbox@toukach.ru

K.F. Aoki-Kinoshita (ed.), *A Practical Guide to Using Glycomics Databases*,
DOI 10.1007/978-4-431-56454-6_5

5.1 Introduction

Carbohydrate Structure Databases (CSDB) provide structural, bibliographic, taxonomic, NMR spectroscopic, and other related information on natural carbohydrate structures assigned to all taxonomic domains except animals (including humans). For animals, especially mammals, other glycan databases exist (Lütteke 2012). CSDB main features are:

1. Coverage above 80 % (for bacterial and archaeal carbohydrates published up to now; for plant and fungal carbohydrates full coverage is a question of future efforts; about 600 structures from 400 publications are added every year, the time lag between the date of publication and date of deposition to the database being 6–18 months)
2. High data consistency (above 90 % of error-free records, according to automated and manual expert estimation)
3. Presence of manually verified bibliographic, NMR spectroscopic, and taxonomic annotations

CSDB includes glycans and glycoconjugates found in prokaryotes, plants, and fungi, notably, re-annotated structures from CarbBank (Doubet et al. 1989) associated with organisms belonging to these domains plus structures manually extracted from the publications indexed in the Web of Science (https://apps.webofknowledge.com) and NCBI PubMed (http://www.ncbi.nlm.nih.gov/pubmed) with a carbohydrate-related term and taxon name from any of the mentioned domains in the keywords, title, or abstract. In this project, a "carbohydrate" means a structure composed of any residues linked via glycosidic, ester, amide, phospho- or sulphodiester, and other bonds, in which at least one residue is a sugar or its derivative. In contrast to other carbohydrate databases, in CSDB, a monovalent substituent attached with the removal of H_2O or other hydride is considered as a separate residue. The historical exception is amino groups in amino sugars. For example, N-acetylglucosamine is considered as two residues, GlcN and Ac, linked via the 1–2 bond, not as a single residue GlcNAc. This approach significantly reduces the number of monomers required.

CSDB is freely available on the Internet as a web service at http://csdb.glycoscience.ru/. As of 2016, CSDB covers ca. 17700 structures from 8100 organisms presented in 6700 publications and 7800 NMR spectra. Domain-restricted parts of CSDB are also available as separate databases: bacterial CSDB (BCSDB, http://csdb.glycoscience.ru/bacterial, ca. 12300 structures) for bacterial and archaeal data only and plant and fungal CSDB (PFCSDB, http://csdb.glycoscience.ru/plant_fungal, ca. 6100 structures) for plant and fungal data only.

CSDB has a web front end, with user operations listed in the main menu of the project. The menu consists of several subtitles. "Search" provides various modes of data search using record, compound, publication or organism IDs, structural fragments, composition, taxonomical information, bibliography, keywords, or NMR data. "Help" provides documentation, examples of usage, structure encoding rules, and other supplementary information. "Extras" provide additional operations, such as data submission, structure translation between different formats, NMR tools, and various statistical instruments. "Maintenance" is for the CSDB staff.

For more details on CSDB, including its principles, comparison with other projects, and details on the database architecture, please refer to dedicated publications (Toukach 2011; Toukach and Egorova 2015).

To illustrate the daily scientific usage of CSDB, we present several examples of solving model problems. These simple tasks do not claim for real scientific significance but are designed to demonstrate various CSDB features. The queries are performed in the merged CSDB; however, they can be performed from the bacterial database or from the plant and fungal database in accordance to the query content. Please note that the number of the found records, as well as nonpersistent IDs, may differ from those stated in this chapter, since the database is continuously updated. The first example is provided with maximum details, whereas the others imply the previous examples have been studied.

5.2 Search Tools

CSDB provides several search modes using IDs, structural fragments and annotations, monomeric composition and taxonomic, bibliographic, or NMR spectroscopic data. In the following six examples, we demonstrate how CSDB search features can be used to solve problems that a glycoscientist may face in daily practice.

5.2.1 Example 1. *Study How Introduction of an Amino Group Affects NMR Chemical Shifts of the Lactose Fragment*

For this purpose, we will search for all records containing the disaccharide 4-O-β-D-galactopyranosyl-2-amino-β-D-glucopyranose (both N-acetylated and non-N-acetylated) OR 4-O-β-D-galactopyranosyl-β-D-glucopyranose for which NMR data are deposited in CSDB.

To perform a search, first of all we should select the search type in the main menu of CSDB. The problem specified requires the (sub)structure search, the form for which is shown in Fig. 5.1. Here we fill in search terms (3, 4), select scope (6) and additional parameters (7), and run the query (8). The structural fragment being searched is previewed (2) in the graphic form according to the Essentials of Glycobiology v. 3 standard: (Symbol Nomenclature For Glycans, SNFG) (Varki et al. 2015), which extends the format known as "CFG notation." If you edit the search term manually and your query cannot be parsed, area (2) displays the parsing errors.

Every search form except "ID search" has a selector identifying the search scope (6). This feature allows refining the queries by their intersection or combination with

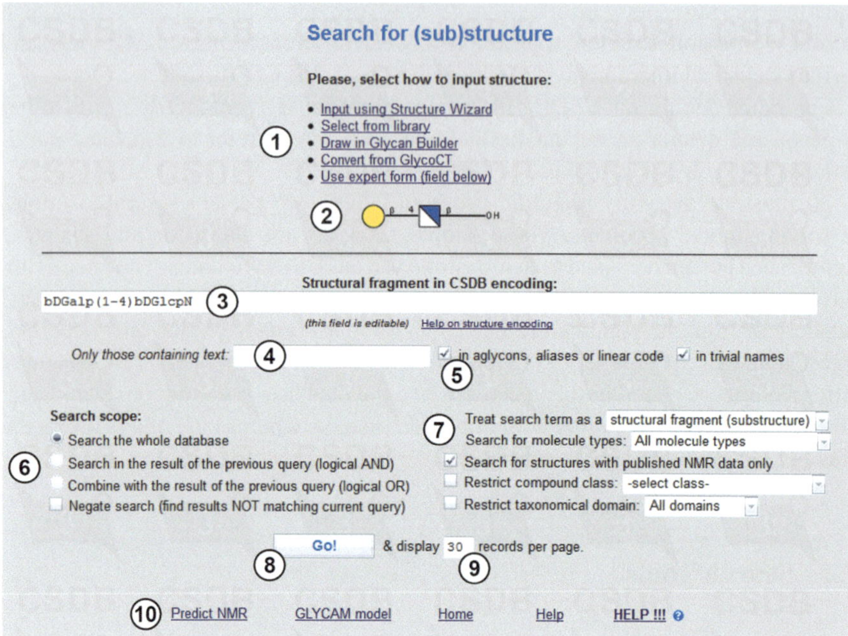

Fig. 5.1 "Search for (sub)structure" page: the query contains the structural fragment 4-O-β-D-galactopyranosyl-2-amino-β-D-glucopyranose (2), (3)

queries of the same or different type. "Search in the result of the previous query" intersects the current and previous queries (logical AND), whereas "Combine with the result of the previous query" unites the current and previous queries (logical OR). "Negate search" negates the current query (logical NOT) and can be used both to get all the results present in the database except those matching the current search criteria and to exclude the results matching the current search criteria from the results of the previous query (when the check box "Search in the result of the previous query" is checked).

The number of found records displayed per page can be specified (9). Links to additional CSDB tools (10) are also available.

To query the database for structures, we should enter a structure of interest into the search field (3). CSDB allows several ways to do that (1):

- "Structure wizard" can be used for visual construction of a structure; it does not require special knowledge except the general nomenclature of carbohydrates, but it has limitations: some specific search queries cannot be constructed via the wizard.
- "Select from library." Here we can select a widespread structure by its trivial name or abbreviation and preview it in the graphical or pseudographical format.
- "Draw in Glycan Builder" allows building and displaying glycan structures in a graphic form (Damerell et al. 2012) in SNFG and other formats.

– "Convert from GlycoCT." Here we can enter a structure in the GlycoCT condensed format (Herget et al. 2008) to convert it to the CSDB linear encoding.
– "Copy from the previous structural query." This option is available if there has already been a structural query within the browser session. It copies the previous structural query to the search term field, so we do not have to reenter the whole structure to make minor changes.
– "Use expert form" allows typing the structure into the search term field manually and demands knowledge of the CSDB structure encoding rules (CSDB linear code, http://csdb.glycoscience.ru/bacterial/index.html?help=rules, Toukach and Egorova 2015).

The disaccharides chosen for this example are simple and can be created without using the CSDB linear code. For exemplary purposes, we will use Glycan Builder to build 4-O-β-D-galactopyranosyl-2-amino-β-D-glucopyranose.

Follow the link "Draw in Glycan Builder" (1) (Fig. 5.1), and the Glycan Builder will open in a new window (Fig. 5.2). This plug-in requires the Java environment (http://java.com) installed on your computer. The menu "View" allows choosing the view mode of the fragment being drawn: the Consortium for Functional Glycomics (CFG) notation (http://www.functionalglycomics.org/), the Oxford (UOXF) notation (Royle et al. 2006), or text only. You can link residues by selecting them from the menu; anomeric configuration (beta, alpha, or unknown) (1), chirality (D, L, or unknown) (3), ring size (pyranose, furanose, open, or unknown) (4), and the linkage between the residues (carbon atom numbers of the donor and the acceptor) (2) can be specified. The resulting structure is shown in field (5). Pressing link (6) returns

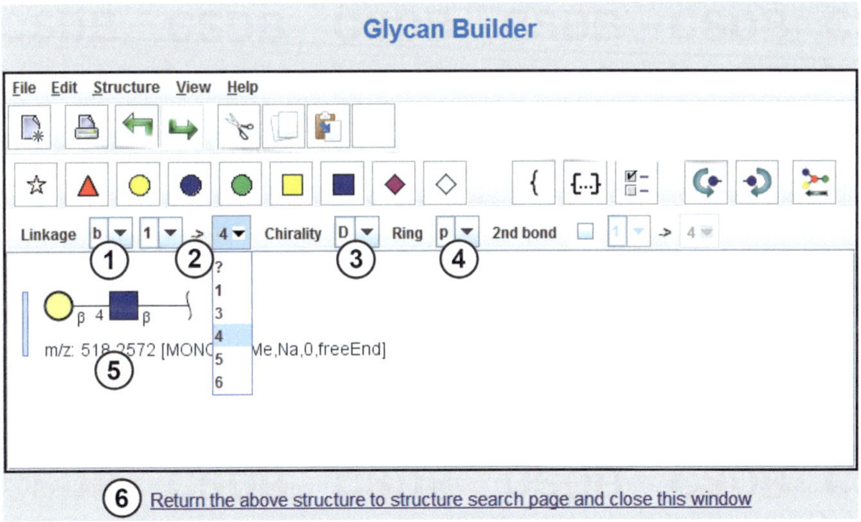

Fig. 5.2 Drawing a structure by using Glycan Builder. The CFG notation mode is used. See explanation in the text

the structure to the "(Sub)structure search" page and closes the Glycan Builder. More information on the Glycan Builder has been published elsewhere (Damerell et al. 2012).

The menu contains no GlcN residue; therefore, we will use a widespread GlcNAc residue (which becomes two residues in the CSDB notation, GlcN and Ac), and the returned structure will be 4-O-β-D-galactopyranosyl-2-deoxy-2-acetamido-β-D-glucopyranose written in the CSDB encoding: bDGalp(1–4)[Ac(1–2)]bDGlcpN. To get the target structure, we must "deacetylate" the N-acetylglucosamine residue, i.e., manually delete the [Ac(1–2)] part from the structure field. The page with the resulting structure (3) and its visualization in the SNFG format (2) is shown in Fig. 5.1. This query will find structures with non-acetylated, as well as with acetylated amino groups.

We will conduct the search with the following restrictions: the search will include molecules of all types (monomers, oligomers, repeating units, cyclic compounds, etc.) (7) and will be conducted through the whole database (6), with no additional restrictions on text content (4) present in aglycons, aliases, linear code or trivial names (5), structure completeness, or compound class (7). Since we are interested only in the records containing the NMR data, the "Search for structures with published NMR data only" checkbox should be checked (7).

Pressing the "Go!" button (8) starts the search.

The search results in 44 structures containing the target fragment. Now we will extend our query by searching for all the structures containing the 4-O-β-D-galactopyranosyl-β-D-glucopyranose disaccharide. By pressing the "New query" link (at the bottom of the result page) or through the main menu, we return to the "Search for (sub)structure" form. Since 4-O-β-D-galactopyranosyl-β-D-glucopyranose is widespread and has the trivial name "lactose," we can select it from the structure library by clicking "Select from library," which opens a new window (Fig. 5.3a). Here we choose "lactose" (2) from "Named saccharides" (1), and the corresponding structure is previewed in the SweetDB pseudographic (Loss et al. 2002) (3) and in the SNFG graphic (Varki et al. 2015) (4) forms. Clicking on link (5) returns the structure to the structure search page and closes the library.

The query form with the lactose disaccharide, bDGalp(1-4)bDGlcp (7), is shown in Fig. 5.3b. Note that we could also copy and manually edit the structure from the previous structural query (6).

Now we are ready for the final step – combining the results of both queries. By choosing "Combine with the result of the previous query" (8) (the number of these results is shown on the page together with the link to their ID list (9)), we define the search scope as the database records matching the current structural fragment, 4-O-β-D-galactopyranosyl-β-D-glucopyranose, plus the records matching the previous one, 4-O-β-D-galactopyranosyl-2-amino-β-D-glucopyranose, with published NMR data only (10). Pressing (11) starts the search. The combined result includes 109 structures containing any of the target disaccharides for which NMR data are present in the database (Fig. 5.4).

On the result page, all compounds are displayed in the collapsed form (Fig. 5.4) showing the most important data only. The header contains the number of structures found, a link to the next page of results, a link expanding all records and a switch

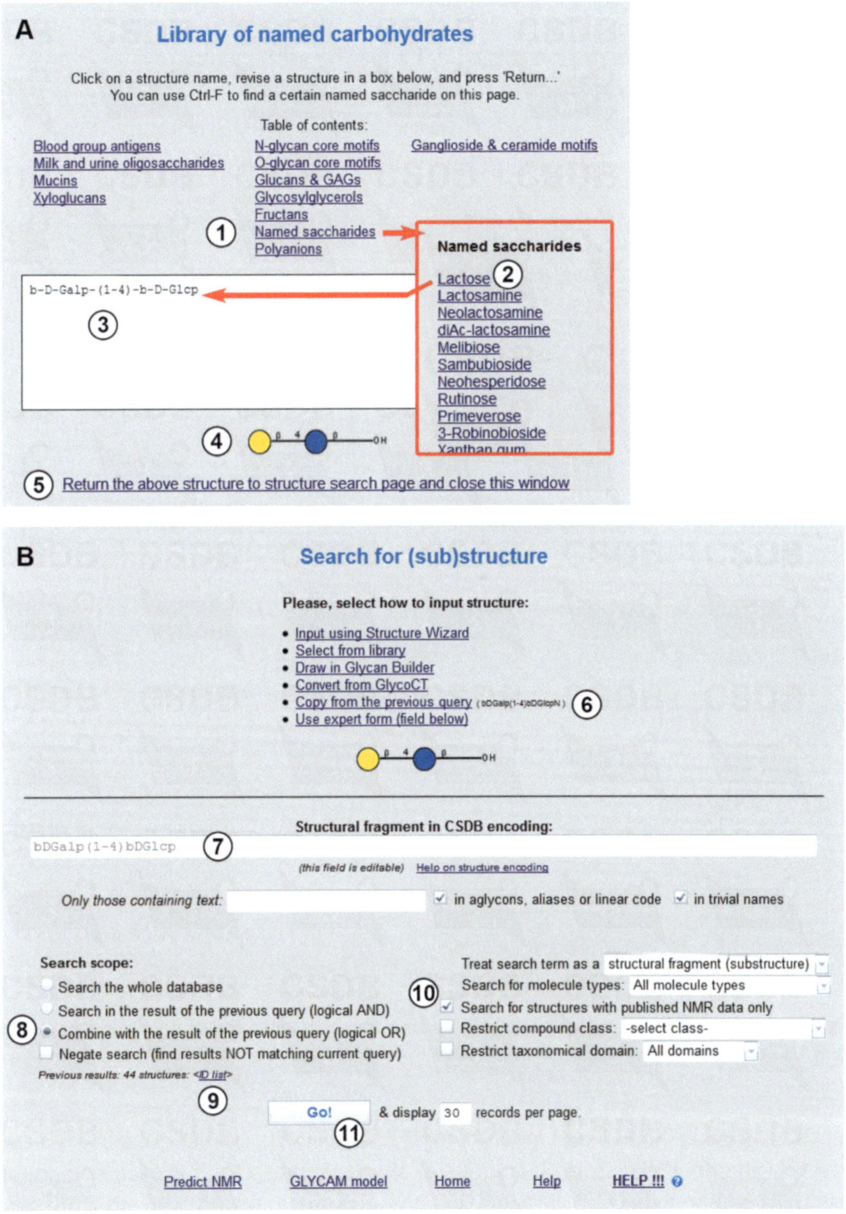

Fig. 5.3 Selecting the target structure from the CSDB library. (**a**) Library of named carbohydrates; (**b**) search page with the structure returned from (**a**) combining the results with those of the previous query

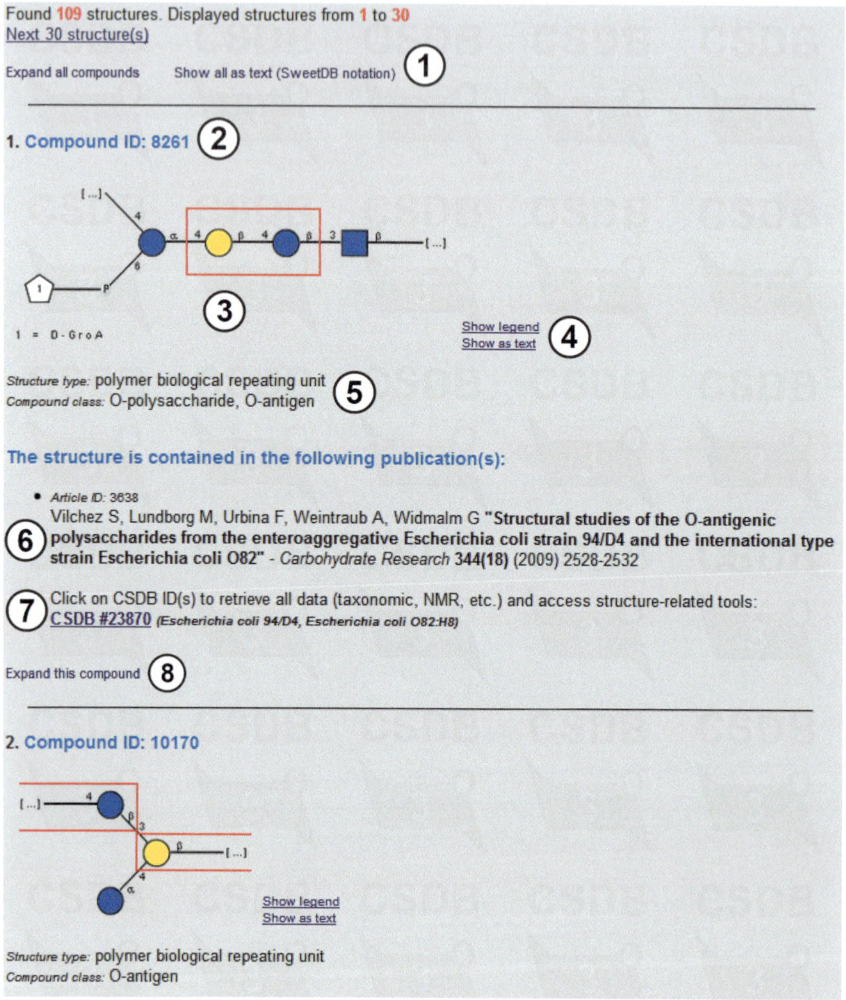

Fig. 5.4 (Sub)structure search result page (only the first record and a part of the second record are shown). The search term is highlighted for clarity

between graphic SNFG and pseudographic (SweetDB, similar to the extended IUPAC (Sharon 1988)) representations of the structures (1). For each compound found, its compound ID (2), structure (3), visualization tools (4), and structure type and compound class (if present) (5) are displayed. Visualization tools include a link to a residue symbol legend and a format switch (graphic vs. pseudographic). Each compound has a list of publications in which it was published (6); in the depicted case it contains only one item. The combination of an article and a compound is called a record and has a persistent CSDB record ID. To access the record, press link (7) that retrieves all the record data, such as NMR spectra and other data. Link (8) expands the entry showing more data associated with the compound.

Figure 5.5 shows record 27282 for the second compound from Fig. 5.4 in the expanded form. The following information is provided: a CSDB ID for the record (1) (it can be used later to access a certain record by its ID); authors and imprint of the paper containing the compound (2); a graphic (SNFG, Varki et al. 2015) or pseudographic (SweetDB, Loss et al. 2002) representation of the structure (3,4); taxonomic data on the organism(s) from which this compound was obtained in this publication, together with taxon renaming information and links to the NCBI Taxonomy database (Acland et al. 2014, http://www.ncbi.nlm.nih.gov/Taxonomy) using name or TaxID (5,12); links to the NCBI PubMed (Acland et al. 2014, http://www.ncbi.nlm.nih.gov/pubmed) and DOI (http://doi.org) resources, as well as information on publisher, affiliation, and contacts of the authors (6); abstract (7) and keywords (8) of the paper; structure type, compound class and location of the structure in the paper (9); methods used in the paper (10); links to the related records within CSDB and other databases if available (11); NMR conditions (temperature and solvent) (13), ^{1}H NMR (14), and ^{13}C NMR (15) signal assignment tables of the compound, as well as an overview of its ^{13}C NMR spectrum (16); and links to additional CSDB tools (17). Link (18) collapses the expanded record.

To see the effect of the presence of an amino group on chemical shifts, we should look at the NMR tables for the compounds found. Figure 5.6 shows NMR data for CSDB records 27282 and 29784: the former contains 4-O-β-D-galactopyranosyl-β-D-glucopyranose and the latter contains both 4-O-β-D-galactopyranosyl-β-D-glucopyranose and 4-O-β-D-galactopyranosyl-2-amino-β-D-glucopyranose. Chemical shifts that differ significantly between the two structures are shown in red (glucose) and green (glucosamine). The highlighted values of chemical shifts account for alpha- and beta-amination effects of galactose within lactose.

5.2.2 Example 2. *Find Bacterial Glycans Containing a Galacturonic Acid Residue and at Least One More Hexose, Published After 2005 in Relation to Antigens*

This problem can be solved by using the Composition search available from the main menu of CSDB. The search query is shown in Fig. 5.7.

Here we can select a partial structural composition for our search. In Fig. 5.7, the composition includes two units, one unspecified hexose, and one galacturonic acid. Drop-down list (1) allows selection of a residue from the list of the most common ones. If a residue of interest is missing from the list, select the first or last entry “show all residues” to look through all residues present in the database. Drop-down list (2) indicates the minimal number of the residue instances in the structures to search for, which is also reflected in the composition preview area (3). Buttons (4) add or remove residues from the composition.

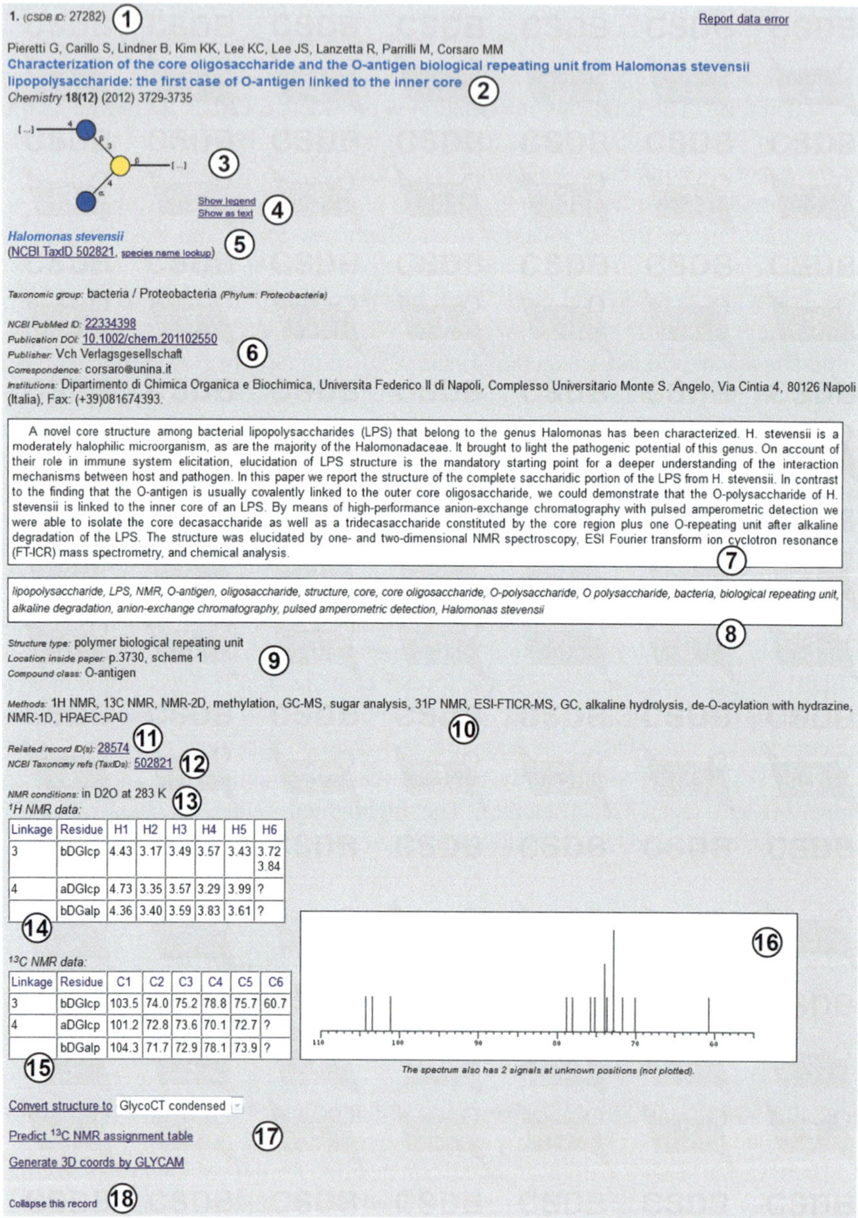

Fig. 5.5 Data referenced from the second record in Fig. 5.4 in the expanded view

In this query, we will include all types of molecules stored in CSDB (5), and the search will be carried out through the whole database (6), with no restriction on the composition completeness or compound class. Since we are interested in

(BCSDB ID: 27282)

```
         a-D-Glcp-(1-4)-+
                        |
-4)-b-D-Glcp-(1-3)-b-D-Galp-(1-
```

NMR conditions: in D2O at 283 K

^{1}H NMR data:

Linkage	Residue	H1	H2	H3	H4	H5	H6
3	bDGlcp	4.43	3.17	3.49	3.57	3.43	3.72 3.84
4	aDGlcp	4.73	3.35	3.57	3.29	3.99	?
	bDGalp	4.36	3.40	3.59	3.83	3.61	?

^{13}C NMR data:

Linkage	Residue	C1	C2	C3	C4	C5	C6
3	bDGlcp	103.5	74.0	75.2	78.8	75.7	60.7
4	aDGlcp	101.2	72.8	73.6	70.1	72.7	?
	bDGalp	104.3	71.7	72.9	78.1	73.9	?

(BCSDB ID: 29784)

```
b-D-Galp-(1-4)-b-D-GlcpNAc-(1-3)-+
                                 |
                   -6)-b-D-Galp-(1-3)-b-D-Galp-(1-4)-b-D-Glcp-(1-
```

NMR conditions: in D2O at 333 K

^{1}H NMR data:

Linkage	Residue	H1	H2	H3	H4	H5	H6
4,3,3,4	bDGalp	4.48	3.55	3.66	3.93	3.72	3.77
4,3,3,2	Ac		2.03				
4,3,3	bDGlcpN	4.75	3.77	3.76	3.71	3.59	3.96 3.83
4,3	bDGalp	4.62	3.67	3.73	4.14	3.86	4.02 3.89
4	bDGalp	4.50	3.71	3.81	4.18	3.73	3.79 3.77
	bDGlcp	4.50	3.33	3.64	3.64	3.59	3.98 3.80

^{13}C NMR data:

Linkage	Residue	C1	C2	C3	C4	C5	C6
4,3,3,4	bDGalp	105.63	73.73	75.39	71.32	78.02	63.65
4,3,3,2	Ac	177.45	24.96				
4,3,3	bDGlcpN	105.18	58.08	74.93	81.48	77.36	62.95
4,3	bDGalp	106.76	72.89	84.48	71.06	75.99	71.62
4	bDGalp	105.31	72.87	84.70	71.06	77.66	63.67
	bDGlcp	105.31	75.56	77.18	81.36	77.49	63.06

Fig. 5.6 Structures and NMR data for CSDB records 27282 and 29784. Chemical shifts that differ significantly between the two structures are shown in *red* (glucose) and *green* (glucosamine)

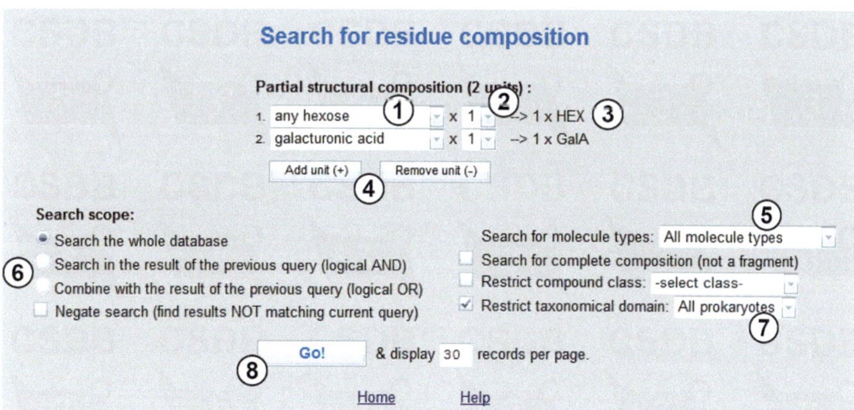

Fig. 5.7 Residue composition search

bacterial glycans in this example, we limit the result to the structures associated with prokaryotes by choosing the taxonomical domain restriction of "All Prokaryotes" (7). The query will return all bacterial or archaeal structures containing at least one hexose residue and one galacturonic acid residue. Pressing the "Go!" button (8) starts the search resulting in 738 compounds.

Now we will refine the results by choosing only structures from papers published after 2005 and containing various word forms of the term "antigen" in the article title or keyword list. For this purpose, we select "Bibliography" from the main menu (Fig. 5.8).

In the "Bibliography" search mode, you can compose queries using the imprint and metadata of publications. We can specify authors (1) directly or from index (2), full or partial title (3) (with or without abstract; see "search also in abstract" checkbox (4)), keywords (5) (with or without title terms; see "search also in title" checkbox (6)); select a journal from list (7) and specify the publication year (8) (exact (=), newer than (>), or older than (<)); as well as volume or pages (9). Term fields (1), (3), and (5) support queries with logical operations, term grouping, and wildcards; specific national symbols are available (1). The search is case-insensitive and accent independent.

To search for papers published after 2005 and containing keywords or titles which include "antigen," we enter *antigen* in field (5) and check checkbox (6).

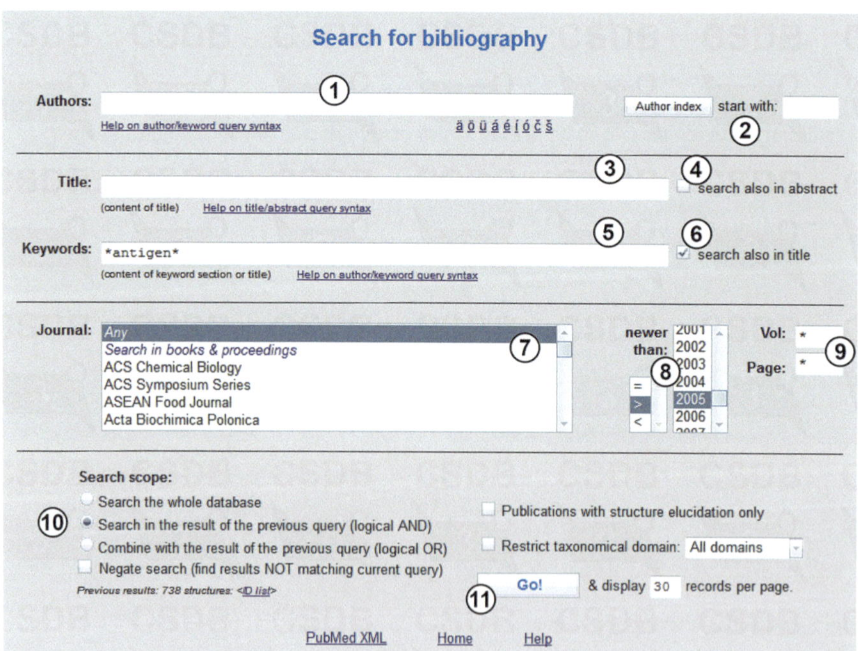

Fig. 5.8 Bibliography search

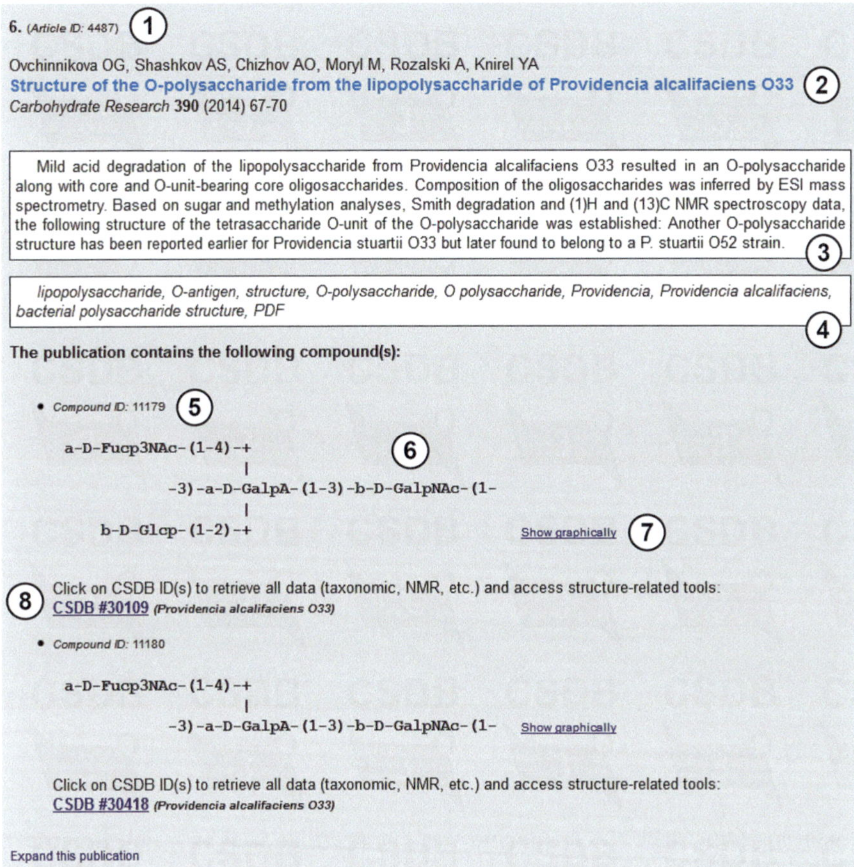

Fig. 5.9 Bibliography search result page (only one publication entry is shown; the structures were chosen to be presented in the pseudographic format)

As the star character is a wildcard for any number of characters, this query will process all terms like “antigens,” “antigenic,” “O-antigen,” etc. To start, we select the year span “newer than 2005” (8), check “Search in the results of the previous query” (10), and press the “Go!” button (11).

The results include 73 publications. One of them is shown in Fig. 5.9.

In contrast to the (sub)structure search, which returns compounds, the bibliography search results in a list of publications. In the case of our combined query, each of these publications includes one or several structures at least one of which matches the composition query and contains at least one galacturonic acid residue and one more hexose residue. For each publication, its article ID (1), imprint data (2), abstract (3), and keywords (4) are shown, and the list of compounds present in the paper is displayed, with the corresponding compound ID (5), pseudographic representation of the structure in the SNFG or SweetDB format (6), and the

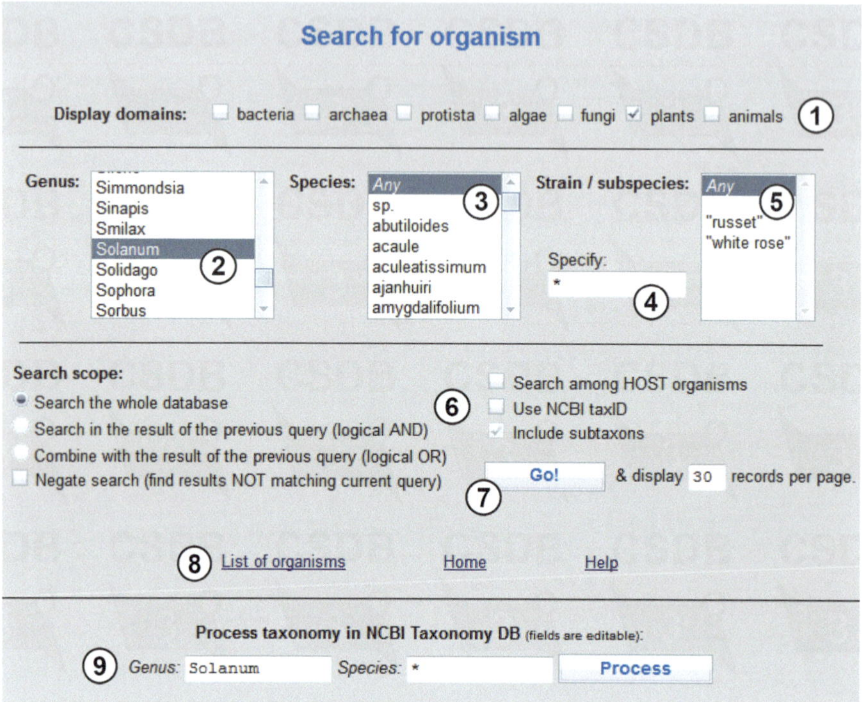

Fig. 5.10 Taxonomy search form

organism from which the compound was extracted, as well as a link to retrieve all the data (8) available in the record. Switch (7) allows selecting the structure representation format (graphic vs. pseudographic). In Fig. 5.9, the structure display format was chosen as SweetDB (pseudographic).

Figure 5.9 shows that in this example all the compounds present in the paper match the initial composition query (1 × HEX + 1 × GalA + …). However, if at least one of the structures in the paper matches the query, such paper is included into the search results, and therefore, compounds from the same paper with composition different from that specified are also shown.

5.2.3 Example 3. *Find All Compounds Extracted from the Plants of the Genus* Solanum *Which Contain a Solanidine Constituent*

To search for compounds obtained from organisms belonging to a certain genus, we will use the Taxonomy search from the main menu (Fig. 5.10).

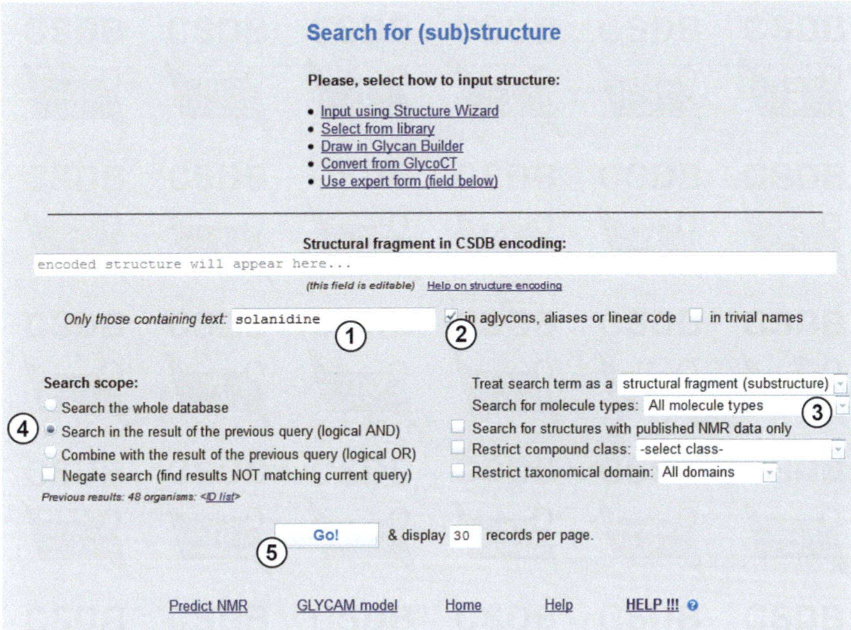

Fig. 5.11 Search for solanidine in aglycons

The "Display groups" option (1) allows specifying a target domain(s) of organisms; in our case it is plants only. By using lists of genera (2) and species (3), we specify *Solanum* as genus and "Any" as species implying that all species belonging to this genus will be searched for, including *Solanum* organisms with undetermined species. For fast navigation in the list of genera, start quick typing the genus name. Fields (4) and (5) specify subspecies by selection from those available for the current genus (5) or by direct input of a character pattern (4) (fragment of name or "*" for no limitations). The search will be conducted through the whole database, with taxonomic children included (6): all organisms belonging to the specified taxon and its subtaxa will be returned; if we use the selection lists, the taxonomic children are always included. Other available options are "Search among HOST organisms" (returns an organism specified and structures found in microorganisms or parasites infecting the target organism or extracted from it) and "Use NCBI TaxID" (the taxon selection lists disappear, and we can indicate a position of taxon on the tree of life by its NCBI TaxID) (Acland et al. 2014, http://www.ncbi.nlm.nih.gov/Taxonomy/). We can access the full list of organisms present in the database (8) or process the taxonomic request in the NCBI Taxonomy database (9) (retrieves taxonomic lineage and other data for the genus and species specified). Pressing button (7) starts the search resulting in 48 organisms.

To find only those compounds from the *Solanum* genus that contain the steroidal alkaloid solanidine, we will use the (sub)structure search (Fig. 5.11).

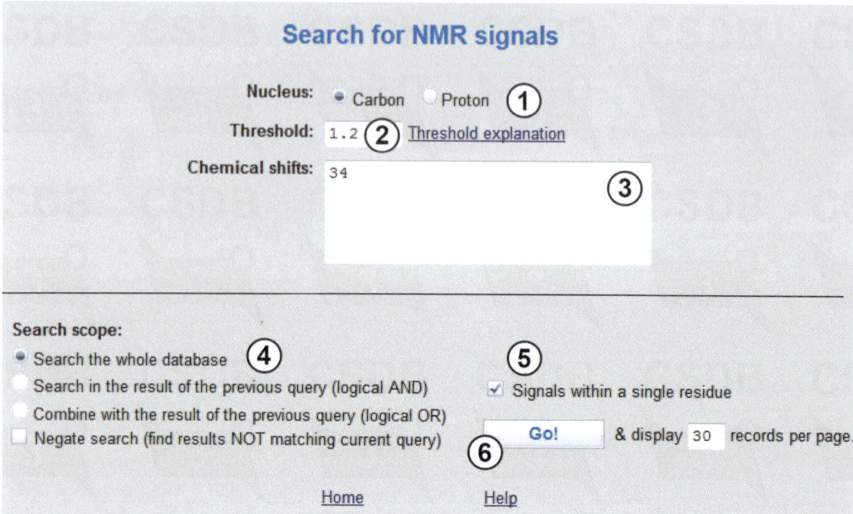

Fig. 5.12 Search for NMR signal form

Solanidine is not a carbohydrate and has no reserved residue name in CSDB; therefore, it can be present only as aglycon or inside the alias explanation. To search through aglycons and aliases, we will enter "solanidine" in field (1), with the "in aglycons, aliases or linear code" checkbox checked (2). The search will include all molecule types (3) and will be carried out in the results of the previous query (4). Pressing button (5) starts the search resulting in 16 compounds each of which contains solanidine as aglycon and is extracted from the *Solanum* plants. The layout of the result page is close to that described in *Example 1*.

5.2.4 Example 4. *Find All Carbohydrate Structures Having a Signal Close to 34 ppm in the ^{13}C NMR Spectrum, Except Those Containing Any Octose*

This problem addresses residues with the specified feature in the NMR spectra. However, an overabundance of NMR data for structures containing 3-deoxy-D-manno-oct-2-ulosonic acid (Kdo) would obscure the results, so we will search only for those compounds, which do NOT contain octoses. We will start this search from the "NMR signals" in the CSDB menu (Fig. 5.12).

Here we specify the nucleus (1) and chemical shift(s) (3) (numerals and decimal dot are allowed). The signals can be separated with spaces or new-line characters; sorting is not required. The "Threshold" field (2) determines how accurate the correspondence of signals to the specified values should be. It filters the results by spectra similarity returning only those with similarity higher than the threshold. To calculate the similarity between the specified and stored spectra, the CSDB engine forms all possible subspectra of the larger NMR spectrum with the number of signals equal to that found in the smaller spectrum, and the best-fitting subspectrum is used to calculate the similarity value. Similarity is an inverse average deviation between the signals normalized by the number of signals in the smaller spectrum. 0 is reserved for no similarity, and 1000 is for full similarity (exact match of chemical shifts). Good similarity values are 1 and above for carbon spectra and 5 and above for proton spectra.

We will carry the search out through the whole database (4). Note that the "Signals within a single residue" checkbox is checked by default (5), but since we are searching for a single signal, it is of no importance in our case. In other cases, the demand for all the specified signals to belong to the same residue narrows the search and makes it faster. Pressing button (6) starts the search resulting in 133 compounds sorted by spectra similarity.

Figure 5.13 shows one of the found compounds and its NMR spectrum.

For each compound found, its ID (1), graphic or pseudographic representation of the structure (2), visualization tools (3), and structure type are displayed. The average similarity between the ^{13}C NMR spectra of this compound and the search term is shown (4), together with the spectrum itself (7, 8), experimental conditions (6), and a link to the paper in which it was published (5). Signals matching the search term are highlighted in yellow. Information about the paper is also given (9), and a link to the full data on the record is present, together with the organism from which this compound was extracted (10).

To refine the search results to the compounds that contain no octose residues, we will use the Composition search mode (Fig. 5.14).

We choose a partial structural composition of one unit, octose (1), and include all molecule types (2). To perform logical exclusion (AND NOT operation) we define the search scope as "Search in the result of the previous query" and check the "Negate search" checkbox (3). Pressing the "Go!" button (4) will find all compounds, which contain no octose residues and spectra of which contain at least one signal close to 34 ppm. The final result list includes 71 compounds, in which this signal originates from non-sugar residues and deoxy-sugars other than Kdo (3-deoxy-D-manno-oct-2-ulosonic acid), Ko (D-glycero-D-talo-oct-2-ulosonic acid) or other octoses.

Since every search form provides more than one search criteria, you should be careful about the usage of negation. For example, specifying the negation together with two criteria (octose-containing and prokaryote domain) will hardly make sense because [NOT (octose-containing AND prokaryotic)] is the same Boolean logic as [(NOT octose-containing) OR (NOT prokaryotic)]. If you wish to filter the results to prokaryotic structures only and you still need a negation in the complex query, you should use the structure search once more, specify structure=ANY and domain=Prokaryotes, and intersect it with the previous results.

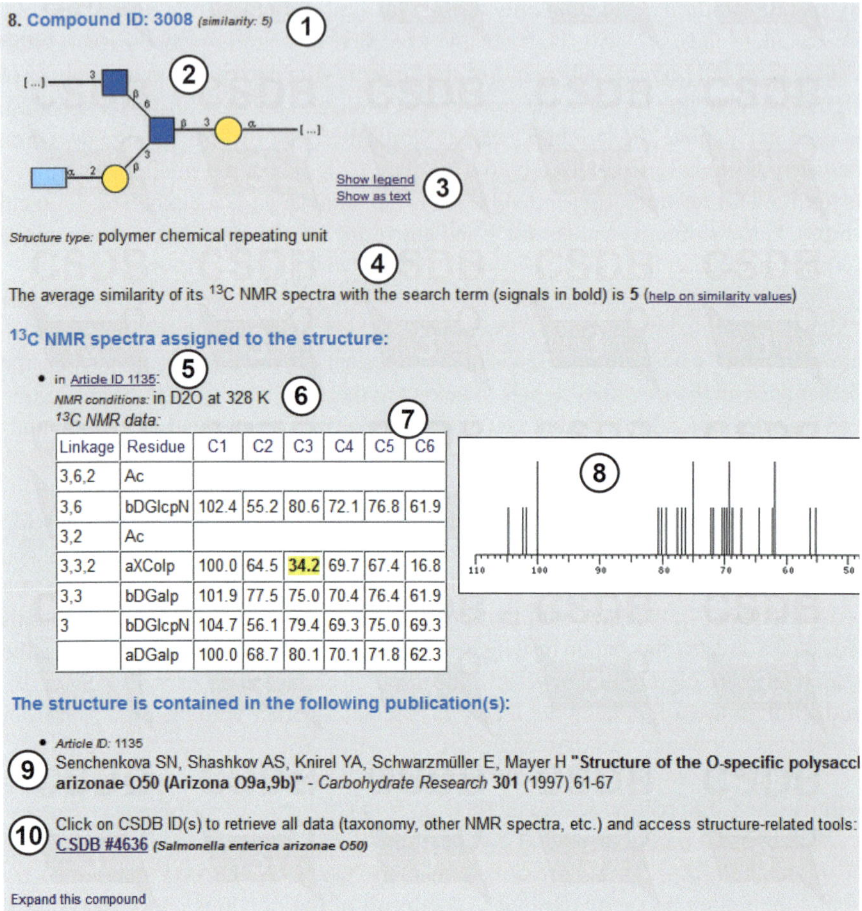

Linkage	Residue	C1	C2	C3	C4	C5	C6
3,6,2	Ac						
3,6	bDGlcpN	102.4	55.2	80.6	72.1	76.8	61.9
3,2	Ac						
3,3,2	aXColp	100.0	64.5	34.2	69.7	67.4	16.8
3,3	bDGalp	101.9	77.5	75.0	70.4	76.4	61.9
3	bDGlcpN	104.7	56.1	79.4	69.3	75.0	69.3
	aDGalp	100.0	68.7	80.1	70.1	71.8	62.3

Fig. 5.13 Search for NMR signal result page (only one record is shown)

5.2.5 Example 5. *Find All Papers by Knirel or Shashkov AS on Bacterial Structures Containing Quinovose-4-Amine Amidated by Any N-Acetylated Amino Acid*

First we will find papers written by Knirel or Shashkov AS using the Bibliography search mode (Fig. 5.15).

The author query ‘Knirel OR “Shashkov AS”’ (1) will find all papers of authors which include at least one of the names specified, Knirel (any initials) or Shashkov AS (explicit initials). To insert a specific author name and initials from the index, type the starting characters in field (2), press “Author index”, and select an author. Quotes allow inclusion of blank spaces in the search term, e.g., the author last name and initials. The author index (2) can be used for picking up the author names. You

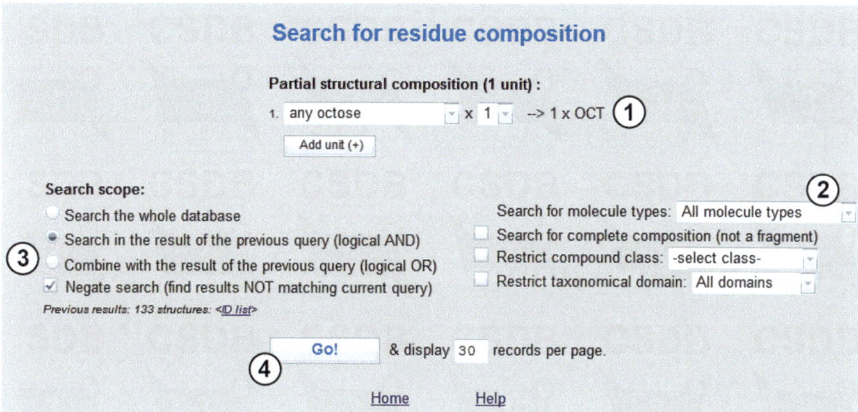

Fig. 5.14 Search for structures NOT containing octose

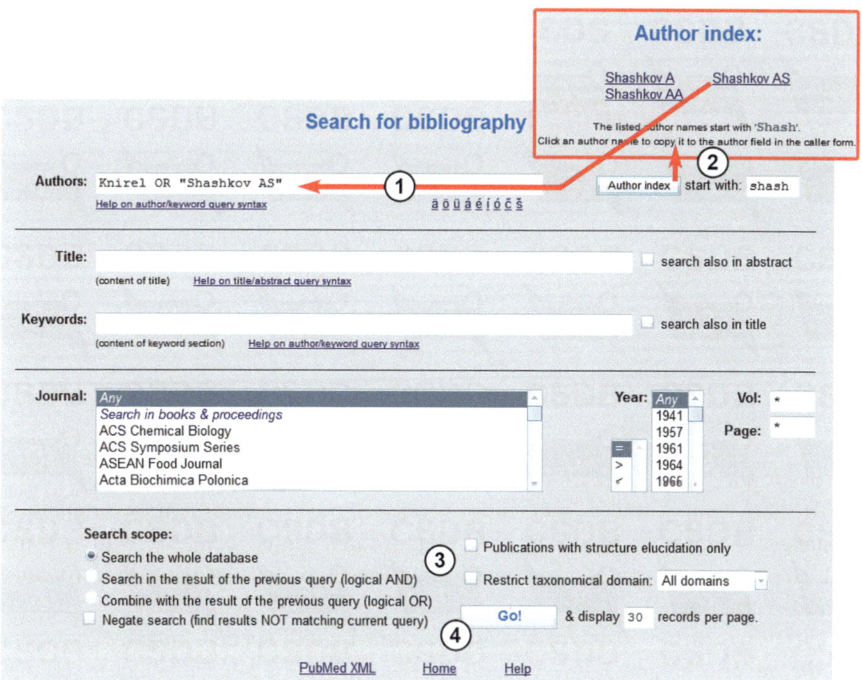

Fig. 5.15 Search for papers written by Knirel or Shashkov AS

should type at least two characters to display the list of author names starting with these characters. The search will be carried out through the whole database (3). For other bibliography search options available, see *Example 2*. Pressing the "Go!" button (4) starts the search resulting in 770 publications.

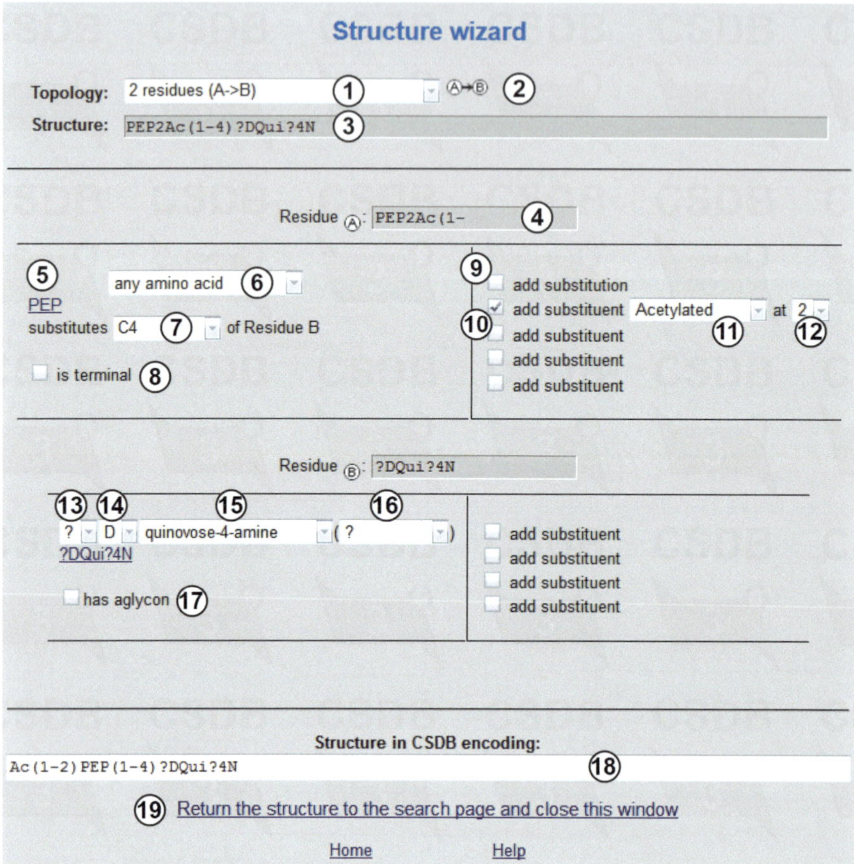

Fig. 5.16 Structure wizard

To find among these publications only those with structures containing quinovose-4-amine amidated by any N-acetylated amino acid, we will use the (sub)structure search mode and will enter the disaccharide using the structure wizard (Fig. 5.16).

The usage of the structure wizard does not require special knowledge except for the general nomenclature of carbohydrates. First of all, we select a suitable topology from drop-down list (1) (topologies of up to four residues are supported), and the graphic representation of the selected topology is displayed (2). When the topology is selected, the corresponding number of residue sections appears below (two in our case). Structure (3) shows the target structure in the IUPAC condensed form (Sharon 1988, http://www.chem.qmul.ac.uk/iupac/2carb/38.html#384). Residue section header (4) shows the position of this residue within the selected topology (e.g., Residue A) and the residue name with configurations and substitutions.

The left part of the residue section includes the drop-down list of residue names (6, 15). Only the most common residues are listed; if a residue of interest is missing, select the first or the last entry "show all residues." In our case, the superclass "any amino acid" is selected for Residue A (6), and "quinovose-4-amine" is selected for residue B (15). If applicable to the selected residue, configuration options are available: anomeric configuration (alpha, beta, or ? for any) (13), absolute configuration (D, L, R, S, or ? for any) (14), and residue ring form (pyranose, furanose, open chain, alditol, or ? for any form) (16). Link (5) shows the resulting residue name without substitutions and leads to the complete residue list window. Linkage control (7) specifies the position in the acceptor residue substituted by the current residue (in this case, C4, which means that Residue A is linked to C4 of Residue B). The position of the current residue by which it substitutes the acceptor is C1 for aldo sugars and non-sugar residues and C2 for keto sugars. The linkage control does not apply to the residue at the reducing end of the fragment. The "is terminal" checkbox (8) indicates that this residue should not be substituted by anything except monovalent substituents and is visible only for residues occupying terminal positions in the selected topology. The residue at the reducing end (in the last residue section or, in this case, Residue B) has the aglycon checkbox (17) and a corresponding drop-down list (hidden in Fig. 5.16) allowing to select an aglycon.

The right half of the residue section allows adding substituents. Usually this feature is used to indicate the position of attachment of another repeating unit (9) to the leftmost residue in the polymer repeat or positions of monovalent substituents like acetyl groups (10–12). Thus, the amino acid residue in our structure is acetylated (11) at position 2 (12).

The constructed structure in the CSDB encoding is displayed at the bottom of the page (18). Pressing the "Return the structure to the structure search page" link (19) transfers the query into the (sub)structure search form (Fig. 5.17) and closes the wizard.

Figure 5.17 shows the (sub)structure search form with the disaccharide fragment (1) returned by the structure wizard. The search will include all molecule types (2) and will be conducted in the results of the previous query (3). Pressing the "Go!" button (4) results in 11 compounds containing the specified fragment published in the papers written by Knirel or Shashkov AS. The layout of results is similar to that in *Example 1*.

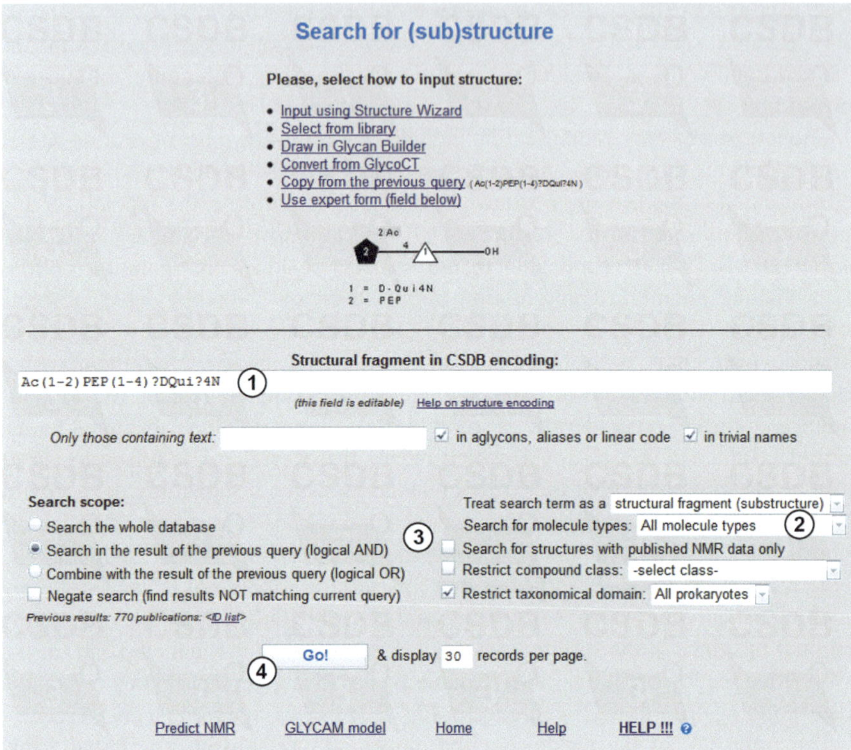

Fig. 5.17 (Sub)structure search form with quinovose-4-amine amidated by any N-acetylated amino acid constructed in the structure wizard

5.2.6 Example 6. *Find All Bacterial Nonose Monosaccharide Structures (Monomers or Homopolymers)*

This simple query demonstrates how the "Search for complete composition" option can be used. We will perform the search in the composition search mode. Figure 5.18 shows the query.

To find all structures containing only one nonose monosaccharide, we select the complete structural composition (3): any nonose (1) occurring once within the structure (2). Then we choose all molecule types (4), check the "Search for complete composition (not a fragment)" checkbox (5), and restrict the taxonomical domain to "All prokaryotes." Pressing the "Go!" button (6) returns seven compounds each of which is a prokaryotic nonose monosaccharide or a prokaryotic nonose homopolymer without monovalent modifications.

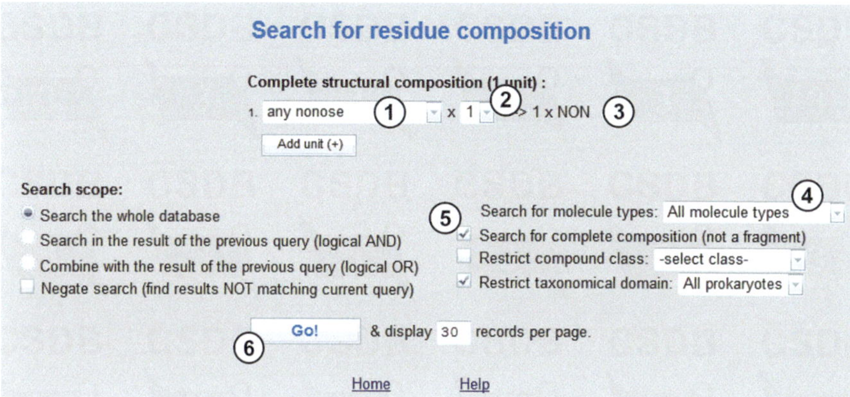

Fig. 5.18 Search for complete composition

5.3 NMR Tools

NMR spectroscopy is one of the principal methods for carbohydrate structure elucidation. To assist NMR structural studies, CSDB provides two NMR services: the NMR simulation tool, which predicts ^{13}C and ^{1}H NMR chemical shifts for a specified compound, and the NMR-based structure ranking tool, which generates all possible structures matching the given constraints and matches them against an experimental ^{13}C NMR spectrum. The key point of the CSDB NMR predictor is the ability to process almost all structural features occurring in natural glycans, including atypical and noncarbohydrate moieties.

In this section, we provide two examples of usage of these tools. They are available under the "Extras" section of the main menu.

5.3.1 Example 7. *Predict ^{13}C NMR Spectrum of 3-O-α-Abequosyl-6-deoxy-β-*D*-mannoheptopyranosyl-(*D*-ribitol-1)-phosphate in Water Solution and Explore Credibility of Chemical Shifts Simulated with Lowest Reported Trustworthiness*

CSDB contains ca. 7800 NMR spectra recorded in water, facilitating statistical simulation of the NMR data in water solutions. The distribution of solvents vs. records can be displayed by clicking on the "Coverage" link (6). The "NMR simulation" link in "Extras" from the main menu opens the spectrum simulation form (Fig. 5.19). As the options for structure input (1) and visualization (2) have been described in the previous examples, here we will use the expert form

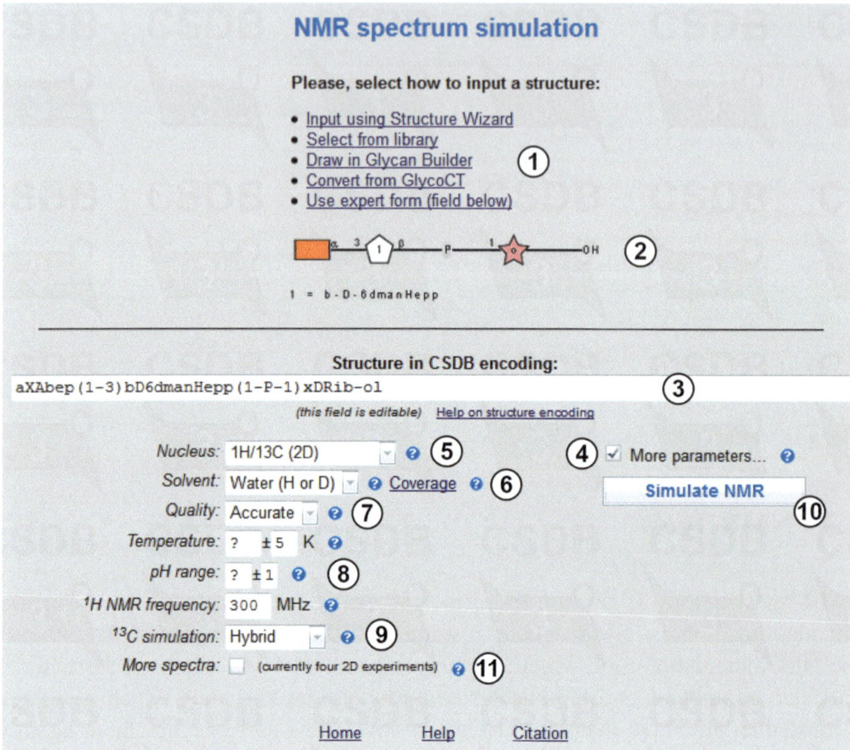

Fig. 5.19 NMR spectrum simulation form

and type the target structure in the CSDB encoding in field (3): aXAbep(1-3)bD6dmanHepp(1-P-1)xDRib-ol.

By default, only the nucleus (5) and solvent (6) parameters are displayed, however checking "More parameters" (4) shows all of them, as in the figure. We will select ^{1}H/^{13}C for nucleus (5) to simulate 1D and 2D NMR spectra, select *water* as a solvent (6), and use the *accurate* quality mode (7). There are three modes of simulation, depending on quality vs. speed: *fast*, *accurate* (default), and *extreme*. If you see unpredicted signals (question marks) in the assignment, set this option to higher quality and press button (10) once again. Note that in the *extreme* quality mode, the calculation may take up to 15 min. Together with the solvent, additional simulation parameters (8), such as pH or temperature range, allow limiting the database scope to the data obtained under certain experimental conditions.

For carbon chemical shifts, both empirical and statistical simulation approaches will be employed to obtain a hybrid result (9). The empirical simulation is based on an incremental scheme with steric correction and utilizes internal databases of reference chemical shifts of mono-, di-, and trimeric fragments and substitution effects. The algorithm was adopted from earlier work (Toukach and Shashkov

2001). The statistical simulation is based on sequential generalization of the atomic surrounding of the predicted atom until enough structurally similar fragments are found in CSDB, with subsequent outlier detection and data averaging (Kapaev et al. 2014; Kapaev and Toukach 2015). The hybrid approach combines the results of empirical and statistical simulations in accordance with trustworthiness reported by both approaches. It becomes available if the hybrid mode is selected for carbon chemical shifts, and the solvent is water or unrestricted. Proton chemical shifts are always simulated statistically.

Pressing button (10) simulates the NMR spectra. Assignment tables and schematic ^{13}C NMR plots predicted by three different approaches (empirical, statistical, and hybrid) are displayed. A part of the result page (statistical assignment tables, full output of hybrid approach, and selected 2D plots) is shown in Fig. 5.20.

Every residue is represented by a single row in the assignment table. Each row is identified by linkage data (the linkage path to the residue from the oligomer reducing end or from the rightmost residue in the repeating unit of a polymer) (1) and a residue name in the CSDB linear code (2). Every row also contains a color square (in the Linkage column (1)) identifying the color code of signals in the 2D NMR plots. C1–C9 columns (4) list the chemical shifts and supplementary data. In the empirical assignment table (not shown), these data include chemical shifts of unsubstituted residues and substitution effects. Figure 5.20a shows the results of the statistical ^{13}C NMR simulation. In the assignment table, there are expected simulation error (in ppm) and trustworthiness metric (in %) for every signal, the number of database records used to obtain the averaged data and a link to the processing report. Clicking on the number of records below each chemical shift opens a list of references (6). These references allow tracking the source of data to CSDB records and corresponding original publications (7). The data used in the simulation are highlighted in yellow in these records.

The trust column (3) provides the trustworthiness averaged from all atoms in the residue. Trustworthiness values vary from 0 to 100 % and are color coded (uniformly from red to green according to the value). Overall prediction trustworthiness is shown below the table (5). The correlation between the trustworthiness level and chemical shift calculation accuracy was established by linear regression (Kapaev et al. 2014; Kapaev and Toukach 2015). It is used to predict the expected simulation error in ppm, which is displayed below each chemical shift.

Figure 5.20b shows the results of ^{13}C simulation via the hybrid approach. It implies heuristic mixing of data from both approaches according to their deviation, trustworthiness, dataset size, and other parameters. As it benefits from both simulation methods, it usually provides the most accurate results. The hybrid supplementary data contain the trustworthiness metric and inter-approach deviation (Δ) for each atom. The reference data from both approaches (red marks = empirical, blue marks = statistical) are added to the spectrum plot (10). The header reports the overall trustworthiness metrics and lists all signals in sorted order (9) for copy and pasting.

For more data on the NMR simulation algorithms utilized, including evaluation of accuracy and trustworthiness of predictions, and hybridization details, please

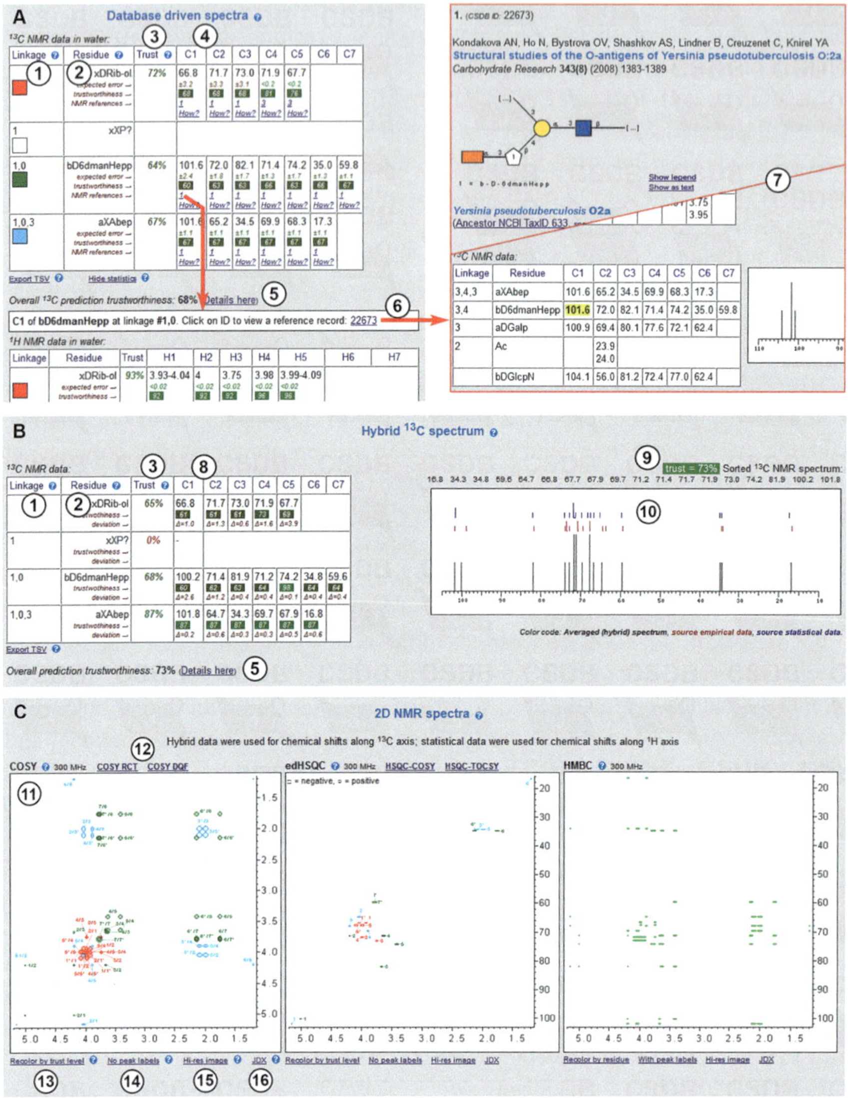

Fig. 5.20 Simulated NMR spectra. (**a**) Statistical assignment table and part of reference record; (**b**) results of hybrid simulation; (**c**) simulated 2D NMR spectra (three selected plots are shown for clarity). COSY and edHSQC spectra are displayed in the assignment mode; HMBC spectrum is displayed in the trustworthiness level mode

refer to dedicated publications (Toukach and Ananikov 2013; Kapaev et al. 2014; Kapaev and Toukach 2016). A brief review is also given in the Help section of the project (http://csdb.glycoscience.ru/bacterial/index.html?help=nmr).

In our example, both empirical (not shown) and statistical (Fig. 5.20a) approaches reported relatively low trustworthiness (60 %) for C1 of β-6-deoxy-mannoheptopyranose. Statistical data were taken from a single database record. By clicking on "1" (the number of used records) in the C1 column in the bD6dmanHepp row, we can explore where the data came from and see that the structure has undergone strong permutations during the generalization process, which explains the low trustworthiness metric. The list of permutations applied to fit records present in the database is individual for each atom and is available under the "How?" link.

If ^{1}H or ^{1}H/^{13}C is specified as a nucleus, 2D NMR spectra are visualized using the predicted chemical shifts (Fig. 5.20c). Depending on the nucleus and the state of the "More NMR experiments" checkbox, two to eight 2D spectra are plotted. These experiments cover most of the proton and carbon spin correlations commonly used in glycobiology (COSY, TOCSY, HSQC, HMBC, and derived experiments). NOE correlations are currently not supported. As an example, COSY, edHSQC, and HMBC spectra are shown in the figure (11). Links (12) lead to the experiments related to that displayed. Links below the spectrum switch useful display options: color mode, signal labels, and image resolution.

The color mode switch (13) determines how the signals are colored; it has two states: signal assignment and trustworthiness level. The former colors all the signals according to the residue color code, as displayed in the first column of the assignment table (exemplified in COSY and edHSQC). The latter colors all the signals in the range from red to green reflecting how accurate the simulation of the signal was (exemplified in HMBC).

The peak label switch (14) hides or shows the numbers beside signals. In the assignment color mode, these numbers correspond to the order of carbon atoms in the structure. The combination of the color (=residue) and the label (=position in a residue) identifies every atom. NMR spectra, which provide non-direct correlations, may have complex labels, including bicolored ones to identify inter-residue cross peaks. In the trustworthiness color mode, numbers are the trustworthiness metrics. In Fig. 5.20c, HMBC is displayed without labels in the trustworthiness color mode, while COSY and edHSQC are displayed with labels in the assignment color mode.

The "Hi-res image" link (15) or clicking on the spectrum displays a larger image in a separate window for copy and pasting. If some of the signals could not be predicted (they have a question mark as a chemical shift in the assignment table), they are listed below the spectra in which they should have appeared. The JDX link (16) exports the spectrum in the JCAMP-DX format, which can be further processed online using "Live view NMR" tool at cheminfo.org or downloaded and opened in the dedicated NMR software, such as Mestrelab MestreNova, ACD/Labs NMR viewer, or Bruker TopSpin.

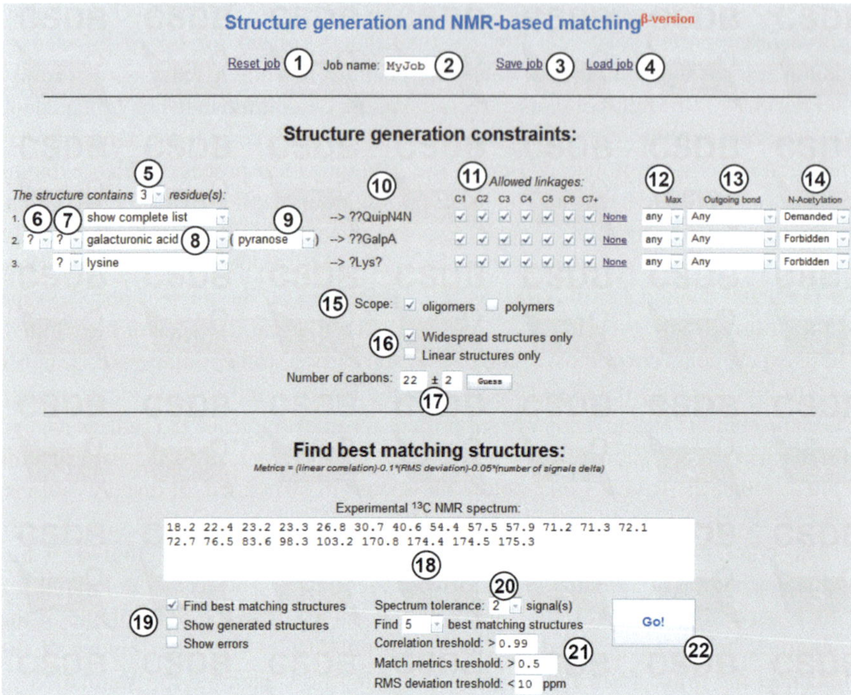

Fig. 5.21 Input form for NMR-based structure prediction

5.3.2 Example 8. *Rank Structural Hypotheses for an Unelucidated Oligomer Conforming to an Experimental ^{13}C NMR Spectrum and Containing Bacillosamine, Galacturonic Acid, and Lysine Residues*

This problem can be solved by using the NMR-based prediction tool available from "Extras" in the main menu. Figure 5.21 shows the input form. Assume that we know the monomeric composition from chromatographic methods and know that the compound contains no furanoses (no characteristic signals are observed in the ^{13}C NMR spectrum), but anomeric configurations of sugars, substitution positions, and the sequence of residues are undetermined.

The tool allows resetting (1), saving (3), and loading (4) a job with the name starting with specified characters (2) from a job list. Structure generation constraints are used to define the scope of the target structures. Drop-down list (5) selects the number of residues, and the corresponding number of residue sections appears below (three in our case). Every residue section has a residue name selector (8), allowing the selection of certain residues, or superclasses, or ANY residue. This drop-down list contains only the most common residues; if a residue of interest is missing, as in the case of bacillosamine (2,4-diamino-2,4,6-trideoxyglucose), it can be selected from the last entry ("show all residues"). Depending on the selected residue, the section may possess the following fields, where applicable: anomeric configuration (6), absolute configuration (7), and residue ring form (9). Each of these fields can remain undefined. If a certain residue is known to have multiple instances in the structure, it should be specified several times. An overview of the specified limitations on the monomeric composition is presented in area (10).

Checkboxes in the "Allowed linkages" area (11) indicate which positions in the residue can be substituted during the structural permutations. Only chemically possible structures will be generated, implying the bonds between the residues can only be formed with elimination of water or ammonia. "C7+" refers to any carbon positions higher than C6. The outgoing linkage position (usually C1 for aldo sugars or C2 for keto sugar) should also be checked unless a residue is at the reducing end. The "Max" drop-down list (12) defines the maximal number of substituents that can be attached to the residue, except glycosidic bond acceptors. The default value "any" means no limitations, whereas "term" means that a residue can be in the terminal position only. "Outgoing bond" (13) restrains the type of an acceptor attached to the selected residue. The "reducing end" option indicates that the residue is forbidden to form an outgoing bond. The "N-acetylation" drop-down list (14) defines whether amino groups of the residue are acetylated ("demanded") or free ("forbidden"); "allowed" means that both variants are possible. To allow N-acetylation, the corresponding linkage position must be checked in area (11).

Although we do not know absolute configurations of the residues, we will assume they fit those occurring in nature. These common values will be used automatically in "Widespread" mode (16), which also excludes rarely occurring structural features, such as atypical residues (if exact monomeric composition is not known) and ring sizes, ether and amide bonds between sugars, highly branched patterns, and large side chains in polymers. We usually know whether the structure is oligo- or polymeric; in this case, we will check "oligomers" (15) only. Field (17) limits the search to the structures containing the given number of carbons (plus-minus the specified delta) in the oligomeric molecule or per repeating unit. The "Guess" button fills in field (17), according to the experimental spectrum provided.

We enter the experimental ^{13}C NMR spectrum in field (18). We will use default ranking parameters (21) for the prediction. The "Spectrum tolerance" field (20), which is 2 by default, defines the maximal allowed difference between the number of signals in the simulated vs. experimental spectrum.

Pressing the "Go!" button (22) runs the empirical ^{13}C NMR simulation. Depending on the constraints, the calculation may take a long time. For example, calculation of hypotheses for 5-6 strictly defined residues or 2-3 unconstrained residues takes from 0.5 to 2 h. The number of structures already processed (in thousands) is continuously updated during the calculation, and the iterated structures are shown if specified in (19). The result contains a list of the best-fitting structures (Fig. 5.22, only the first two are shown for clarity).

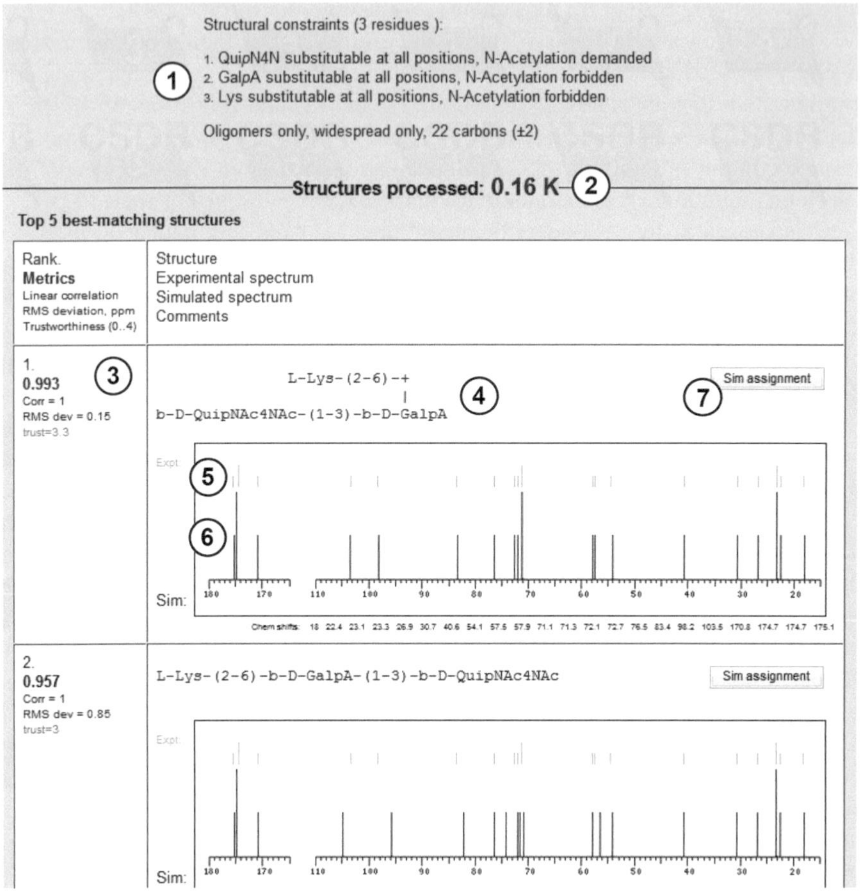

Fig. 5.22 Prediction results (only first two predicted structures are shown)

The result is shown in the same window below the task form and includes the query overview (1) as well as the number of structures processed (2). In our case, only 164 structures could be generated due to strict constraints; the query took less than 2 s. The prediction results are shown as a table containing the best-matching structures, one per row. The rows are sorted according to the similarity between the experimental and simulated NMR spectra. The first column contains the information on spectrum similarity (3): structure rank in the top list of structural hypotheses, similarity metric (shown in bold), linear correlation factor, root-mean-square deviation in ppm, and color-coded spectrum simulation trustworthiness level from 0 (red) to 4 (green) returned by the simulation engine. The similarity metric is calculated in regard to the linear correlation factor, RMS deviation, and the number of signals required to equalize the size of the experimental and simulated spectra (see more details in the tool help at http://csdb.glycoscience.ru/bacterial/index.html?help=nmr).

The second column shows the predicted structures in the pseudographical SweetDB format (Loss et al. 2002) (4) and their NMR spectra. The experimental spectrum (in gray) (5) is displayed above the simulated one (6) for easy visual comparison. The sorted simulated chemical shifts (as well as other comments, e.g., warnings about missing signals) are listed below the spectrum. The "Sim assignment" button (7) runs NMR simulation for this structure and returns the signal assignment tables.

Please note the limitations:

1. This example was specially chosen for demonstration purposes. Real examples may show lower deviation between the best-fitting and the second structural hypotheses.
2. The more constraints you specify, the lower number of structures is iterated, the higher the deviation between them and more reliable the result. A typical tool usage is determination of a residue sequence and anomeric configurations when monomeric composition, absolute configurations, and most of substitution positions are known from chromatographic methods and methylation analysis.
3. Spectral effects at some glycosidic bonds are insensitive to some structural parameters; for example, absolute configurations of galactose and of its substituent at C4 do not affect carbon substitution effects at bonded atoms (Lipkind et al. 1988). In such cases, hypotheses occupying the neighboring positions in the results may differ very slightly. They can hardly be distinguished from the NMR data and thus have almost equal metrics. Common sense or additional experiments are required for further verification.
4. This version of the tool is still in the testing phase and has been proven to work with a relatively small number of generated structures (up to 10,000) as a proof of concept, assuming that tasks contain 2–3 unconstrained or 5–6 highly constrained residues. Implementation of a postponed execution and login-based provision of results, which should eliminate the performance limitation, is our prospect for the next version of the structure predictor (release is planned in 2017).

5.4 Statistical Tools

An overview of database content often presents valuable information for researches. To assist cumulative and statistical studies, CSDB provides several dedicated tools. Links to them can be found in the "Extras" section of the main menu. The "Fragment abundance" tool generates abundance tables of monomers and dimers present in carbohydrates from specified taxonomic groups, whereas the "Coverage statistics" tool produces statistics on the database coverage for specified taxonomic groups. The following three examples demonstrate the usage of these instruments. For the novel glycome-based taxon clustering tool, please refer to the dedicated publication (Egorova et al. 2015).

5.4.1 Example 9. *Study Monomeric Composition of Two Fungal Species, Aspergillus oryzae and Aspergillus fumigatus, and Reveal Which Monomers Occupy the Termini of Side Chains*

Fragments at the termini of side chains of O-antigens often determine the immune response in higher organisms. Terminal residues present in the structures from human-pathogenic *Aspergillus fumigatus* are potential targets for probing how their presence affects the immunospecificity of species.

The "Fragment abundance" link from the main menu opens the "Monomer and dimer abundance" tool (Fig. 5.23).

First we should select the target taxonomic rank, in our case – species, from drop-down list (1). We can select one or more taxa of this rank later. Other available ranks are domain, phylum, class, genus, and strain/subspecies. Since we are interested only in species from a single fungal genus, the specified display group is "fungi" (2). List (3) selects the genus of interest (*Aspergillus*) from all fungal genera present

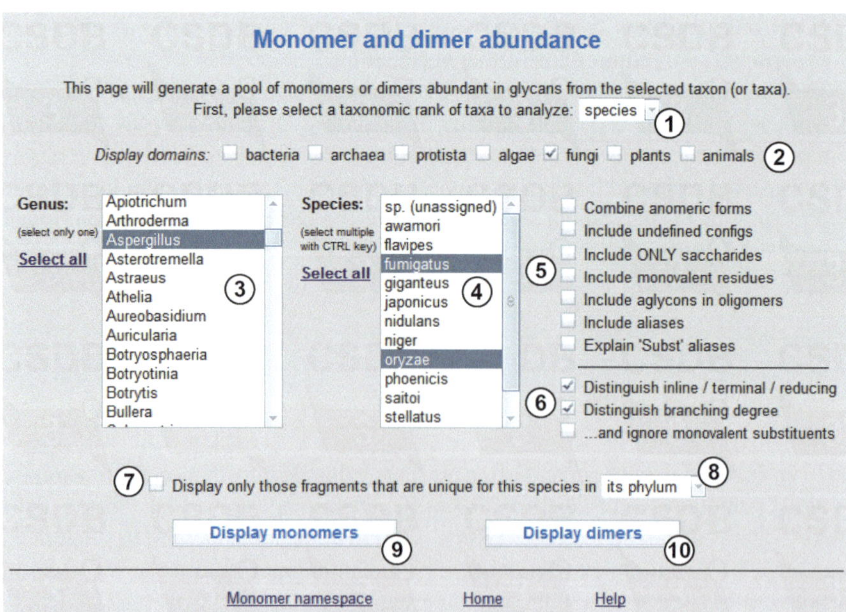

Fig. 5.23 Fragment abundance form

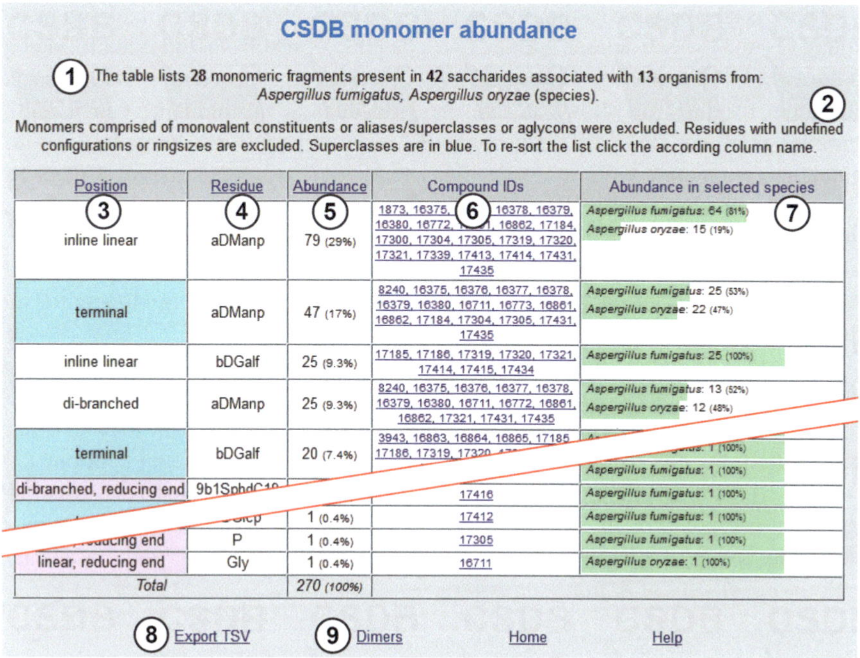

Fig. 5.24 Monomeric composition for *A. oryzae* and *A. fumigatus*

in the database. For fast navigation, we can focus on the list and start quick typing the genus name.

When the genus is selected, a list of its species appears in field (4). This list contains only those species whose compounds are present in the database. From the list, we select two species, *A. fumigatus* and *A. oryzae*.

Checkbox groups (5) and (6) define the search scope. Not checking "Combine anomeric forms" will treat different anomers as separate residues rather than combine them in a single entity. The "Include undefined configs" option allows processing of residues with undetermined anomeric, absolute, or ring size configurations, which are otherwise excluded from the statistics. We will not include them, as well as aglycons, aliases, and monovalent residues (such as methyl and acetyl substituents) for clarity (5). Checkboxes (6) determine how to display the position of a residue in the structure and residue branching degree (the number of substituents, including monovalent ones; to ignore monovalent substituents, check the corresponding checkbox).

Checkbox (7) allows displaying only fragments unique for this taxon (in our case – species) as compared to all biota, its kingdom or phylum (8). Buttons (9) and (10) run statistics on monomers and dimers, correspondingly. Pressing button (9) displays the table of monomers present in glycans and glycoconjugates from *A. fumigatus* and *A. oryzae* (Fig. 5.24).

There is an overview stating the number of fragments, structures, and organisms found (1) and the restrictions applied (2). The table of results includes the following columns: position of the residue in the structure if it was checked to be distinguished (terminal residues are shown in cyan and residues at the reducing end in pink; the branching degree is also indicated) (3), residue names and configurations (4), abundance (how many times a particular residue occurs in the structures matching the query) (5), compound IDs (links to the corresponding compounds) (6), and abundance in the selected taxa (7). The page also contains accessory links, e.g., export of results to tab-separated values (which can be copied and pasted into Microsoft Excel or other spreadsheet software) (8) and statistics on dimers for the current query (9).

According to the results, CSDB contains 42 saccharides from the *A. fumigatus* and *A. oryzae* strains, and these saccharides are comprised of 28 monomeric residues, α-D-mannopyranose being the most abundant. The columns may be sorted by position, residue name, or abundance by clicking on the column captions (3), (4), or (5). The results show that rarely occurring β-D-galactofuranose occupies the terminal position in 20 times in 13 distinct structures from *A. fumigatus*; thus, the presence of this residue is a potential candidate for probing for the immunospecificity of *A. fumigatus* fungi.

5.4.2 Example 10. *Find Which Dimeric Fragments (Including Sugars, Aglycons, and Other Residues) of Higher Plant Carbohydrates Are Specific to Lupins*

This problem addresses the unique features of the genus *Lupinus* in terms of biosynthesis of glycans and allows prediction of unique lupin glycosyltransferases for further search in proteomic databases.

This query can be performed using the "Fragment abundance" tool available from the main menu. The query form is shown in Fig. 5.25.

Here we select the taxonomic rank as genus (1) and the display group as plants (2) and then choose the *Lupinus* genus from the corresponding list (3). To see dimers built not only of monosaccharides but also of monovalent residues, aglycons, and aliases, the corresponding checkboxes should be checked (4). If the "Explain 'Subst' aliases" checkbox is unchecked, all substituents for which there are no reserved residue names in the database will be displayed as "Subst" and treated together. Since the study of dimeric fragments is the first step for revealing transferases, the "Include undefined configs" checkbox is unchecked, and the result will contain no fragments with underdetermined entities or unknown linkage positions.

To find only fragments extracted from this genus but not from any other organism belonging to the *Streptophyta* phylum (higher plants), we check checkbox (5) and select *phylum* from drop-down list (6). Pressing button (7) processes the query.

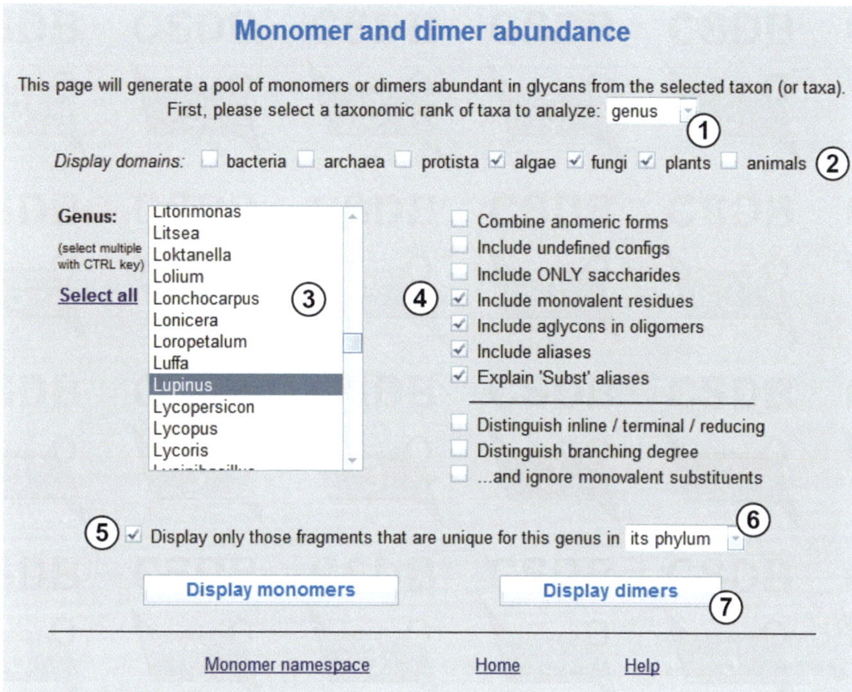

Fig. 5.25 Fragment abundance form

The resulting table is shown in Fig. 5.26. There is an overview stating the number of fragments, structures, and organisms found (1). The table consists of the following columns: donor (2), linkage (3), and acceptor (4) (reflecting two residues in the dimeric fragment and the linkage between them, respectively), abundance (how many times a particular dimer occurs in the structures matching the query) (5), compound IDs (links to the corresponding compounds) (6), and abundance in the selected taxa (may be less than 100 % if multiple taxa were selected for comparison; in our case, there is a single taxon, *Lupinus*) (7). This table can be exported to tab-separated values for further processing in spreadsheet software (8). Link (9) displays the monomeric fragment statistics for the same query. We can see that there are seven lupin-specific dimers containing a characteristic aglycon moiety at the reducing end, of which 21-O-β-D-xylopyranosyl-soyasapogenol A is the most frequent. Of those compounds deposited in the database, lupins possess a single disaccharide unique in higher plants, namely, 4-O-α-L-arabinofuranosyl-β-D-rhamnopyranose.

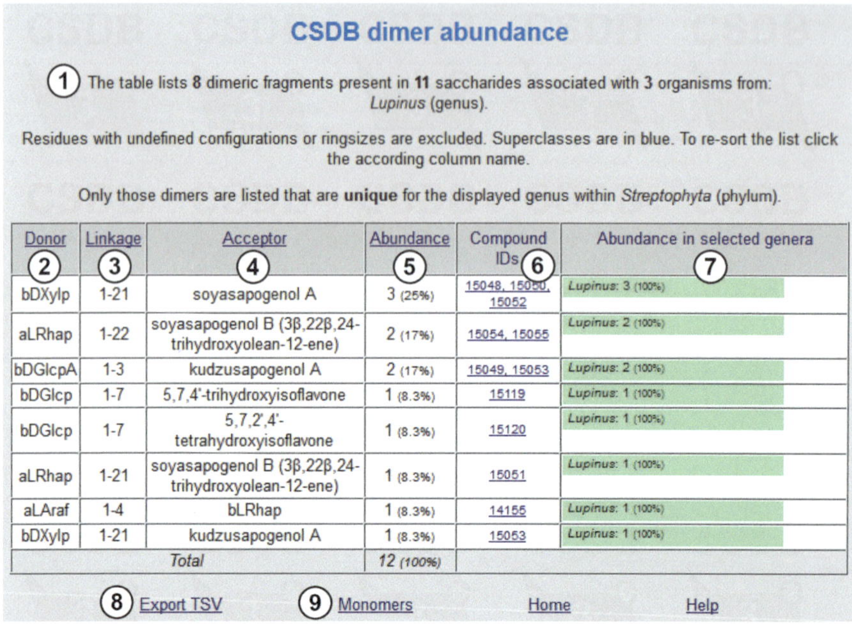

CSDB dimer abundance

(1) The table lists **8** dimeric fragments present in **11** saccharides associated with **3** organisms from: *Lupinus* (genus).

Residues with undefined configurations or ringsizes are excluded. Superclasses are in blue. To re-sort the list click the according column name.

Only those dimers are listed that are **unique** for the displayed genus within *Streptophyta* (phylum).

Donor (2)	Linkage (3)	Acceptor (4)	Abundance (5)	Compound IDs (6)	Abundance in selected genera (7)
bDXylp	1-21	soyasapogenol A	3 (25%)	15048, 15050, 15052	*Lupinus*: 3 (100%)
aLRhap	1-22	soyasapogenol B (3β,22β,24-trihydroxyolean-12-ene)	2 (17%)	15054, 15055	*Lupinus*: 2 (100%)
bDGlcpA	1-3	kudzusapogenol A	2 (17%)	15049, 15053	*Lupinus*: 2 (100%)
bDGlcp	1-7	5,7,4'-trihydroxyisoflavone	1 (8.3%)	15119	*Lupinus*: 1 (100%)
bDGlcp	1-7	5,7,2',4'-tetrahydroxyisoflavone	1 (8.3%)	15120	*Lupinus*: 1 (100%)
aLRhap	1-21	soyasapogenol B (3β,22β,24-trihydroxyolean-12-ene)	1 (8.3%)	15051	*Lupinus*: 1 (100%)
aLAraf	1-4	bLRhap	1 (8.3%)	14155	*Lupinus*: 1 (100%)
bDXylp	1-21	kudzusapogenol A	1 (8.3%)	15053	*Lupinus*: 1 (100%)
		Total	*12 (100%)*		

(8) Export TSV (9) Monomers Home Help

Fig. 5.26 Unique dimeric fragments in plant compounds from the *Lupinus* genus

5.4.3 Example 11. *Study Coverage Statistics of* Proteobacteria

To review coverage for a particular taxonomic group(s), we select the "Coverage stats" link from the main menu. Figure 5.27 shows the "Coverage statistics" form. Here we select a taxonomic rank, in our case – phylum (1) – and the display group, bacteria (2). Then we select "Proteobacteria" from drop-down list (3). Publication year span (4) and structure type (5) filters are available. The "Display coverage" button (6) processes the query.

The results are shown in Fig. 5.28. The table includes the following columns: selected taxon(s) (1) (in our case, a single phylum was selected, so all the rows in this column are the same); subtaxa of the selected taxon(s) (2) (in our case, classes comprising the *Proteobacteria* phylum); structures (number of structures for the corresponding subtaxon found in the database, together with their part of the total number of structures found for the whole taxon) (3); publications (number of publications in which these structures are present) (4); organisms (number of taxonomically distinct organisms or groups of organisms from which these structures were obtained) (5); and NMR spectra (number of NMR spectra for these structures present in the database) (6). The cumulative values are shown in the last row (7).

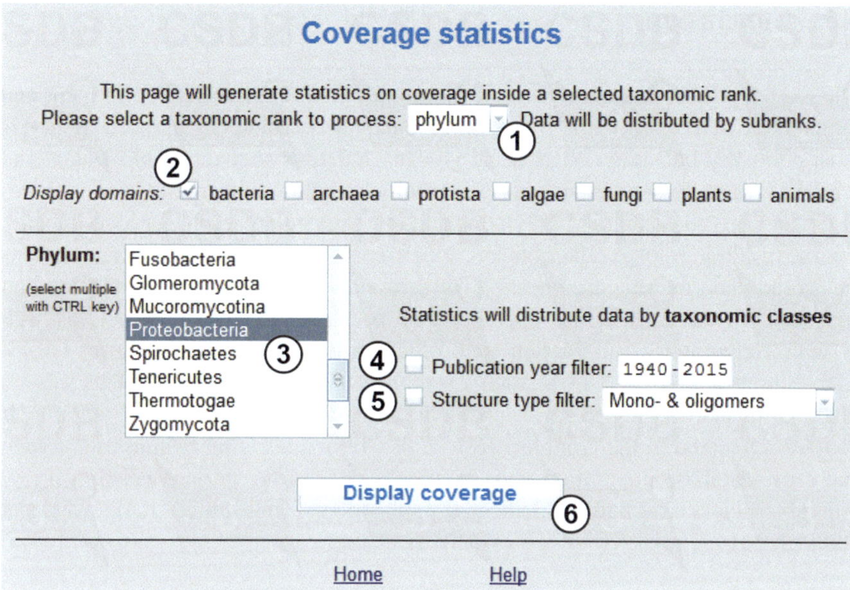

Fig. 5.27 Coverage statistics form

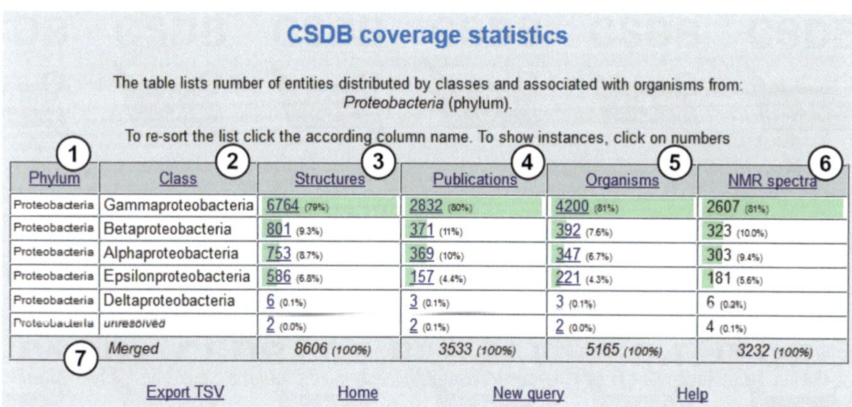

Phylum	Class	Structures	Publications	Organisms	NMR spectra
Proteobacteria	Gammaproteobacteria	6764 (79%)	2832 (80%)	4200 (81%)	2607 (81%)
Proteobacteria	Betaproteobacteria	801 (9.3%)	371 (11%)	392 (7.6%)	323 (10.0%)
Proteobacteria	Alphaproteobacteria	753 (8.7%)	369 (10%)	347 (6.7%)	303 (9.4%)
Proteobacteria	Epsilonproteobacteria	586 (6.8%)	157 (4.4%)	221 (4.3%)	181 (5.6%)
Proteobacteria	Deltaproteobacteria	6 (0.1%)	3 (0.1%)	3 (0.1%)	6 (0.2%)
Proteobacteria	*unresolved*	2 (0.0%)	2 (0.1%)	2 (0.0%)	4 (0.1%)
	Merged	*8606 (100%)*	*3533 (100%)*	*5165 (100%)*	*3232 (100%)*

Fig. 5.28 Coverage statistics for the *Proteobacteria* phylum

The numbers of structures, publications, and organisms are links to lists of the corresponding compounds, articles, and organisms. The table can be sorted by clicking column captions (1)–(6).

5.5 Conclusion

The Carbohydrate Structure Database combines bacterial (BCSDB) and plant and fungal (PFCSDB) parts. It is freely accessible and continuously updated. It covers most published bacterial and archaeal glycans and aims at covering most plant and fungal glycans. Besides structures of carbohydrate and noncarbohydrate moieties of glycans, glyco-containing polymers, and low molecular weight glycoconjugates, chemical, bibliographic, taxonomical, NMR spectroscopic, and other annotations are stored. The records are checked for errors and undergo manual curation and approval. The database is equipped with a user-friendly web interface, as well as with endpoints for interaction with other projects in glycoinformatics (Aoki-Kinoshita et al. 2013). This chapter exemplifies application of CSDB to solving several typical everyday scientific problems and gives basic CSDB user features. If you are interested in the complete feature list, principles lying behind the project, and cross database integration options, please refer to the project website (http://csdb.glycoscience.ru) and the dedicated publications (Toukach 2011; Egorova and Toukach 2014; Toukach and Egorova 2016).

References

Acland A, Agarwala R, Barrett T, Beck J, Benson DA, Bollin C, Bolton E, Bryant SH, Canese K, Church DM, Clark K, DiCuccio M, Dondoshansky I, Federhen S, Feolo M, Geer LY, Gorelenkov V, Hoeppner M, Johnson M, Kelly C, Khotomlianski V, Kimchi A, Kimelman M, Kitts P, Krasnov S, Kuznetsov A, Landsman D, Lipman DJ, Lu Z, Madden TL, Madej T, Maglott DR, Marchler-Bauer A, Karsch-Mizrachi I, Murphy T, Ostell J, O'Sullivan C, Panchenko A, Phan L, Pruitt DP, Rubinstein W, Sayers EW, Schneider V, Schuler GD, Sequeira E, Sherry ST, Shumway M, Sirotkin K, Siyan K, Slotta D, Soboleva A, Soussov V, Starchenko G, Tatusova TA, Trawick BW, Vakatov D, Wang Y, Ward M, John Wilbur W, Yaschenko E, Zbicz K (2014) Database resources of the National Center for Biotechnology Information. Nucleic Acids Res 42:D7–D17

Aoki-Kinoshita KF, Bolleman J, Campbell MP, Kawano S, Kim JD, Lütteke T, Matsubara M, Okuda S, Ranzinger R, Sawaki H, Shikanai T, Shinmachi D, Suzuki Y, Toukach P, Yamada I, Packer NH, Narimatsu H (2013) Introducing glycomics data into the Semantic Web. J Biomed Semant 4:39

Damerell D, Ceroni A, Maass K, Ranzinger R, Dell A, Haslam SM (2012) The GlycanBuilder and GlycoWorkbench glycoinformatics tools: updates and new developments. Biol Chem 393:1357–1362

Doubet S, Bock K, Smith D, Darvill A, Albersheim P (1989) The complex carbohydrate structure database. Trends Biochem Sci 14:475–477

Egorova KS, Toukach PhV (2014) Expansion of coverage of Carbohydrate Structure Database (CSDB). Carbohydr Res 389:112–114

Egorova KS, Kondakova AN, Toukach PhV (2015) Carbohydrate structure database: tools for statistical analysis of bacterial, plant and fungal glycomes. Database, (article ID bav073):1–22. doi: 10.1093/database/bav073

Herget S, Ranzinger R, Maass K, von der Lieth CW (2008) GlycoCT—a unifying sequence format for carbohydrates. Carbohydr Res 343:2162–2171

Kapaev RR, Toukach PhV (2015) Improved carbohydrate structure generalization scheme for ^{1}H and ^{13}C NMR simulations. Anal Chem 87(14):7006–7010

Kapaev RR, Egorova KS, Toukach PhV (2014) Carbohydrate structure generalization scheme for database-driven simulation of experimental observables, such as NMR chemical shifts. J Chem Inf Model 54:2594–2611

Kapaev RR, Toukach PhV (2016) Simulation of 2D NMR Spectra of Carbohydrates Using GODESS Software. J Chem Inf Model 56:1100–1104

Lipkind GM, Shashkov AS, Knirel YA, Vinogradov EV, Kochetkov NK (1988) A computer-assisted structural analysis of regular polysaccharides on the basis of ^{13}C-n.m.r. data. Carbohydr Res 175:59–75

Loss A, Bunsmann P, Bohne A, Loss A, Schwarzer E, Lang E, von der Lieth CW (2002) SWEET-DB: an attempt to create annotated data collections for carbohydrates. Nucleic Acids Res 30:405–408

Lütteke T (2012) The use of glycoinformatics in glycochemistry. Beilstein J Org Chem 8:915–929

Royle L, Dwek A, Rudd M (2006) Determining the structure of oligosaccharides N- and O-linked to glycoproteins. In: Current protocols in protein science. Wiley, Newyork, pp 12.6.1–12.6.45

Sharon N (1988) Nomenclature of glycoproteins, glycopeptides and peptidoglycans. Pure Appl Chem 60:1389–1394

Toukach PhV (2011) Bacterial carbohydrate structure database 3: principles and realization. J Chem Inf Model 51:159–170

Toukach FV, Ananikov VP (2013) Recent advances in computational predictions of NMR parameters for the structure elucidation of carbohydrates: methods and limitations. Chem Soc Rev 42:8376–8415

Toukach PhV, Egorova KS (2015) Bacterial, plant, and fungal carbohydrate structure databases: daily usage. In: Lütteke T, Frank M (eds) Glycoinformatics, vol 1273, Methods in molecular biology. Springer, New York, pp 55–85, ch.5

Toukach PhV, Egorova KS (2016) Carbohydrate Structure Database merged from bacterial, archaeal, plant and fungal parts. Nucleic Acids Res 44(D1):D1229–D1236

Toukach FV, Shashkov AS (2001) Computer-assisted structural analysis of regular glycopolymers on the basis of ^{13}C NMR data. Carbohydr Res 335:101–114

Varki A, Cummings RD, Aebi M, Packer NH, Seeberger PH, Esko JD, Stanley P, Hart G, Darvill A, Kinoshita T, Prestegard JJ, Schnaar RL, Freeze HH, Marth JD, Bertozzi CR, Etzler ME, Frank M, Vliegenthart JF, Lütteke T, Perez S, Bolton E, Rudd P, Paulson J, Kanehisa M, Toukach P, Aoki-Kinoshita KF, Dell A, Narimatsu H, York W, Taniguchi N, Kornfeld S (2015) Symbol nomenclature for graphical representations of glycans. Glycobiology 25(12):1323–1324

Part III
Glyco-related Genes and Proteins

Chapter 6
The CAZy Database/the Carbohydrate-Active Enzyme (CAZy) Database: Principles and Usage Guidelines

Nicolas Terrapon, Vincent Lombard, Elodie Drula, Pedro M. Coutinho, and Bernard Henrissat

Keywords Carbohydrate-active enzymes • Protein domains • Family classification • Structural and functional annotation • Sequence • genome • and metagenome analysis • Polysaccharide Utilization Loci

Carbohydrate-Active enZymes (CAZymes) assemble, breakdown, and modify glycans and glycoconjugates using their catalytic and binding modules (functional protein domains). The CAZy database offers since 1998 an online and continuously updated classification of CAZyme modules (Lombard et al. 2014). Each module family in the CAZy classification has been created based on experimentally characterized protein modules from the literature, and the families are populated by related module sequences from public protein sequence databases. Since no universal threshold allows the systematic classification of the various CAZyme families, CAZy annotations result from an expert combination of module modeling/calibration and human curation. CAZy annotations are made publicly available for all proteins released by GenBank (Benson et al. 2012), Swiss-Prot (Boutet et al. 2016) and the Protein Data Bank (PDB; http://www.rcsb.org; (Berman et al. 2000)).

N. Terrapon • V. Lombard • P.M. Coutinho
Aix-Marseille Université, AFMB UMR 7257, 163 Avenue de Luminy, 13288 Marseille, France

Centre National de la Recherche Scientifique, AFMB UMR 7257, 163 Avenue de Luminy, 13288 Marseille, France

E. Drula
Aix-Marseille Université, AFMB UMR 7257, 163 Avenue de Luminy, 13288 Marseille, France

Institut National de la Recherche Agronomique, BBF UMR 1163, Polytech Marseille, 163 Avenue de Luminy, 13288 Marseille, France

B. Henrissat (✉)
Centre National de la Recherche Scientifique, AFMB UMR 7257, 163 Avenue de Luminy, 13288 Marseille, France

Department of Biological Sciences, King Abdulaziz University, Jeddah, Saudi Arabia
e-mail: Bernard.Henrissat@afmb.univ-mrs.fr

K.F. Aoki-Kinoshita (ed.), *A Practical Guide to Using Glycomics Databases*,
DOI 10.1007/978-4-431-56454-6_6

Further, functional and 3-D structural information, curated from the literature on a regular basis, constitute essential added values to the CAZy annotation. In this spirit, the display of ligand information from crystallographic complexes has been recently developed (Lombard et al. 2014). This chapter will guide the reader through the usage of CAZy to search enzyme annotations. It will also answer frequent questions such as (i) how to obtain CAZy annotations for a specific protein, a genome, or a metagenome, (ii) how to have a newly characterized family included in the CAZy classification scheme, (iii) why CAZy does not cover all protein families related to glycans/glycoconjugates, and (iv) why CAZy does not transfer functional annotation to similar sequences. Finally, we present here a recent CAZy-associated tool, namely, the Polysaccharide Utilization Loci (PUL) predictor and database in *Bacteroidetes* species (Terrapon et al. 2015).

6.1 Classes of CAZy Modules

The CAZy classification covers sequences from all taxonomic groups and provides the ground for common nomenclature for CAZymes across many glycobiologists, often specialized in some preferred taxa. Among the large diversity of proteins acting on glycoconjugates, poly- and oligosaccharides, the CAZy classification covers several enzyme classes that catalyze their assembly, breakdown, or modifications.

- Glycosyltransferases (GTs) represent the unique class in charge of glycan assembly, forming glycosidic bonds from phospho-activated sugar donors by either inverting or retaining the anomeric configuration (Campbell et al. 1997; Coutinho et al. 2003).
- Glycoside hydrolases (GHs) and polysaccharide lyases (PLs) are responsible for the cleavage of glycans (Lombard et al. 2010). GHs hydrolyze or transglycosylate glycosidic bonds, while PLs cleave the glycosidic bonds of uronic acid-containing polysaccharides by a β-elimination mechanism. Because of their widespread importance for biotechnological and biomedical applications, GHs and PLs constitute so far the best biochemically characterized set of enzymes present in the CAZy database. Interestingly, while GH-coding genes are abundant and present in the vast majority of genomes corresponding to almost half of the enzymes classified in CAZy, PLs only represent a very small proportion (Table 6.1).
- Because lignin is invariably found together with polysaccharides in the plant cell wall, CAZy recently integrated enzyme families known to be involved in lignin degradation along with lytic polysaccharide monooxygenases in a new class termed auxiliary activities (AAs) to accommodate the large range of mechanisms and substrates (Levasseur et al. 2013).
- Carbohydrate esterases (CEs) are enzymes that remove O- or N-acyl substituents on glycans (Coutinho 1999) and thereby often facilitate the action of GHs and PLs on complex polysaccharides. However, as the specificity barrier between

Table 6.1 Statistics of the CAZy website content in April 2016

CAZy class	No. of families	No. of modules in a family	No. of nonclassified	Experimentally characterized	With a 3-D structure
GH	135	289,722	4366	12,377	5065
GT	98	238,679	4449	2205	889
CBM	73	70,049	358	925	911
PL	24	7119	466	406	186
CE	16	32,843	929	311	209
AA	13	12,205	253	457	233

For each module class (first column), we indicate the number of distinct families in CAZy classification scheme (second column), the number of module occurrences classified within a specific CAZy family (distinct protein IDs, possibly having multiple module occurrences – third column), and the number of occurrences nonclassified into any family (fourth column). The numbers of corresponding proteins that have been experimentally characterized in the literature (fifth column) or have been structurally described (sixth column) are provided

carbohydrate esterases and other esterase activities is thin, it is likely that the CAZy sequence-based classification incorporates some enzymes that may act on noncarbohydrate esters (as illustrated by the high proportion of CEs falling in the "Nonclassified" category – see below).

- Carbohydrate-binding modules (CBMs) have no enzymatic activity per se but are known to potentiate the activity of many enzyme activities described above by targeting to and promoting a prolonged interaction with the substrate. If CBMs can occasionally exist in isolated or tandem forms, they are usually combined with catalytic modules within enzymes (Boraston et al. 2004). For this reason CBMs are set apart from other non-catalytic sugar-binding proteins (such as lectins and sugar transporters – see Sect. 6.6) and integrated in the CAZy classification scheme (Coutinho 1999).

The CAZy module classes are subdivided into families (see Table 6.1) based on amino acid sequence similarity, which almost invariably involves similar mechanisms. Families are designated using a simple formula including the class and a number referring to the order of family creation within the class, such as GT1 or GH130. However, the occurrence of enzymes that act on different substrates within a single family prevents the direct functional annotation of CAZymes based on family assignment. Phylogenetic analyses can frequently improve the correlation between sequence and specificity by defining subfamilies as was done for families GH5 (Aspeborg et al. 2012), GH13 (Stam et al. 2006), GH30 (St John et al. 2010), GH43 (Mewis et al. 2016), and all PL families (Lombard et al. 2010). More subfamilies are currently under development internally in CAZy and could be released when in-depth analyses confirm the stability of the subfamilies when the number of sequences increases. Finally, some of most remote homologs, for which sequence similarity is still detectable but cannot guarantee anymore any level of functional similarity nor family assignment, are also reported but without family assignment in a "Nonclassified modules" list, for each CAZy class (see

Table 6.1). The "Nonclassified modules" list is not a family per se but gathers many heterogeneous remote sequences, some that might give rise to distinct CAZy families in the future.

6.2 Browsing the CAZy Website

The homepage of the CAZy website includes a banner with several links to browse the CAZy annotation, either by CAZyme class (tabs labeled "enzyme classes" and "associated modules") or by genome (see Fig. 6.1).

6.2.1 Browsing by CAZy Class and Families

6.2.1.1 CAZy Class Webpages

The webpage dedicated to each CAZy class, illustrated in Fig. 6.2, starts with an introduction to the module function, completed by some details about the catalytic mechanisms for GHs, PLs, and GTs. Further, some statistics are given about the number of occurrence of modules in one family of this class and about the most distant homologs assigned to this class but not into a family, referred to as "Nonclassified modules." Finally, it provides the user with access to all families created in this class – links to individual webpages – in two tables: a simple ordered enumeration of existing families and a functionally oriented table that lists the different families by EC number. Please note that due to the modular nature of CAZymes, these EC numbers may not be directly associated with the family but simply borne by adjacent modules. Hence, enzymatic families with more than one known activity are repeated along this table.

6.2.1.2 CAZy Family Webpages

Each webpage dedicated to a CAZy family, illustrated in Figs. 6.3 and 6.4, contains a synthetic and updated report with all known activities (EC numbers and activity names) in the family. It should be noted that contrary to the class webpage, the activities that are listed in the header of the families correspond to the actual modules of the family and not the activity of adjacent modules. The report also

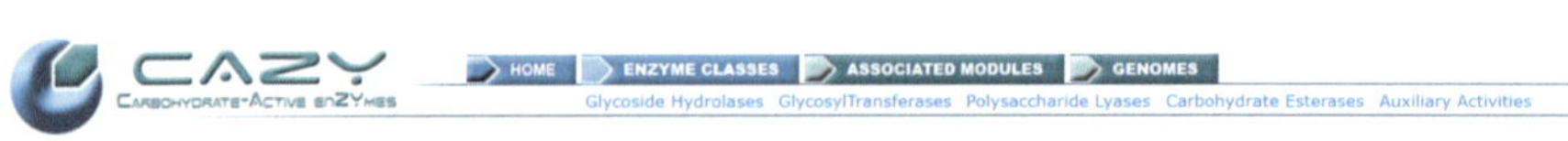

Fig. 6.1 Banner of the CAZy website where the user can choose to browse the data by CAZyme module class/family or search for a specific genome annotation

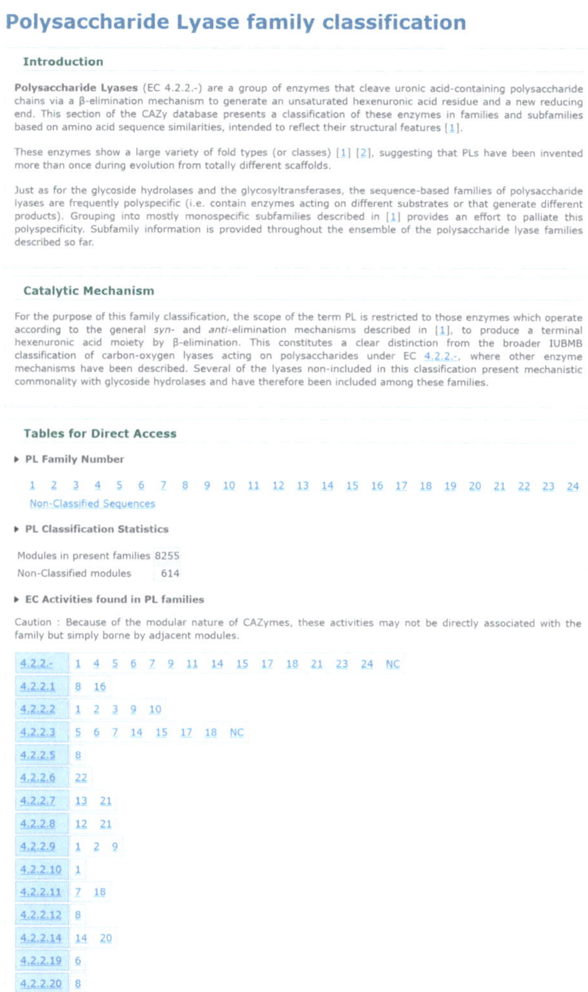

Polysaccharide Lyase family classification

Introduction

Polysaccharide Lyases (EC 4.2.2.-) are a group of enzymes that cleave uronic acid-containing polysaccharide chains via a β-elimination mechanism to generate an unsaturated hexenuronic acid residue and a new reducing end. This section of the CAZy database presents a classification of these enzymes in families and subfamilies based on amino acid sequence similarities, intended to reflect their structural features [1].

These enzymes show a large variety of fold types (or classes) [1] [2], suggesting that PLs have been invented more than once during evolution from totally different scaffolds.

Just as for the glycoside hydrolases and the glycosyltransferases, the sequence-based families of polysaccharide lyases are frequently polyspecific (i.e. contain enzymes acting on different substrates or that generate different products). Grouping into mostly monospecific subfamilies described in [1] provides an effort to palliate this polyspecificity. Subfamily information is provided throughout the ensemble of the polysaccharide lyase families described so far.

Catalytic Mechanism

For the purpose of this family classification, the scope of the term PL is restricted to those enzymes which operate according to the general *syn*- and *anti*-elimination mechanisms described in [1], to produce a terminal hexenuronic acid moiety by β-elimination. This constitutes a clear distinction from the broader IUBMB classification of carbon-oxygen lyases acting on polysaccharides under EC 4.2.2.-, where other enzyme mechanisms have been described. Several of the lyases non-included in this classification present mechanistic commonality with glycoside hydrolases and have therefore been included among these families.

Tables for Direct Access

▸ **PL Family Number**

1 2 3 4 5 6 7 8 9 10 11 12 13 14 15 16 17 18 19 20 21 22 23 24
Non-Classified Sequences

▸ **PL Classification Statistics**

Modules in present families	8255
Non-Classified modules	614

▸ **EC Activities found in PL families**

Caution : Because of the modular nature of CAZymes, these activities may not be directly associated with the family but simply borne by adjacent modules.

4.2.2.-	1 4 5 6 7 9 11 14 15 17 18 21 23 24 NC
4.2.2.1	8 16
4.2.2.2	1 2 3 9 10
4.2.2.3	5 6 7 14 15 17 18 NC
4.2.2.5	8
4.2.2.6	22
4.2.2.7	13 21
4.2.2.8	12 21
4.2.2.9	1 2 9
4.2.2.10	1
4.2.2.11	7 18
4.2.2.12	8
4.2.2.14	14 20
4.2.2.19	6
4.2.2.20	8

Fig. 6.2 Screenshot of the CAZy webpage that describes the PL class. Following an introductory description, statistics data and direct access to individual families are provided in different tables

specifies the mechanism (e.g., inverting or retaining), structural fold, catalytic residues, etc. where known or appropriate. More extensive encyclopedic knowledge of the biology/chemistry of some families can be obtained through links to the CAZypedia resource (see Sect. 6.8). CAZy also provides statistics about the number of known modules in each family, the number of members with a 3-D structure, and the number of functionally characterized enzymes. Finally, the complete list of modules can be browsed with a tab subdivision to see either all or restricted to a specific kingdom of life or to structurally/experimentally characterized cases. Almost all tabs present modules as lines containing the protein name, EC numbers

Glycoside Hydrolase Family 5

Known Activities	endo-β-1,4-glucanase / cellulase (EC 3.2.1.4); endo-β-1,4-xylanase (EC 3.2.1.8); β-glucosidase (EC 3.2.1.21); β-mannosidase (EC 3.2.1.25); β-glucosylceramidase (EC 3.2.1.45); glucan β-1,3-glucosidase (EC 3.2.1.58); licheninase (EC 3.2.1.73); exo-β-1,4-glucanase / cellodextrinase (EC 3.2.1.74); glucan endo-1,6-β-glucosidase (EC 3.2.1.75); mannan endo-β-1,4-mannosidase (EC 3.2.1.78); cellulose β-1,4-cellobiosidase (EC 3.2.1.91); steryl β-glucosidase (EC 3.2.1.104); endoglycoceramidase (EC 3.2.1.123); chitosanase (EC 3.2.1.132); β-primeverosidase (EC 3.2.1.149); xyloglucan-specific endo-β-1,4-glucanase (EC 3.2.1.151); endo-β-1,6-galactanase (EC 3.2.1.164); hesperidin 6-O-α-L-rhamnosyl-β-glucosidase (EC 3.2.1.168); β-1,3-mannanase (EC 3.2.1.-); arabinoxylan-specific endo-β-1,4-xylanase (EC 3.2.1.-); mannan transglycosylase (EC 2.4.1.-)
Mechanism	Retaining
Clan	GH-A
3D Structure Status	(β / α) 8
Catalytic Nucleophile/Base	Glu (experimental)
Catalytic Proton Donor	Glu (experimental)
Note	Once known as cellulase family A; many members have been assigned to subfamilies as described by Aspeborg et al. (2012) BMC Evol Biol. 12(1):186 (PMID: 22992189).
External resources	CAZypedia; HOMSTRAD; PROSITE;
Commercial Enzyme Provider(s)	MEGAZYME; NZYTech; PROZOMIX;
Statistics	GenBank accession (9499); Uniprot accession (1927); PDB accession (193); 3D entries (67); cryst (0)

Summary | All (8139) | Archaea (63) | Bacteria (6093) | Eukaryota (1865) | Viruses (3) | unclassified (115) | Structure (67) | Characterized (540) | Subfamilies (7294)

< | 1 | 2 | 3 | 4 | 5 | 6 | 7 | 8 | >

Archaea

Protein Name	EC#	Organism	GenBank	Uniprot	PDB/3D	Subf
Igag_0570		Ignisphaera aggregans DSM 17230	ADM27405.1	E0SSD3		1
Igag_0224		Ignisphaera aggregans DSM 17230	ADM27073.1	E0SQC3		1
endo-β-1,4-glucanase (EglB;TCel4;PAB0632)	3.2.1.4	Pyrococcus abyssi GE5	CAB49854.1 AEG76944.1 NP_126623.1	Q9V052		1
endo-β-1,4-glucanase (EGPh;EglB;TCel3;PH1171)	3.2.1.4	Pyrococcus horikoshii OT3	AAQ31833.1 AAQ31832.1 BAA30271.1 NP_143072.1	O58925	2ZUM[A] 2ZUN[A,B,C] 3AXX[A,B,C] 3QHM[A,B,C] 3QHN[A,B,C] 3QHO[A,B,C] 3VVG[A,B,C] 3W6L[A,B,C] 3W6M[A,B,C] 4DM1[A,B,C] 4DM2[A,B,C]	1
Py04_1787		Pyrococcus sp. ST04	AFK23355.1			1
Shell_0798		Staphylothermus hellenicus DSM 12710	ADI31911.1	D7D815		1
CHITON_2145		Thermococcus chitonophagus	CUX78924.1			1
ADU37_CDS22600		Thermococcus sp. 2319x1	ALV63957.1			1
Sequence 2 from patent US 6329187 (fragment)		unidentified archaeon AEPII1a	AAQ77850.1			1
Huta_2396		Halorhabdus utahensis DSM 12940	ACV12562.1	C7NMC3		7
Htur_4673		Haloterrigena turkmenica DSM 5511	ADB63456.1	D2S259		7
L593_05380		Salinarchaeum sp. Harcht-Bsk1	AGN01025.1			7
QT06_C0001G0664		archaeon GW2011_AR15	AJF61504.1			13
ASAC_0734		Acidilobus saccharovorans 345-15	ADL19140.1	D9Q1F2		19
SE86_04550 (fragment)		Acidilobus sp. 7A	AMD30698.1			19
SE86_03690 (fragment)		Acidilobus sp. 7A	AMD30586.1			19
SE86_03695 (fragment)		Acidilobus sp. 7A	AMD30587.1			19
Cmaq_0936		Caldivirga maquilingensis IC-167	ABW01768.1	A8MDB2		19
Igag_0312		Ignisphaera aggregans DSM 17230	ADM27158.1	E0SQT4		19
PTO1452		Picrophilus torridus DSM 9790	AAT44037.1	Q6KZ15		19
SiH_2292		Sulfolobus islandicus HVE10/4	ADX83632.1			19
LD85_2652		Sulfolobus islandicus L.D.8.5	ADB88268.1	D2PGC8		19
LS215_2522		Sulfolobus islandicus L.S.2.15	ACP36477.1	C3ML75		19
SiL_2199		Sulfolobus islandicus LAL14/1	AGJ63637.1			19
M1425_2350		Sulfolobus islandicus M.14.25	ACP39078.1	C3MSD5		19
M1627_2428		Sulfolobus islandicus M.16.27	ACP56280.1	C3N1Q3		19
M164_2357		Sulfolobus islandicus M.16.4	ACR42955.1	C4KL74		19

Fig. 6.3 Screenshot of the CAZy webpage that describes the GH5 family information with the specific details on protein sequences attributed to subfamilies in the rightmost column ("Subfamilies" tab)

if any, the organism, the GenBank accessions (one reference in bold, and redundant ones below), and the UniProt and PDB identifiers (crystals not yet solved/deposited labeled as "cryst"). Further, for families with subfamily division, a tab at the very right shows the subfamily number (see Fig. 6.3). Finally, the "Structure" tab is the special case (see Fig. 6.4): it does not contain GenBank nor UniProt accessions but instead displays more detailed information from the PDB files. For each PDB file, we extract and display the resolution when the structure was solved by x-ray

Fig. 6.4 Screenshot of the CAZy webpage corresponding to the GT3 family with the specific display (at the *bottom*) of structurally characterized proteins and related information ("Structure" tab)

crystallography (otherwise we indicate the method: powder diffraction or nuclear magnetic resonance).

6.2.1.3 Recent Addition of Carbohydrate Ligands

The PDB does not provide any option to perform a comprehensive search for carbohydrate structures found in CAZyme binding sites, and, unlike proteins or nucleic acids, the nomenclature for carbohydrate residues within PDB files is not yet standardized. Significantly, the information on how the isolated carbohydrate residues are linked to each other is not described in PDB files. For each PDB file, we thus extract the carbohydrate ligand information using PDB-care (www.glycosciences.de/tools/pdb-care/; (Lütteke and Von Der Lieth 2004)). These ligands are filtered for display as follows. N- and O-glycans covalently linked to Asn or Ser/Thr residues are discarded as they correspond to posttranslational modifications of the protein structure, and generally not directly linked to enzyme function. The remaining carbohydrate ligands are retained as they should describe functional recognition in catalytic or other binding sites in CAZymes and are displayed in the structure pages of CAZy (see Fig. 6.4) following IUPAC nomenclature (Lombard et al. 2014). Not all carbohydrate structures are susceptible to automated description by PDB-care. In a number of cases, we must manually curate and provide IUPAC descriptions for structures that are unsuitable to PDB-care such as (i) nonreducing glycans (cyclodextrins, sucrose and sucrose derivatives, trehalose, kestose, raffinose, nystose, etc.), (ii) ligands that are made of both carbohydrate and noncarbohydrate moieties such as acarbose, (iii) thio-oligosaccharides, (iv) fluorine-containing carbohydrates, and (v) oligosaccharides containing 3,6-anhydro bridges. In addition, automated scripts have been devised to handle close to 200 carbohydrate analogues that we denote <carb_like_*ligandref*> where *ligandref* corresponds to the three-letter ligand name given by the PDB. For instance, the carbohydrate-like inhibitor 1-deoxynojirimycin appears as <carb_like_NOJ>. Significantly, nearly half (45 %) of the approx. 7500 PDB structures present in CAZy as of April 2016 bear a glycan-containing ligand or a glycan analog revealing enzyme-glycan interactions.

6.2.2 Browsing by Genome

The collection of carbohydrate-active enzymes encoded by the genome of an organism, hereafter referred to as "CAZome," provides an insight into the nature and extent of the metabolism of complex carbohydrates of the species. The CAZomes of free-living organisms typically correspond to 1–5 % of the predicted coding genes. Because of the massive chemical, structural, and functional variability of carbohydrates, CAZome comparisons can highlight the adaptation of the CAZymes repertoire of species to their carbohydrate environment.

The CAZy website allows to browse CAZomes by kingdom of life, where species are presented in alphabetically ordered tabs. For each organism, the complete list of CAZymes is displayed in addition to the family distribution, as illustrated in Fig. 6.5. As of April 2016, CAZy is close to 5000 public genomes, with more than 4000 bacterial genomes but less than 200 eukaryotes. This is due to the fact that the CAZomes listed in the CAZy website correspond to protein models of *finished* genomes from the daily releases of GenBank. In just a few rare cases, genomes with protein models not released as finished entries in GenBank, but publicly available, have been analyzed and are presented in CAZy. However, for these few cases, the display only shows the number of proteins in each family, but does not feature the actual list of proteins with database accessions. The taxonomical lineage of the genome is directly extracted and updated from the NCBI Taxonomy database.

6.3 Retrieving Information from the Search Form/Engine in the CAZy Website

To facilitate search of specific information, the CAZy website includes a search tool, which appears at the top right of every page. The search form is composed of a text area with a magnifying glass to enter the required query and a drop-down list to indicate the field of searched information (see Fig. 6.6, with "Site" option to search in every field). Main fields notably allow the user to search by CAZy family, organism (name, even partial, or taxonomy id), protein name or accessions (GenBank, UniProt, and PDB), ligand (indicating a sugar-like compound, a part of a chain, or the catalytic residue to which the ligand is attached, e.g., "GLU"), activity (EC number/name), etc. The result of the search either indicates the modularity of the protein or provides direct links to the relevant genome and family webpages.

Porphyromonas gingivalis ATCC 33277

Taxonomy ID : 431947
Lineage: cellular organisms; Bacteria; FCB group; Bacteroidetes/Chlorobi group; Bacteroidetes; Bacteroidia; Bacteroidales; Porphyromonadaceae; Porphyromonas; Porphyromonas gingivalis

Glycoside Hydrolase Family	2	3	13	20	23	24	27	29	33	57	73	77	92	108	109
Number of sequences	3	1	2	1	2	2	1	2	1	1	1	1	4	1	1

GlycosylTransferase Family	2	3	4	5	9	19	26	28	30	35	51	83
Number of sequences	12	1	7	1	1	1	1	1	1	1	1	1

Carbohydrate Esterase Family	4	11
Number of sequences	1	1

Carbohydrate-Binding Module Family	20	50
Number of sequences	2	3

List Of Proteins

Protein Name	Family	Reference Accession
PGN_0009	GH3	BAG32528.1
PGN_0030	GH2	BAG32549.1
PGN_0039	GH20	BAG32558.1
PGN_0055	GH24	BAG32574.1
PGN_0193	CE4	BAG32712.1
PGN_0206	GT19	BAG32725.1
PGN_0225	GT2	BAG32744.1
PGN_0227	GT4	BAG32746.1
PGN_0232	GT2	BAG32751.1
PGN_0233	GT26	BAG32752.1
PGN_0242	GT4	BAG32761.1
PGN_0252	GH23,CBM50	BAG32771.1
PGN_0361	GT2	BAG32880.1
PGN_0405	GH92	BAG32924.1
PGN_0406	GH92	BAG32925.1
PGN_0427	GH57	BAG32946.1
PGN_0428	GT4	BAG32947.1
PGN_0544	GT30	BAG33063.1
PGN_0627	GT28	BAG33146.1
PGN_0700	GH109	BAG33219.1
PGN_0701	GH2	BAG33220.1
PGN_0733	GT35	BAG33252.1
PGN_0777	GT2	BAG33296.1
PGN_0793	CBM20,CBM20,GH77	BAG33312.1
PGN_0817	GT51	BAG33336.1
PGN_0980	GH92	BAG33499.1
PGN_1026	GT2	BAG33545.1
PGN_1039	GH92	BAG33558.1
PGN_1044	GH13	BAG33563.1
PGN_1045	GH2	BAG33564.1
PGN_1134	GT4	BAG33653.1
PGN_1135	GT4	BAG33654.1
PGN_1239	GT2	BAG33758.1
PGN_1240	GT4	BAG33759.1
PGN_1251	GT4	BAG33770.1
PGN_1255	GT9	BAG33774.1
PGN_1286	GH24	BAG33805.1
PGN_1310	GT3	BAG33829.1
PGN_1362	GH23	BAG33881.1
PGN_1627	GT83	BAG34146.1
PGN_1628	GT2	BAG34147.1
PGN_1651	GT2	BAG34170.1
PGN_1668	GT2	BAG34187.1
PGN_1670	GH108	BAG34189.1
PGN_1690	GH29	BAG34209.1
PGN_1724	GT2	BAG34243.1
PGN_1736 (fragment)	GT5	BAG34255.1
PGN_1772	GH13	BAG34291.1
PGN_1807	GT2	BAG34326.1
PGN_1811	GH29	BAG34330.1
PGN_2019	CE11	BAG34537.1
PGN_2024	GH73,CBM50,CBM50	BAG34542.1
PGN_2032	GH27	BAG34550.1
PGN_2087 (probable fragment)	GT2	BAG34605.1
sialidase (PGN_1608)	GH33	BAG34127.1

Fig. 6.5 Screenshot of the CAZy webpage for the genome of *Porphyromonas gingivalis ATCC 33277*. The CAZy family distribution (*top*) is followed by the list of all identified CAZymes with their modularity (where relevant)

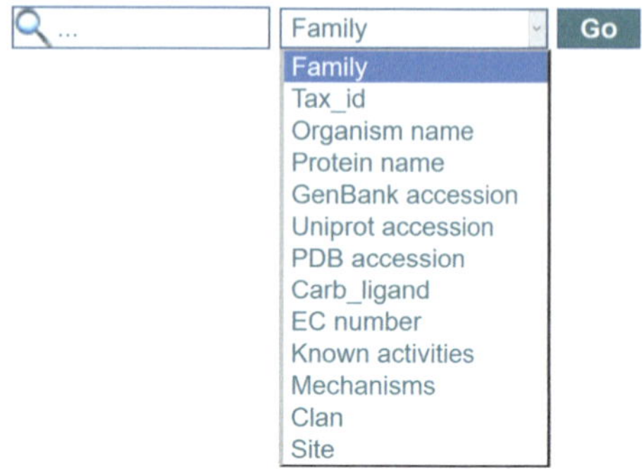

Fig. 6.6 Search tools and fields that accept queries in the CAZy database

6.4 How to Get CAZy to Annotate Your Studied Protein, Genome, or Metagenomic Sample?

The most straightforward way to obtain a CAZy annotation for a genome is to submit your sequence(s) to the NCBI with the "finished" status or by contacting us (cazy@afmb.univ-mrs.fr) to request our analysis as part of a collaborative effort. Every day, the internal CAZy tool for the semiautomatic modular assignment runs on the protein sequences from the daily release of NCBI GenBank, and our computational equipment makes it possible to perform several large-scale analyses such as the annotation of CAZyme repertoire in genomic and metagenomic investigations (the latter can be in the form of DNA or protein sequences). These putative assignments are thus manually validated (or rejected) by expert curators. Subsequent family comparisons provide insights into how similar or different might be the newly sequenced organisms compared to closely related species or how metagenomic samples differ relative to each other. Differences in the relative family size, for example, can reflect the relative diversity or complexity of the inherent biological processes and, therefore, the biology of the compared species/samples.

Automatic tools, freely available on the web, attempt to emulate the CAZy classification scheme. Our experience is that these fully automatic methods provide results that can be substantially different from actual CAZy assignments. Further, these tools sometimes include outdated module families and automatic subfamilies that are not curated. And finally (and most importantly), automatic predictors are dependent on the user's parameters for detection threshold, generally applied to e-value statistics. The issues with an e-value threshold is that (i) the e-value varies with the length of the aligned sequences for identical sequence similarity percentage, (ii) such threshold completely bypasses curation to distinguish possibly functionally related homologs from locally shared secondary structures, (iii) a unique threshold is not appropriate for families of unequal diversity, and (iv) low/significant e-values do not guarantee the completeness of modules since all detection tools (BLAST- or HMM-based) are local by nature.

6.5 What to Do If You Obtain a New Activity or the 3-D Structure for a CAZyme or If You Characterize a New CAZy Family?

We cordially invite biologists with experimental results to contact us (with appropriate material such as a peer-reviewed preprint of the work), to reduce the number of reports that was missed during our bibliography surveillance. If the subject protein has not yet been submitted to GenBank/PDB, we strongly encourage you to do so. This will allow the automatic capture and display of the annotation on the CAZy website as follows:

6.5.1 *Novel Activity, 3-D Structure, or New Chemical Information for an Enzyme in an Existing CAZy Family*

If the studied enzyme is already assigned to a numbered family in the CAZy website, we will complete the family records (content and description) with the new information. The new information will thus be displayed on the webpage dedicated to the family, in the corresponding sections as described in Sect. 6.2.1. The accumulation of experimental evidence will notably help in refining the classification system by the population of subfamilies based on phylogenetic analyses (see Table 6.1).

Warning If your enzyme appears in the CAZy website but in the "Nonclassified modules" listed for each CAZy class, it has to be considered as new regarding the CAZy classification (see below).

6.5.2 *Novel Family in the CAZy Classification*

If the newly characterized enzyme does not belong to any known CAZy family, or belongs to the "Nonclassified modules" of a CAZy class, we will create a new CAZy family, as follows. Starting from the subject sequence, we first collect the most similar homologs in GenBank by BLAST. Then, we iteratively gather more distant family members using HMMs, which capture the family diversity (flexible and constrained positions in the multiple sequence alignment and corresponding structure). The delineation of the module boundaries is guided by family conservation and is generally facilitated or refined when a 3-D structure becomes available. The creation and analysis of a new CAZy family remains private until notification by the original requestor or until publication.

6.6 Why Doesn't CAZy Extend Its Classification Scheme to Other Classes of Enzymes?

Even though the CAZy families do not always coincide with a precise substrate specificity, family assignment often gives clues on what the broad substrate category might be. And when the relatedness to a functionally characterized enzyme is high, typically at the subfamily level, then the functional predictions for CAZymes can be very good. In any case, this is substantially more informative than most other families of enzymes (kinases, proteases, esterases, etc.) whose substrates are difficult or impossible to derive from their sequence alone. Due to our limited number of expert curators and to the poorer relationship between family and function in other enzyme categories, we prefer to stay within our field of competence and do not expand the scope of CAZy beyond what it is.

6.7 Why Doesn't CAZy Propagate Experimentally Established Function to Similar Sequences?

All too often during a protein/genome study, the functional annotations automatically inferred by computational methods contain a significant amount of low-quality and even erroneous information that are then propagated to the next projects. For example, the transfer of Gene Ontology (GO) terms based on Pfam modules usually assigns excessively general terms. This can be explained by the stringent policy of module annotation that links a module solely to the GO terms common to all proteins having this module, whatever the diversity of the possible module combination and associated functions. Other widely used tools are also prone to overprediction by transferring annotation from a demonstrated example to distant homologs or by creating annotation based on hypothesis devoid of any experimental evidence, as, for example, with the BACON domain (Pfam ID PF13004) which stands for *Bacteroidetes*-Associated Carbohydrate-binding Often N-terminal based on a conjecture-only publication. This conjecture has been recently challenged in a publication showing that the BACON domain of BACOVA_02653 protein of *Bacteroides ovatus ATCC 8483* does not have any carbohydrate-binding activity (Larsbrink et al. 2014). As a consequence CAZy did not create a new CBM family for such modules. More generally, to avoid problems linked to annotation transfer, CAZy policy is to display EC numbers only for the experimentally characterized enzymes.

6.8 Links and Announcements on the CAZy Website

In addition to multiple links to essential enzymatic and glycogenomic resources, CAZy contains many cross-links to the CAZypedia resource. CAZypedia is a community-driven encyclopedic resource meant to be the logical extension of the CAZy classification. It contains extensive information about CAZy families with especial emphasis on GHs, but the other CAZy families are now being filled progressively. The CAZy website also offers an opportunity for commercial enzyme providers to present their products which follow the CAZy nomenclature and to announce scientific meetings and opened job positions related to CAZymes.

6.9 What Is the PULDB Database?

PULDB is a recent addition to CAZy that describes Polysaccharide Utilization Loci (PULs) experimentally characterized in the literature and our automated PUL predictions in *Bacteroidetes* species (Terrapon et al. 2015). A PUL is a set of physically linked genes organized around a *susCD* gene pair. Named according to the prototypic starch utilization system, *susC* is a characteristic membrane transporter, and *susD* encodes an outer membrane-binding protein (Shipman et al. 2000). PULs are prevalent in the *Bacteroidetes* phylum, with species encoding dozens of PULs, each tailored to degrade a particular glycan structure. PULs provide an evolutionary advantage to these gram-negative species by orchestrating the breakdown of complex glycans, thanks to the encoded CAZymes, and by sequestrating these nutrients away from competitors (Terrapon and Henrissat 2014). PULDB offers a query engine to search PULs by species, by (combination of) CAZy modules, and by locus tags. It also contains a JBrowse engine (Skinner et al. 2009) to visualize the genomic context of CAZymes and PULs for all the integrated genomes (source: IMG HMP project at the JGI (Markowitz et al. 2012)) as illustrated in Fig. 6.7.

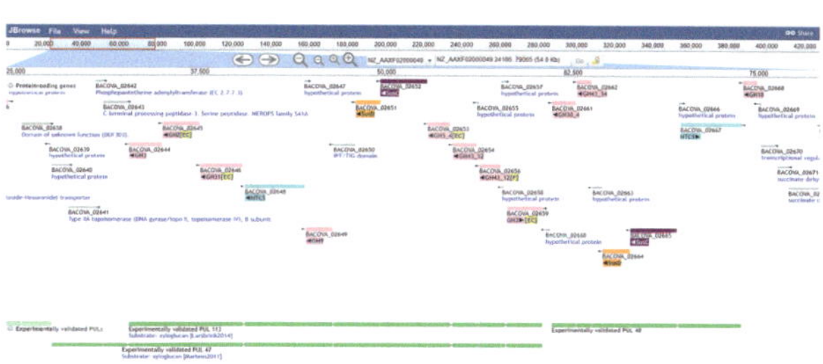

Fig. 6.7 Screenshot of the PULDB website. JBrowse visualization of a xyloglucan PUL in *Bacteroides ovatus ATCC 8483*

6.10 Conclusion

The CAZy database is based on family classification schemes that were established in the 1990s, before any genome had been completely sequenced. A key feature of the success of CAZy is the stability of its underlying classification system. The earliest GH families have survived a >500 times expansion since their creation in 1991. Other key features of CAZy are the integration of the variable modular architecture of CAZymes and its panel of expert curators to capture structural and functional data from the literature. In the near future, however, high-throughput enzymology will deliver more data in 1 year than what has accumulated during the last 50 years. Without a mechanism to capture functional information reliably, a large amount of experimental data will remain buried and underexploited.

References

Aspeborg H, Coutinho PM, Wang Y, Brumer H, Henrissat B (2012) Evolution, substrate specificity and subfamily classification of glycoside hydrolase family 5 (GH5). BMC Evol Biol 12(1):186

Benson DA, Cavanaugh M, Clark K, Karsch-Mizrachi I, Lipman DJ, Ostell J, Sayers EW (2012) GenBank. Nucleic Acids Res:gks1195

Berman HM, Westbrook J, Feng Z, Gilliland G, Bhat T, Weissig H, Shindyalov IN, Bourne PE (2000) The protein data bank. Nucleic Acids Res 28(1):235–242

Boraston A, Bolam D, Gilbert H, Davies G (2004) Carbohydrate-binding modules: fine-tuning polysaccharide recognition. Biochem J 382:769–781

Boutet E, Lieberherr D, Tognolli M, Schneider M, Bansal P, Bridge AJ, Poux S, Bougueleret L, Xenarios I (2016) UniProtKB/Swiss-Prot, the manually annotated section of the UniProt KnowledgeBase: how to use the entry view. Plant Bioinf Methods Protocols:23–54

Campbell JA, Davies GJ, Bulone V, Henrissat B (1997) A classification of nucleotide-diphospho-sugar glycosyltransferases based on amino acid sequence similarities. Biochem J 326(Pt 3):929

Coutinho P (1999) Carbohydrate-active enzymes: an integrated database approach. In Recent advances in carbohydrate bioengineering

Coutinho PM, Deleury E, Davies GJ, Henrissat B (2003) An evolving hierarchical family classification for glycosyltransferases. J Mol Biol 328(2):307–317

Larsbrink J, Rogers TE, Hemsworth GR, McKee LS, Tauzin AS, Spadiut O, Klinter S, Pudlo NA, Urs K, Koropatkin NM, Creagh AL, Haynes CA, Kelly AG, Cederholm SN, Davies GJ, Martens EC, Brumer H (2014) A discrete genetic locus confers xyloglucan metabolism in select human gut Bacteroidetes. Nature 506(7489):498–502

Levasseur A, Drula E, Lombard V, Coutinho PM, Henrissat B (2013) Expansion of the enzymatic repertoire of the CAZy database to integrate auxiliary redox enzymes. Biotechnol Biofuels 6(1):41

Lombard V, Bernard T, Rancurel C, Brumer H, Coutinho P, Henrissat B (2010) A hierarchical classification of polysaccharide lyases for glycogenomics. Biochem J 432:437–444

Lombard V, Golaconda Ramulu H, Drula E, Coutinho PM, Henrissat B (2014) The carbohydrate-active enzymes database (CAZy) in 2013. Nucleic Acids Res 42 (Database issue):D490–D495

Lütteke T, Von Der Lieth CW (2004) pdb-care (PDB carbohydrate residue check): a program to support annotation of complex carbohydrate structures in PDB files. BMC Bioinf 5(1):69

Markowitz VM, Chen IMA, Palaniappan K, Chu K, Szeto E, Grechkin Y, Ratner A, Jacob B, Huang J, Williams P (2012) IMG: the integrated microbial genomes database and comparative analysis system. Nucleic Acids Res 40(D1):D115–D122

Mewis K, Lenfant N, Lombard V, Henrissat B (2016) Dividing the large glycoside hydrolase family 43 into subfamilies: a motivation for detailed enzyme characterization. Appl Environ Microbiol AEM. 03453–03415

Shipman JA, Berleman JE, Salyers AA (2000) Characterization of four outer membrane proteins involved in binding starch to the cell surface of Bacteroides thetaiotaomicron. J Bacteriol 182(19):5365–5372

Skinner ME, Uzilov AV, Stein LD, Mungall CJ, Holmes IH (2009) JBrowse: a next-generation genome browser. Genome Res 19(9):1630–1638

St John FJ, González JM, Pozharski E (2010) Consolidation of glycosyl hydrolase family 30: a dual domain 4/7 hydrolase family consisting of two structurally distinct groups. FEBS Lett 584(21):4435–4441

Stam MR, Danchin EG, Rancurel C, Coutinho PM, Henrissat B (2006) Dividing the large glycoside hydrolase family 13 into subfamilies: towards improved functional annotations of α-amylase-related proteins. Protein Eng Des Sel 19(12):555–562

Terrapon N, Henrissat B (2014) How do gut microbes break down dietary fiber? Trends Biochem Sci 39(4):156–158

Terrapon N, Lombard V, Gilbert HJ, Henrissat B (2015) Automatic prediction of polysaccharide utilization loci in Bacteroidetes species. Bioinformatics 31(5):647–655

Chapter 7
Glyco3D: A Suite of Interlinked Databases of 3D Structures of Complex Carbohydrates, Lectins, Antibodies, and Glycosyltransferases

Serge Pérez, Anita Sarkar, Alain Rivet, Sophie Drouillard, Christelle Breton, and Anne Imberty

Abstract Glyco3D is a portal for structural glycobiology of several interlinked databases that is covering the three-dimensional features of monosaccharides, disaccharides, oligosaccharides, polysaccharides, glycosyltransferases, lectins, monoclonal antibodies, and glycosaminoglycan-binding proteins. Collection of annotated NMR data of bioactive oligosaccharides is also available. A common nomenclature has been adopted for the structural encoding of the carbohydrates. Each individual database stands by itself as it covers a particular family of either complex carbohydrates or carbohydrate-binding proteins. A unique search engine is available that scans the full content of all the databases for queries related to sequential information of the carbohydrates. The interconnection of these databases provides a unique opportunity to characterize the three-dimensional features that a given oligosaccharide molecule can take in different environments, i.e., vacuum, crystalline state, or interacting with different proteins having different biological function. The databases, which have been manually curated, were developed with nonproprietary software. They are web-based platform and are freely available to the scientific community at http://glyco3d.cermav.cnrs.fr.

Keywords Monosaccharides • Disaccharides • Oligosaccharides • Polysaccharides • Glycosyltransferases • Lectins • Antibodies • Glycosaminoglycan-binding proteins

S. Pérez (✉)
Département de Pharmacochimie, UMR5063 CNRS-Université Grenoble Alpes, BP53, 38041 Grenoble cédex 09, France
e-mail: spsergeperez@gmail.com

A. Sarkar • A. Rivet • S. Drouillard • C. Breton • A. Imberty (✉)
Centre de Recherches sur les Macromolécules Végétales, UPR5301 CNRS (affiliated with Université Joseph Fourier and ICMG), BP53, 38041 Grenoble cédex 09, France
e-mail: imberty@cermav.cnrs.fr

K.F. Aoki-Kinoshita (ed.), *A Practical Guide to Using Glycomics Databases*,
DOI 10.1007/978-4-431-56454-6_7

7.1 Introduction

Major advances in structural elucidation methods are benefitting glycobiology at large. Progress arising from the use of synchrotron radiation sources along with major advances in high-resolution NMR spectrometry and electron microscopy contributes strongly to these advances. In conjunction with these experimental sources of structural investigations, computational and molecular modeling methods are providing complementary information (Demarco and Woods 2008; Perez and Tvaroška 2014). Several distinct repositories hold 3D structural information which has been experimentally and theoretically determined for carbohydrates and carbohydrate-containing molecules (Perez and Mulloy 2005).

Among the 5000 entries related to carbohydrates in the Cambridge Structural Database (Allen 2002), only a small fraction is relevant to the field of glycobiology since most glyco-related entries relate to monosaccharides or to substituted intermediates for synthetic pathways (Perez 2007; Perez et al. 2000). As for polysaccharide structures, although a large amount of 3D structural models has accumulated over time, the effort to collect, curate, and disseminate this data electronically and freely to the scientific community has been limited when compared to other initiatives dealing with biomacromolecules.

In the past 45 years, more than 100,000 atomic-resolution structures have been deposited into the Protein Data Bank (PDB) (Berman et al. 2003). An increasing number of crystal structures have been reported for glycoproteins and protein-carbohydrate complexes. More than 5000 entries for glycoproteins or protein-carbohydrate complexes have been deposited, forming a valuable resource for glycoscientists (Jo and Im 2013). They occur, in their vast majority, as N-glycans or non-covalently bound ligands, O-Glycan chains forming a minority. In total, about 3.5 % of the proteins in the PDB carry covalently bound glycan chains and thus can be classified as glycoproteins. The quality of the data does not always meet the high-quality standards; the structure needs to be curated and annotated. The glycosciences.de web portal (Lutteke et al. 2006) is a valuable utility for searching carbohydrate structures in the PDB.

As regards to theoretical characterizations, the complexity of oligosaccharide and polysaccharide topologies has required the design of dedicated molecular building procedures. These procedures can convert sequence information into reliable 3D models prior to any optimization through molecular mechanics or molecular dynamics methods, either as isolated molecules or in their interactions with proteins. All these constructions are based on the linking of preconstructed 3D molecular templates of monosaccharides (Engelsen et al. 2013; Lutteke et al. 2006; Woods 2014).

Despite their availability, these structural information have not yet gained full utilization in the rational design and engineering of glycan-based multivalent vehicles or glycan-grafted materials in view of their potential implications. These approaches, along with those provided by novel methodologies in the field of click chemistry or in the understanding of multivalency, are now part of the tool box that glycoscientists have in hand to address and develop smart constructions that would exploit the full repertoire of the informational power of glycans.

Indeed, a vast majority of structures deal with those carbohydrate molecules that are referred to as "glycan determinants," i.e., those which are recognized by glycan-binding proteins. Those proteins are lectins, receptors, toxins, antibodies, microbial adhesins, carbohydrate-binding modules, transporters, but also enzymes involved in their synthesis, modification, and degradation. In the present work, only some of these glycan-binding proteins have been selected; they are lectins, monoclonal antibodies, glycosaminoglycan-binding proteins, and glycosyltransferases. As regards to the carbohydrates, the three-dimensional features of monosaccharides, disaccharides, oligosaccharides, and polysaccharides are covered in the form of databases, without claiming exhaustivity. A collection of annotated NMR data of bioactive oligosaccharides is also available. These databases have been developed with nonproprietary software, and they are opened freely to the scientific community. They are accessible throughout a common portal called "Glyco3D" http://glyco3d.cermav.cnrs.fr. Each individual database stands by itself as it covers a particular field of structural glycosciences. Nevertheless, the interconnection of these databases provides a unique opportunity to characterize the three-dimensional features that a given oligosaccharide molecule can take in different environments, i.e., vacuum, crystalline state, or interacting with different proteins having different biological function. To this aim, a common nomenclature has been adopted for the structural encoding of the carbohydrates.

7.2 Representing and Encoding Complex Carbohydrates

Representation in text of the primary structure, or sequence, of complex carbohydrates was first described following the IUPAC-IUBMB terminology in its extended and condensed forms (Mcnaught 1997). These forms are used within the carbohydrate community and are adequate for describing complex sugar sequences. Recommendations apply to the description of polysaccharides and glycoproteins (Mcnaught 1997). Other types of representations have been developed in glycobiology, favoring pictorial representations that facilitate the visualization of the monosaccharides. This is adequate as the number of basic carbohydrate units found in mammals is limited. Extension to the constituents found in bacterial and plant polysaccharides has also been developed and adopted (Varki et al. 2015). Figure 7.1 presents in a non-exhaustive fashion the results of such an extension, which allows the description of some structural descriptors which were not taken into account previously.

From the standpoint of bioinformatics, it is impractical to encode glycans (composed of more than 100 monosaccharide units) into distinct graphical symbols. To establish effective databases that can intercommunicate, a simple representation in a common/standard format is essential. This would facilitate computational

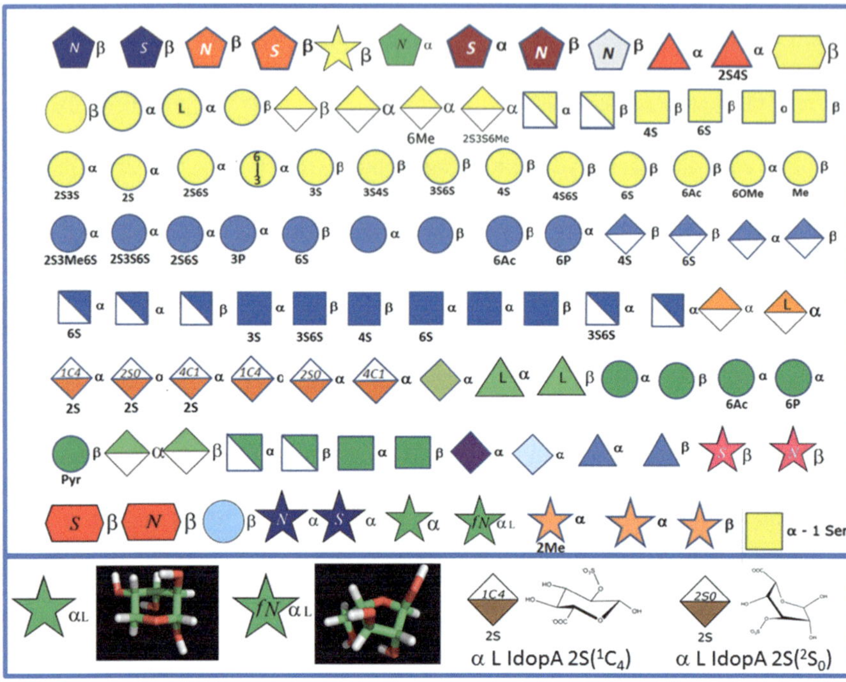

Fig. 7.1 Pictorial representation of some important monosaccharide units

processing and ensure that the data content is non-redundant. Two approaches can be followed to encode a carbohydrate molecule:

Connecting atom sets through chemical bonds is commonly used in chemoinformatics. Chemical file formats like InChi (Mcnaught 1997) and SMILES (Weininger 1988) have been developed to aid storing of molecule information in chemical databases like PubChem (Wang et al. 2010) or ChEBI (Degtyarenko et al. 2008). IUPAC (extended), InChi, and SMILES encoding are computed from the chemical drawing (ring structure), thereby allowing auto-generation of these encodings. There are severe limitations that do not make this type of encoding the favored choice, but InChi and SMILES are the proper formats to exchange data between distinct databases.

Connecting building blocks (monosaccharides) through glycosidic linkages is far more efficient to encode carbohydrates using a residue-based approach (Frank and Schloissnig 2010). As compared to nucleic acids or proteins, there are a far greater numbers of monosaccharides. In addition, carbohydrates are frequently found to have branched structures; most of them are tree-like molecules. The prerequisite for a residue-based encoding format is a controlled vocabulary of its residue names. It makes sense to restrict the number of residues to as low as possible. The lack of clear rules to subscribe atoms of a molecule to one particular monosaccharide, and not to its substituent(s), is the main hurdle in encoding monosaccharide names. The

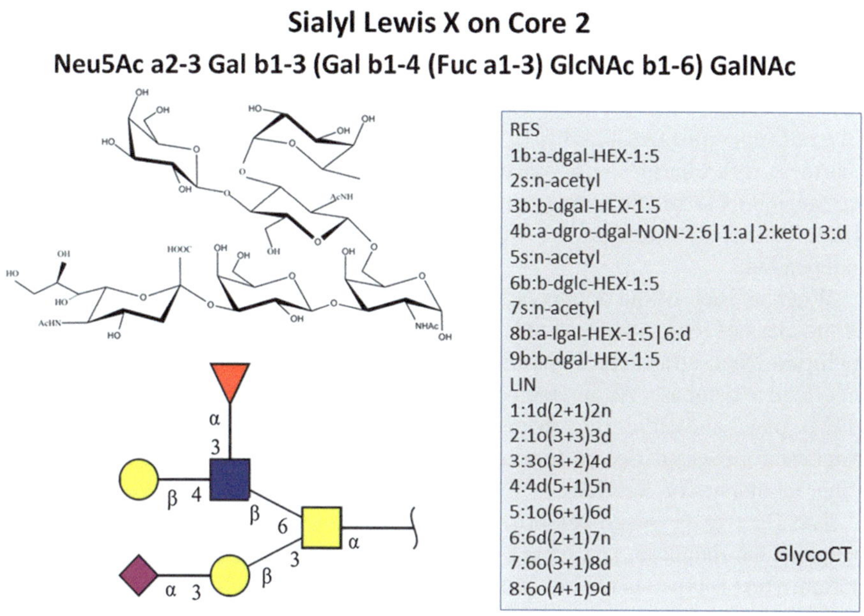

Fig. 7.2 The different levels of glycan encodings. The example used to illustrate the variety of notations in this figure is Sialyl Lewis X on core2

variety in nomenclature and structural representation of glycans makes it complex to decide the best form to use. The choice of notation is frequently based on whether the study is focused on the chemistry or on biology. The information content of each representation may vary or highlight a particular aspect as compared to others. While representing a complex glycan structure, chemists prefer to elucidate the structure that includes information about the anomeric carbon, the chirality of the glycan, the monosaccharides present, and the glycosidic linkages that connect them. For others, it is more interesting to visualize the monosaccharides present, and hence a symbolic/diagrammatic notation is favored (Fig. 7.2).

Due to the independent development of glycan databases in various geographical locations, several formats for representing glycan structures have been created. Chapter 2 provides an overview of the most common representations used in the field.

The three-dimensional depiction of glycans, polysaccharides, and glycoconjugates, coping with the accepted nomenclature and pictorial representation used in carbohydrate chemistry, biochemistry, and glycobiology, is made possible throughout the molecular visualization program Sweet Unity Mol (Perez et al. 2015).

7.3 Glyco3D Portal: An Ensemble of 3D Databases for Glycosciences

Glyco3D encompasses a family of databases covering the 3D features of monosaccharides, disaccharides, oligosaccharides, polysaccharides, glycosyltransferases, lectins, monoclonal antibodies, and glycosaminoglycan-binding proteins (Fig. 7.3). These databases, which have been manually curated, are web based and platform independent.

Whereas each of the databases has been set to account for the specific features of the class of molecules covered, the set of databases in Glyco3D share common facilities. Data retrieval and usability are the primary goals set by the developers of an effective database. An interactive front end was designed for each database with HTML pages and server side scripts that extract data from the tables on the relational database for user queries on *Data Query* and display the retrieved information in a coherent manner on *Results*.

The *Data Query* page comprises primarily two levels of increased complexity to query the database, i.e., *Simple and Advanced* searches. In the *Simple Search* option, a text is typed in the box provided, based upon which a result prompt appears to guide the user in selecting from the "hits" found in the database. An accordion function eases preview of the results. This can be used to expand or minimize the preview of the listed results of the user query for a first glance into the entries matching the request to the database. The *Advanced Search* is a multi-criteria search that can be used together for querying in various combinations as best suits the user's requirement. Both the Simple Search and the Advanced Search options are equipped with an "auto-complete" function, which guides the user while querying

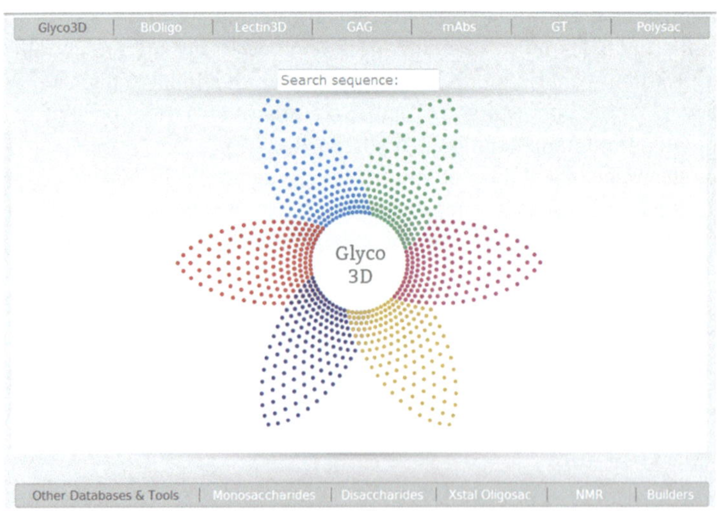

Fig. 7.3 The Glyco3D home page

the database. It comprises two parts: (1) a single field of entered text and (2) the auto-prompt when the data is entered, through which the desired hit in the database can be selected either by scrolling down with the mouse or by using the arrow keys on the keyboard.

The *Results* page details the results which are organized under two tabs, namely, *Molecule Information* and *Display and Download*. The Molecule Information page provides a detailed description of the molecule (or macromolecule) as described throughout the several nomenclature schemes, along with key elements representative of the type of databases (e.g., trivial name, molecular weight, some literature references, etc.). The illustrative representations of the glycans or the glycan-interacting proteins can be viewed through the "Zoombox" feature that was developed by modifying an existing JQuery plug-in that allows the selected image to be zoomed and highlighted. The *Display and Download* tab incorporates the best representatives of the most probable low-energy conformational families. As regards to the display of the results, 3D structures can be viewed over the website via the Jmol application http://jmol.sourceforge.net/. Jmol is an interactive web browser applet that is an open-source, cross-platform 3D Java visualizing tool for viewing chemical and molecular structures. It provides high-performance 3D rendering with standard available hardware. Downloading the atomic coordinates (in pdb format) for further independent use is an option provided for all the databases. Additionally, a GUI has been designed to retrieve, interpret, and display the related information about each entry stored in the back end in the tables of the relational database and to display it interactively to the user. Finally, in an effort to assimilate other relevant resources for sugars, "External Links" are provided that empowers the user to explore more online glycoinformatics resources.

The databases run on an Apache web server (http://www.apache.org/) with the application program Hypertext Preprocessor (PHP) (http://www.php.net/). It has been implemented using the open source MySQL database (http://www.mysql.com/). They have been developed based on a combination of three layers. The underlying layer is the MySQL database system, a relational database management system that stores all the structure-related information in the back end and provides the facility to link two or more tables in the database. The intermediate layer is an Apache-PHP application [Apache 2, PHP] that receives the query from the user and connects to the database to fetch data from the upper layer, which comprises populated HTML pages, to the web browser client. The PHP and Java scripts are embedded in the HTML web pages for this effect and are used as application programs for integrating the back end (MySQL database) to the web pages (HTML). Apache has been used as the web server for building the interface between the web browser and the application programs. PHP was used for writing scripts to query the database, and Javascript (with JQuery plug-in) was used to design the auto-complete function for the user interface. The graphical user interface was developed with HTML (version 5) and CSS (version 3).

7.3.1 Monosaccharides

Database Content This is an annotated database that contains the 3D structural information of about 100 entries of monosaccharides. These monosaccharides constitute the building blocks of the vast majority of oligosaccharides, complex carbohydrates such as "glycan determinants" (blood group antigens, core structures, fucosylated oligosaccharides, sialylated oligosaccharides, Lewis antigens, GPI anchors, N-linked oligosaccharides, globosides, etc.), glycosaminoglycans, plant and algal polysaccharides, as well as some bacterial polysaccharides. For establishing the 3D database, they all have been subjected to systematic conformational sampling to determine their conformational preferences, using molecular mechanics optimization. Whereas most of the monosaccharides exhibit a fairly rigid ring conformation, some cases exist such as in the case of iduronic acid, idose, and all furanosides where several ring shapes can occur. In these cases, the low-energy conformations are available for each entry.

Data Query Upon reaching the search page, two buttons to query the database appear on the left hand panel.

Simple Search A search box is provided, in which the user inputs textual information related to the search. The result is a prompt to guide the user in selecting from the "hits" found in the database, by a simple search engine. A preview of the results is displayed in an accordion fashion. This can be used to expand or minimize the preview of the listed results of the user query for a first glance into the entries matching the request to the database. The preview provides the monosaccharide name, its absolute configuration (D or L), its anomeric configuration, category, and molecular weight to the user to make an informed choice.

Advanced Search Four search boxes appear, each of them offering the choice between criteria to select: trivial name, type of constituent, category, and molecular weight. A slider is provided for assigning a range of values to be queried in the molecular weight of the database entries. It consists of two cursors that can navigate on a bar for specifying the minimum and maximum limit of the search. Two text fields display the values of the current position on the slider bar. The slider cursors auto-adjust themselves when values are entered directly in the text boxes.

Results The detailed results are organized under two tabs: "Molecule Information" and "Display and Download" (Fig. 7.4).

Molecule Information This includes the trivial name of the monosaccharide, the graphical representation of the stereochemical configuration (when available, the symbol notation for carbohydrates of the Consortium for Functional Glycomics (http://www.functionalglycomics.org/)), and the molecular weight. Additional comments and literature references are present if available. The illustrative representations of the monosaccharide can be viewed through the "Zoombox" feature that allows the selected image to be zoomed and highlighted.

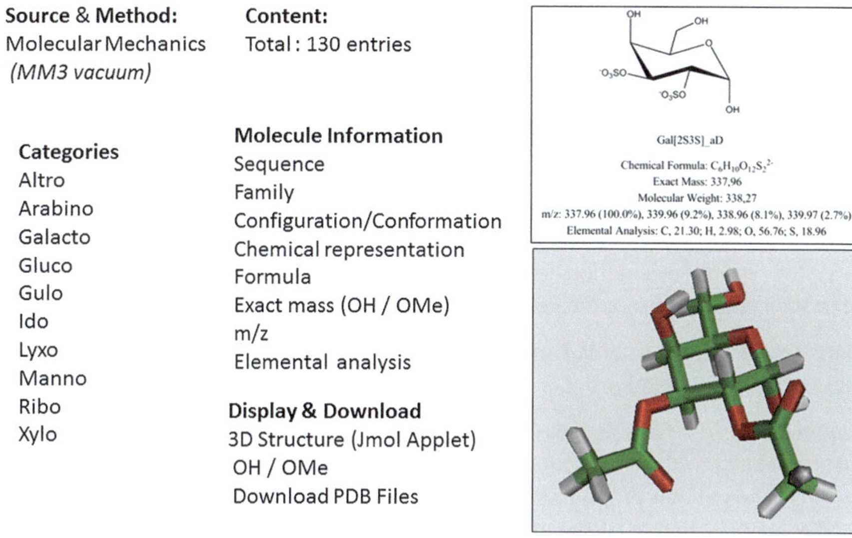

Fig. 7.4 The monosaccharide page

Display and Download This tab incorporates the low-energy conformation of the monosaccharide and its methyl glycoside.

7.3.2 Disaccharides

Database Content This annotated database contains the 3D structural information of about 120 entries of disaccharides. These disaccharides constitute molecules in their own rights, and they constitute the building blocks of the vast majority of oligosaccharides, complex carbohydrates such as "glycan determinants" (blood group antigens, core structures, fucosylated oligosaccharides, sialylated oligosaccharides, Lewis antigens, GPI anchors, N-linked oligosaccharides, globosides, etc.), glycosaminoglycans, plants and algal polysaccharides, and some bacterial polysaccharides.

The relative orientation of two contiguous monosaccharides linked by a glycosidic bond in a disaccharide is characterized by the Φ and Ψ torsion angles. In the so-called Heavy Atom Definition commonly used in crystallography, Φ is the torsion angle $\Phi = \text{O5-C1-O-Cx}$, and Ψ is the torsion angle $\Psi = \text{C1-O1-Cx-Cx+1}$, where x is the number of the carbon atom of the second monosaccharide with which the $1 \rightarrow$ x glycosidic bond is formed. An alternate definition of use in NMR spectroscopy refers to the hydrogen atoms about the glycosidic bond in a way such as $\Phi^{H} = \text{H1-C1-O-Cx}$ and $\Psi^{H} = \text{C1-O-Cx-Hx}$. For two monosaccharides linked by a $1 \rightarrow 6$ linkage, another parameter (ω) is required describing the orientation about the exocyclic bond C5-C6. Its orientation is customarily described by the torsion angles O5-C5-C6-O6 and C4-C5-C6-O6, which combination defines the so-called *gauche-trans* (gt), *gauche-gauche* (gg), and *trans-gauche* (tg) conformations (Marchessault and Perez 1979). For each disaccharide, an exhaustive search was performed using the MM3 molecular mechanics force field. This gave a complete sampling of the conformational space, yielding the construction of a relaxed adiabatic energy map, which is represented as a function of Φ and Ψ torsion angles. In the case of $1\rightarrow 6$ linkages, relaxed adiabatic maps can be established for the three low-energy orientations of the torsion angle ω. Typically, the exploration of each such energy maps indicates the occurrence of 2–4 energy minima.

Data Query Upon reaching the search page, two buttons to query the database appear on the left hand panel.

Simple Search A search box is provided, in which the user inputs textual information related to the search. A preview of the results is displayed in an accordion fashion. The preview provides the disaccharide name, its absolute configuration, the axial-equatorial nature of the glycosidic linkage, and molecular weight to the user to make an informed choice.

Advanced Search Four search boxes appear, each of them offering the choice between criteria to select: trivial name, type of constituent, category, and molecular weight. A slider is provided for assigning a range of values to be queried in the molecular weight of the database entries. It consists of two cursors that can navigate on a bar for specifying the minimum and maximum limit of the search. Two text fields display the values of the current position on the slider bar. The slider cursors auto-adjust themselves when values are entered directly in the text boxes.

Results The detailed results are organized under two tabs: "Molecule Information" and "Display and Download" (Fig. 7.5).

Molecule Information This includes the trivial name of the disaccharide, its sequence, the graphical representation of the stereochemical configuration (when available, the symbol notation for carbohydrates of the Consortium for Functional Glycomics), and the molecular weight. Additional comments and literature references are present if available. The illustrative representations of the disaccharide can be viewed through the "Zoombox" feature that allows the selected image to be zoomed and highlighted.

Display and Download This tab incorporates the best representatives of the families of the most probable low-energy conformation(s) from 1 to 4.

7.3.3 BiOligo

Database Content More than 250 entries of bioactive oligosaccharides are listed in the BiOligo-annotated database, with details about 3D structural information.

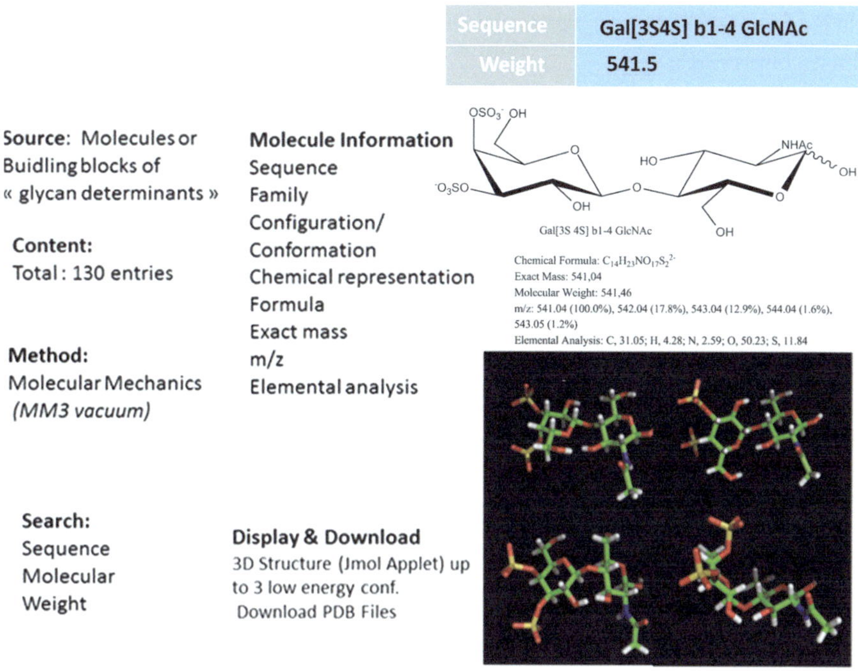

Fig. 7.5 The disaccharide page

The glycan epitopes are complex carbohydrates with their associated substitutions and aglycones, most of them being targets for glycan-binding proteins. The glycan epitopes belong to widely occurring families like the blood group antigens, core structures, fucosylated oligosaccharides, sialylated oligosaccharides, Lewis antigens, GPI anchors, N-linked oligosaccharides, globosides, etc. Table 7.1 gives the classification of glycan determinants in BiOligo.

For establishing the database, the three-dimensional structures of each constituent were generated using a combination of the available carbohydrate molecular builders (Engelsen et al. 2013; Lutteke et al. 2006; Woods 2014) or the building facilities offered by Sybyl (Tripos Inc.) and Chimera (Pettersen et al. 2004). Once constructed, the glycans were subjected to systematic conformational sampling to determine their conformational preferences, using the Shape software (Rosen et al. 2009). In such cases, several low-energy conformations (1–5) are available for each entry. At the present time, the monumental work required to complete the computation work is still ongoing, and the results are being implemented in the database on a regular basis.

Data Query The database is available from Glyco3D portal. Upon reaching the search page, two buttons to query the database appear on the left hand panel.

Simple Search A search box is provided, in which the user inputs textual information related to the search. The result is a prompt to guide the user in selecting

Table 7.1 Classification of glycan determinants in BiOligo

Index	BiOligo category
1	Blood group A antigens
2	Blood group B antigens
3	Blood group H antigens (blood group O)
4	Blood group H antigens (blood group O) and Globo H tetraose
5	Core structures
6	Core structures (type 1 and type 2)
7	Core structures (type 1)
8	Core structures (type 2)
9	Core structures (type 4)
10	Fucosylated oligosaccharides
11	Fucosylated oligosaccharides (3-fucosyllactose core)
12	Fucosylated oligosaccharides (lacto series)
13	GAGs
14	Galα-3Gal oligosaccharides (Galili and xeno antigens)
15	Galα-3Gal oligosaccharides (isogloboseries)
16	Ganglioside sugars
17	Globoside sugars (P antigens) (Forssman antigens)
18	Globoside sugars (P antigens) (Globo series: core structure type 4)
19	Globoside sugars (P antigens) (P blood group antigens and analogues)
20	Globoside sugars (P antigens) (stage-specific embryonic antigens: SSEA-3 and SSEA-4)
21	Glucuronylated oligosaccharides
22	Glycosphingolipid
23	Lewis antigens
24	Miscellaneous
25	Miscellaneous (blood group-related oligosaccharides)
26	Miscellaneous (chitin oligosaccharides)
27	Miscellaneous (fibrinogen-related oligosaccharides)
28	Miscellaneous (LDN-related oligosaccharides)
29	Miscellaneous (Lewis X-related oligosaccharides)
30	Miscellaneous (TF-related oligosaccharides)
31	Miscellaneous (TN-related oligosaccharides)
32	Miscellaneous (Trehalose-like sugars)
33	N-linked oligos
34	Sialylated oligosaccharide (type 1)
35	Sialylated oligosaccharide (type 2)
36	Disaccharides (***GlycoLego***)
37	Monosaccharides (***GlycoLego***)

from the "hits" found in the database, by a simple search engine. A preview of the results is displayed in an accordion fashion. This can be used to expand or minimize the preview of the listed results of the user query for a first glance into

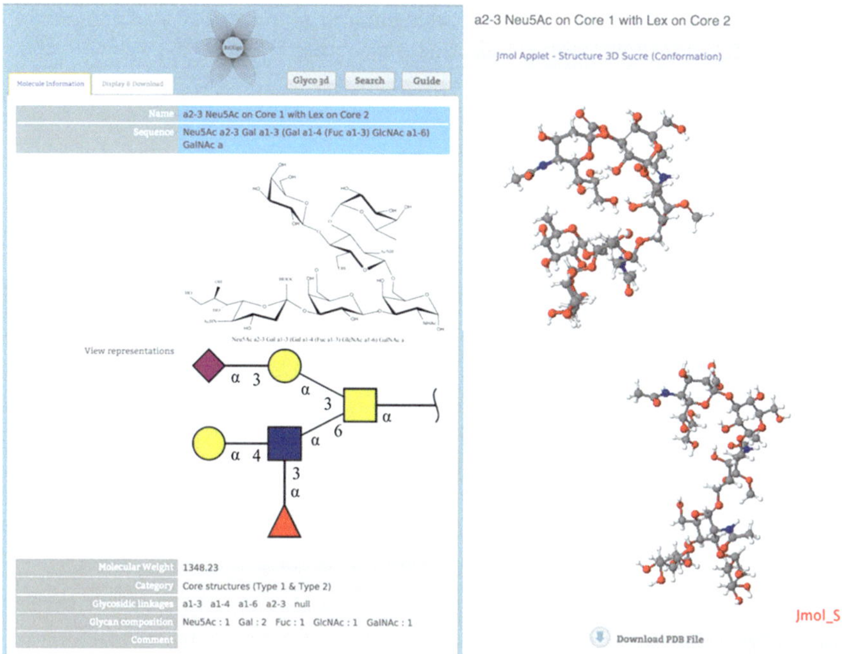

Fig. 7.6 The BiOligo page

the entries matching the request to the database. The preview provides the glycan name, category, and molecular weight to the user to make an informed choice.

Advanced Search Four search boxes appear each of them offering the choice between criteria to select: trivial name, type of constituent, category, and molecular weight. A slider is provided for assigning a range of values to be queried in the molecular weight of the database entries. It consists of two cursors that can navigate on a bar for specifying the minimum and maximum limit of the search. Two text fields display the values of the current position on the slider bar. The slider cursors auto-adjust themselves when values are entered directly in the text boxes.

Results The detailed results are organized under two tabs: "Molecule Information" and "Display and Download" (Fig. 7.6).

Molecule Information This includes the trivial name of the glycan, its sequence, the graphical representation of the stereochemical configuration, the symbol notation for carbohydrates of the Consortium for Functional Glycomics, the molecular weight, the glycan category or family in which it has been classified in the BiOligo Database, the glycan composition (i.e., the comprising glycan type and number of each such glycan), and the glycosidic linkages present in it. Additional comments and literature references are present if available. The illustrative representations of the glycan can be viewed through the "Zoombox" feature that allows the selected image to be zoomed and highlighted.

View and Download This tab incorporates the best representatives of the families of the most probable low-energy conformation(s) from 1 to 4.

7.3.4 NMR: A NMR Database of Bioactive Oligosaccharides

This database contains the NMR structural information of more than 150 entries of bioactive oligosaccharides. This set of glycans determinants is a subset of the group of bioactive oligosaccharides which constitute the core of the BiOligo Database (Sarkar et al. 2015). They have been systematically organized using standard names in the field of glycobiology, into 31 categories and subcategories. The glycan determinants in the NMR Database constitute a subset of the entries of BiOligo. Prior to the establishment of the database, these glycan were synthesized in pure form and in sufficient quantity to be investigated throughout NMR spectroscopy. The synthetic work was conducted using recombinant methodology (Priem et al. 2002). For each of these glycans, the experimental work encompassed the recording and interpretation of ^{1}H and ^{13}C spectra, along with COSY, TOCSY, HMQC, and HMBC correlation spectra.

Data Query The database is available from Glyco3D portal. Upon reaching the search page, two buttons to query the database appear on the left hand panel.

Simple Search A search box is provided, in which the user inputs textual information related to the sequence of the glycan. The result is a prompt to guide the user in selecting from the "hits" found in the database, by a simple search engine. A preview of the results is displayed in an accordion fashion. This can be used to expand or minimize the preview of the listed results of the user query for a first glance into the entries matching the request to the database.

Advanced Search can be performed on different criteria: trivial name, sequence, category, and type of constituents. More complex searches can be made by combining criteria which can be interlaced from up to four search boxes.

Results The detailed results are organized under two tabs: "Molecule Information" and "Display and Download" (Fig. 7.7).

Molecule Information provides the trivial name, the sequence, the graphical representation of the stereochemical configuration, the symbol notation for carbohydrates of the Consortium for Functional Glycomics, the type of constituent, and the glycan category. The experimental conditions used to record the NMR spectra are given, i.e., temperature, solvent, frequency, and concentration.

Display and Download This tab incorporates the representations of the chemical repeat and, in many cases, the ^{1}H and ^{13}C spectra, along with COSY, TOCSY, HMQC, and HMBC correlation spectra.

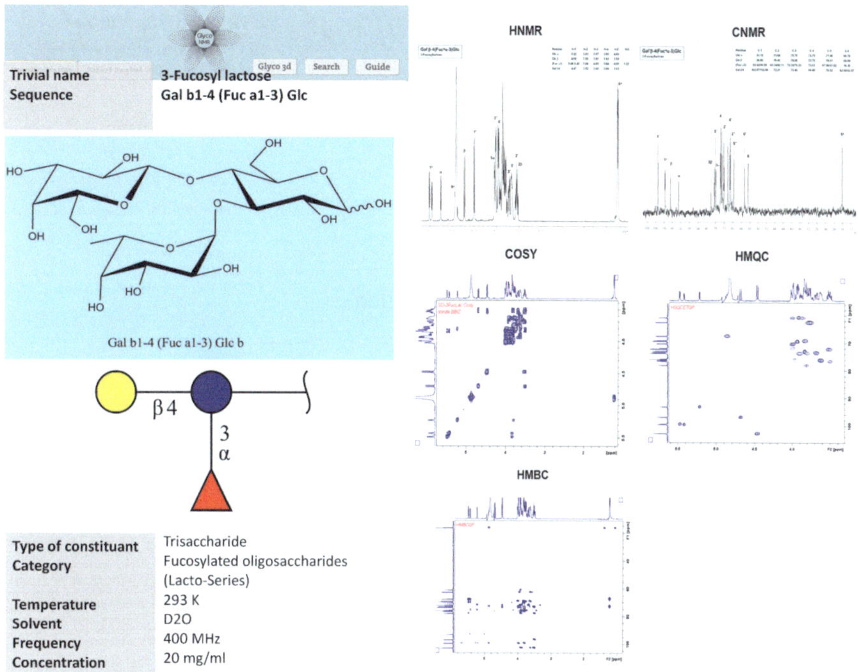

Fig. 7.7 The NMR page

7.3.5 PolySac: A 3D Structural Database of Polysaccharides

Database Content The database contains the 3D structural information of about 140 polysaccharide entries that have been collected from an extensive screening of scientific literature (for review, see Perez (2007)). These were established using various structure determination techniques (fiber X-ray and neutron diffraction, electron diffraction on single crystals, molecular modeling, and high-resolution NMR spectroscopy). The details concerning the construction of the atomic coordinates of polysaccharides have been published previously (Sarkar and Perez 2012). The classification of polysaccharide families present in Polysac3DB is shown in Fig. 7.8.

Data Query The database is available from Glyco3D portal. The polysaccharide data organized in the database can be browed starting from the *search* page. The data can be accessed in two ways.

Simple Search This option searches the database by just entering the name of the polysaccharide of interest. This is available through a drop-down button that enlists all the polysaccharides present in the database and groups all the entries in the database into 18 groups/families to clearly categorize the overall properties

Source: (Literature)
Methods: X-ray fiber diffraction
Neutron fiber diffraction, Electron diffraction
Molecular Modelling, NMR

Categories
Agarose
Alginates
Amylose and Starches
Bacterial Polysaccharides
Carrageenans
Celluloses
Chitins and Chitosans
Curdlans
Glycosaminoglycans (GAG)
Galactoglucans
Galactommans
Glucomannans
Mannans
Pectins
Scleroglucans
Xylans
Nigeran
Others (including inulin)

Content:
Total : 160 entries

Search:
Sequence
Method

Molecule Information
Family
Sub-family
Sequence
Repeat unit (*Chem, CFG*)
PDB Code
Method
Origin
Links (*PDB, Medline, Polysac3DB*)

Display & Download
3D Structure (Jmol Applet)
- Repeat unit
- Chain
- Crystal Packing

Download PDB File

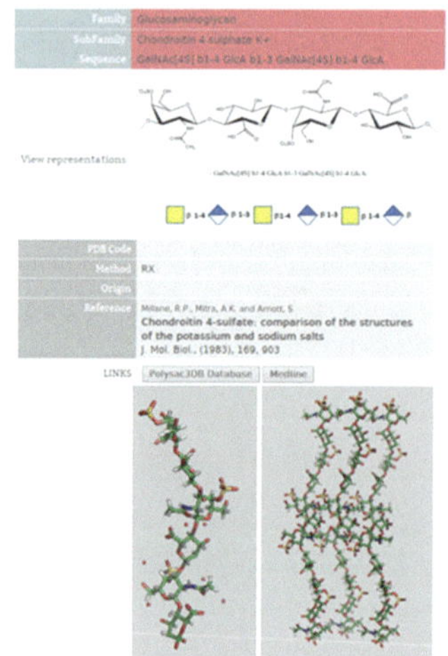

Fig. 7.8 The polysaccharide page

displayed by these polysaccharides. Upon selection of a family, a further drop-down menu offers a list of the polysaccharides belonging to this family for which 3D structural information is available.

Advanced Search offers the choice among two criteria, either the chemical structure of the repeat unit or the method of resolution used to establish the structure.

Results The detailed results are organized under two tabs: "Molecule Information" and "Display and Download" (Fig. 7.8).

Molecule Information This includes the trivial name of the polysaccharide, its family and subfamily, its origin, the sequence of the repeating unit, the graphical representation of the stereochemical configuration, and the symbolic representation. Additional comments and literature references are present if available. The illustrative representations of the glycan can be viewed through the "Zoombox" feature that allows the selected image to be zoomed and highlighted.

View and Download This tab incorporates the molecular representations of the repeat unit, the macromolecular chain, and in some instances some packing features.

7.3.6 GT: A 3D Structural Database of Glycosyltransferases

Database Content Glycosyltransferases (GTs) constitute a ubiquitous group of enzymes that catalyze the synthesis of glycosidic linkages by the transfer of a sugar residue from a donor to an acceptor (Breton et al. 2012). Acceptor substrates are carbohydrates, proteins, lipids, DNA, and numerous small molecules such as antibiotics, flavonol, steroids, etc. Glycosyl donor substrates are mostly sugar nucleotides, such as UDP-GlcNAc, UDP-Gal, and GDP-Man. However, lipid-linked sugars, e.g., dolichol phosphate saccharides and unsubstituted phosphates, are also utilized. Acceptor substrates are carbohydrates, proteins, lipids, DNA, antibiotic, or other small molecules. Sugar-nucleotide-dependent GTs are often referred to as Leloir enzymes. The transfer of saccharides by GTs is regiospecific and stereospecific with two possible stereochemical outcomes resulting in either inversion or retention of the anomeric configuration of the transferred sugar. Glycosyltransferases display low sequence homology, and they have recently been classified into 90 families (CAZY database: http://www.cazy.org). At the present time, more than 140 GT crystal structures are available that have been grouped in 40 families. Surprisingly, the three-dimensional architectures of Leloir-type GTs are remarkably conserved, and their X-ray structures exhibit mostly two general types of folds, termed GT-A and GT-B. As for the GTs that utilize lipid-phosphate donor substrates, different folds have been observed.

Data Query Upon reaching the search page, two buttons to query the database appear on the left hand panel: Simple Search and Advanced Search.

Simple Search The classification of the GT proteins is made based on their origin: (1) animal, (2) archaea, (3) bacteria, (4) plant, (5) virus, and (6) yeast and fungi. Upon selecting one organism, a click opens a new menu that prompts the user to choose among a subclassification based on the fold, i.e., GT-A or GT-B. A further click opens a menu where the GTs are numbered according to the CAZY classification, and upon clicking on a CAZY family, the user can select the requested protein and be brought to the "GT information" page by selecting one PDB access code.

Advanced Search Under the name "Select Criteria," a search box offers to select among the following items: (1) organism, (2) family, (3) PDB, and (4) authors. A search box is provided in which appears either a drop-down button enlisting all the entries corresponding to the selected item, or the user can directly enter a query (i.e., a PDB code). The result is a prompt to guide the user in selecting the "hits" found in the database, by a simple search engine. More complex searches can be made by combining criteria which can be combined from up to four search boxes. A preview of the results is displayed in an accordion fashion, whereby the enzyme name, the organism and type of complex if any, are given. The amount of information provided allows the user to make an informed choice prior going to more information on the selected GT.

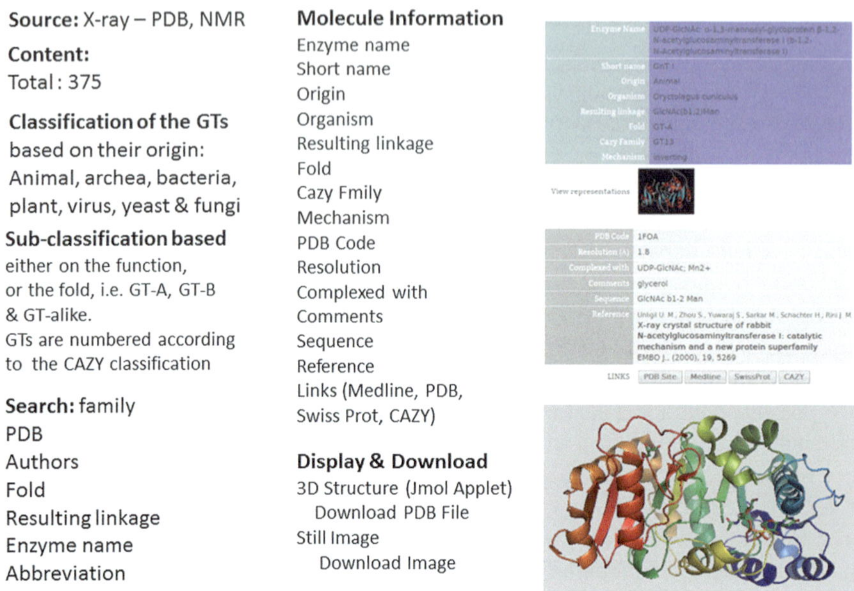

Fig. 7.9 The glycosyltransferases page

Results The detailed results are available under two tabs: "Molecule Information" and "Display and Download" (Fig. 7.9).

Molecule Information Under this button, the following information are provided: enzyme name, short name, origin, organism, resulting linkage, fold, CAZY family, and mechanism. As regards to the crystal structure, the PDB code, the resolution, the nature of the complex (if any), comments, and references are given.

Display and Download On this page will be represented one or more graphical representations of the glycosyltranferase.

7.3.7 mABS: A 3D Structural Database of Monoclonal Antibodies Against Carbohydrates

Database Content Antibodies are glycoproteins belonging to the immunoglobulin superfamily. Three-dimensional structures have been established from X-ray crystallography as listed in http://www.bioinf.org.uk/abs/sacs/ (Allcorn and Martin 2002). Carbohydrate determinants recognized by antibodies are expressed on the cell surface as glycolipids and glycoproteins. In many instances, the minimum carbohydrate epitopes are located at the terminal end of more complex carbohydrate chains, experiencing a wide range of contexts, surface densities, and surroundings. Therefore, antibodies with similar specificities for individual carbohydrate epitopes can exhibit different selective cell profiling depending upon the unique presentation of the carbohydrate on the target cells. As a consequence, anti-carbohydrate antibodies with specificity to oligosaccharides and polysaccharides are of a high importance in immunology and are attractive targets for vaccine design (Pazur 1998). In the present database, the set of high-resolution structures of carbohydrate-antibody complexes is somehow limited. These studies are typically limited to systems involving antibody fragments, such as the antigen-binding fragment (Fab) or variable fragment (Fv), and to small oligosaccharides. Analysis of these complexes reveals general trends about how antibodies recognize different types of carbohydrates. Antibodies which recognize a terminal carbohydrate motif generally feature cavity-like binding sites, where one or more carbohydrate residues are anchored in the cavity by "end-on" extension. Antibodies which recognize an internal carbohydrate motif, as a single repeat of a bacterial polysaccharide, for example, generally exhibit groove-like binding sites or very large cavities which are open at both ends of the site, allowing for "side-on" entry of the antigen.

Data Query Upon reaching the search page, two buttons to query the database appear on the left hand panel: Simple Search and Advanced Search.

Simple Search The first level offers the choice between the natures of the antibody, i.e., human, humanized, or mouse. After selection, a further right click opens a window on "Antibody Information."

Advanced Search Under the name "Select Criteria," a search box offers to select among the following items: (1) nomenclature, (2) antibody, (3) origin, and (4) immunoglobulin type. A search box is provided in which appears a drop-down button enlisting all the entries corresponding to the selected item. The result is a prompt to guide the user in selecting the "hits" found in the database, by a simple search engine. More complex searches can be made by combining criteria which can be combined from up to four search boxes. A preview of the results is displayed. The amount of information provided allows the user to make an informed choice prior going to "Antibody Information."

Results The detailed results are available under two tabs: "Antibody Information" and "Display and Download" (Fig. 7.10).

Molecule Information lists the origin, name of the antibody, nomenclature of the bound carbohydrate, PDB code, resolution, comment, immunoglobulin class, and reference to the original article. Provision is also given to view a still three-dimensional ribbon-type representation of the three-dimensional structure. Links to Medline (http://www.ncbi.nlm.nih.gov/pubmed/) and Protein Data Bank (http://www.rcsb.org/pdb) are also provided.

Display and Download On this page, is given a three-dimensional representation of the three-dimensional structure of the complex which has been constructed from the reported atomic coordinates. In the case of mAbs–carbohydrate crystalline

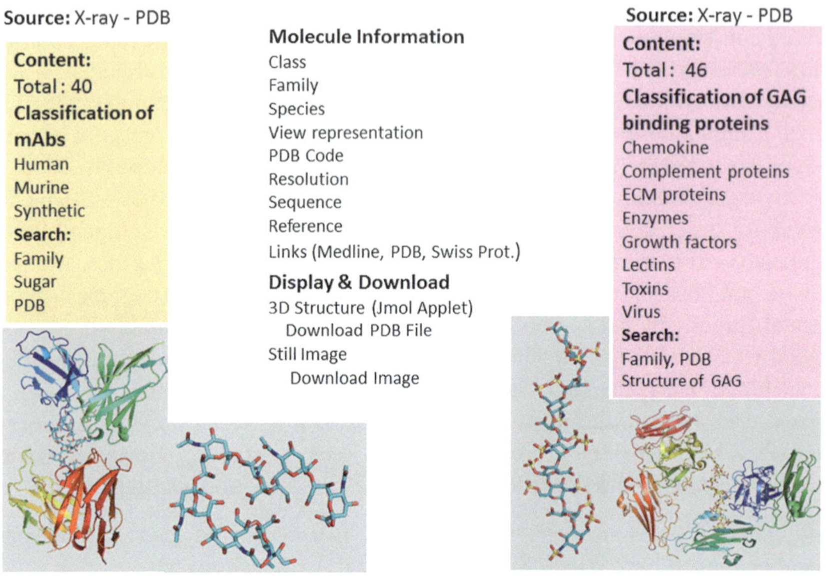

Fig. 7.10 The mAbs and the GAG-binding proteins page

complexes, a particular emphasis is given to indicate the conformation of the bound carbohydrate which can be viewed.

7.3.8 GAG: 3D Structural Database of Glycosaminoglycan-Binding Proteins

Database Content The glycosaminoglycans (GAGs) comprise a class of complex anionic polysaccharides which, through their linkage to a core protein, are part of more complex macromolecules (proteoglycans). There are several GAG families: (1) glycosaminoglycans (heparin and heparan sulfate), (2) galactosylaminoglycans (chondroitin sulfate and dermatan sulfate), and (3) hyaluronic acid and keratan sulfate. GAGs are assembled from repeating disaccharides, and they exhibit diverse patterns of sulfation. Among them, hyaluronic acid is unique as it is not attached covalently to a core protein and it lacks sulfation. In addition to their participation in the physicochemical properties of the extracellular matrix, glycosaminoglycan fragments are specifically recognized by protein receptors, and they play a role in the regulation of many processes, such as hemostasis, growth factor control, anticoagulation, and cell adhesion (Gandhi and Mancera 2008; Imberty et al. 2007; Raman et al. 2005).

Given the importance of protein–GAG interactions, oligosaccharide fragments are prime targets for drug design. The classes of proteins interacting with GAGs are chemokines, complement proteins, components of the extra cellular matrix, enzymes, growth factors, lectins, toxins, and viruses. There is a small of number of crystal structures of complexes available in the Protein Data Bank. Most of the reported structures deal with proteins which have been co-crystallized with heparin oligosaccharides. Enzymes are limited to only two cases: one heparinase and one sulfotransferase illustrating that a large number of protein interacting with heparin sulfate are receptors.

Data Query Upon reaching the search page from Glyco3D, two buttons to query the database appear on the left hand panel: Simple Search and Advanced Search.

Simple Search The classification of the GAG-binding proteins is made based on their biological function: chemokine, complement protein, extracellular matrix (ECM) protein, enzyme, growth factor, lectin, toxin, and virus. Upon selecting one family, a right click opens a new menu that prompts the user to choose among a subclassification. A further right click opens a window on "GAG information."

Advanced Search Under the name "Select Criteria," a search box offers to select among the following items: (1) protein, (2) nature of GAG, and (3) PDB. A search box is provided in which appears a drop-down button enlisting all the entries corresponding to the selected item. The result is a prompt to guide the user in selecting the "hits" found in the database, by a simple search engine. More complex searches can be made by combining criteria which can be interlaced from up to four search boxes. A preview of the results is displayed in an accordion fashion, whereby the classification, the protein name, the GAG type, and the size of the oligosaccharide are given. The amount of information provided allows the user to make an informed choice prior to going to "GAG Information."

Results The detailed results are available under two tabs: "GAG Information" and "Display and Download" (Fig. 7.10).

Molecule Information Under the button, "GAG Information" is given: protein, classification, GAG type, species, PDB code, resolution, length of oligosaccharide, comments, and references. Provision is also given to view an image of the source of the protein, along with a graphical representation of the three-dimensional structure. Links to Medline (http://www.ncbi.nlm.nih.gov/pubmed/), Protein Data Bank (http://www.rcsb.org/pdb), and Uniprot (http://www.uniprot.org/uniprot) are also provided.

Display and Download On this page, is represented a still three-dimensional ribbon-type representation of the three-dimensional structure which has been constructed from the reported atomic coordinates. In the case of protein–GAG crystalline complexes, a particular emphasis is given to indicate the location and conformation of the bound carbohydrate.

7.3.9 LECTIN3D: A 3D Structural Database of Lectins

Database Content Lectins are proteins of nonimmune origin that bind to specific carbohydrates without modifying them. Per current knowledge, they act like molecular readers to decipher sugar-encoded information. They play biologically important roles in recognition processes involved in fertilization, embryogenesis, inflammation, metastasis, and parasite–symbiote recognition in microbes, invertebrates, plants, and vertebrates (Ambrosi et al. 2005; Arnaud et al. 2013, Sharon 2007). In the plant kingdom, lectins have been demonstrated to play a role in defense against pathogens or predators and hypothesized to be involved in establishing symbiosis with mushrooms and bacteria of the Rhizobia species. Among the proteins that interact non-covalently with carbohydrates, lectins bind mono- and oligosaccharides reversibly and specifically.

More than 1400 three-dimensional structures of lectins have been solved and are available in the database as of January 2016. They have been determined by X-ray diffraction, although some neutron diffraction structures are available as well as NMR solution structures or theoretical models. This covers almost 250 different proteins. Most 3D structures have been for the time being, obtained for plant and animal lectins. Nevertheless, the number of structural investigations dealing with viral and bacterial materials increases rapidly. Among these structures, 64 % are complexed with a carbohydrate ligand which can be multiforms; i.e., they are monosaccharides, oligosaccharides, glycoproteins, or synthetic glyco-compounds from organic chemistry (Table 7.2). Accurate determination of carbohydrate–lectin complexes remains a nontrivial problem due to the shallow and multichambered binding sites of many lectins. This is nevertheless a requirement to access information about their binding mechanisms in biologically relevant conditions.

Structures of plant and animal lectins are the most abundant in the database (Fig. 7.11). The number of structures of lectins from bacteria, fungi, and viruses is growing steadily as their biological functions are being recognized (Imberty and Varrot 2008; Varrot et al. 2013). Only one structure of algal lectin is available at the present time (Ziolkowska et al. 2006). As for structural motifs, there is a strong predominance of β-sheets. Two such β-sheets can assemble to form a β-sandwich, an architecture commonly found in more than half of the entries. These β-sandwiches exhibit dissimilarities and the location of binding sites shows many variations. For example, the immunoglobulin-like fold of animal sialo-adhesins is very different from the jelly-roll fold of legume lectins. More complex combinations of β-sheets occur, giving rise to β-propellers and β-prisms. Despite the fact that topologies are very different between families, some interesting structural convergences are nevertheless observed. Intracellular animal lectins which are involved in the quality control of glycoprotein synthesis share the same protein fold with legume lectins, now referred to as L-lectin (Loris 2002).

Data Query Upon reaching the search page form, two buttons to query the database appear on the left hand panel: Simple Search and Advanced Search.

Table 7.2 Classification and distribution of apo/complex 3D structures of lectins

	Lectin	Free	Complex	Lectin	Free	Complex
Plant	L-type (legume)	75	152	Hevein type	13	19
	β-prism I (jacalin)	16	35	β-prism II (monocot)	7	12
	R-type lectin (β-trefoil)	15	20	Cyanovirin-N family	1	0
	β-prism II (monocot)	7	12			
Bacteria	Pili adhesin	9	42	Cytolysin	2	7
	Neurotoxin	21	20	Staphylococcal toxin	2	5
	AB5 toxin	15	20	β-trefoil	1	4
	2-Ca β-sandwich	3	30	Scytovirin	3	0
	Cyanovirin-N family	17	8	Serine-rich repeat	2	1
	1-Ca β-sandwich	2	16	Toxin repetitive domain	1	1
	β-propeller	0	11	TNFα-like	0	1
	Oscillatoria agglutinin	7	3			
Animal	C-type	44	86	Calnexin-calreticulin	5	2
	Galectin	33	103	TIM-lectin	2	6
	Fibrinogen-like	6	15	L-rhamnose binding	4	3
	I-type	5	11	C-type lectin-like	3	1
	P-type	6	9	Malectine	1	2
	R-type (β-trefoil)	3	10	F-type	0	2
	H-type	4	11	Cys-knot	1	0
	Pentraxin	7	3	β-propeller	2	2
	Microneme MAR	2	7	Chitin binding	1	0
	L-type	5	3			
Virus	Hemagglutinin	44	23	Rotavirus spike	4	11
	Norovirus capsid	18	38	Fiber knob	2	12
	Polyomavirus capsid	12	15	Coat protein	0	3
	Phage tailspike	8	8	Coronavirus spike	1	0
Fungi and yeast	Galectin	13	21	L-type 1	7	0
	Actinoporin-like	7	14	7-blades β-propeller	2	6
	β-trefoil	8	20	6-blades β-propeller	1	9
	Yeast adhesin	2	11	Ig-like	1	0
	Cyanovirin-N hom.	7	1			
Algae	Griffithsin	5	8			

Simple Search The classification of the lectins is made based on their origin: (1) algae, (2) animal, (3) bacteria, (4) fungi and yeast, (5) plant, and (6) virus. Upon selecting one family, a right click opens a new menu that prompts the user to choose among a subclassification based on the fold family and then on the species of organisms. A further right click opens a new menu that contains all the three-dimensional structures of the selected lectin either in the apo state or complexed with ligand. A preview of the results is displayed in an accordion fashion, whereby the PDB code, the species, the resolution at which the structure has been solved, and the

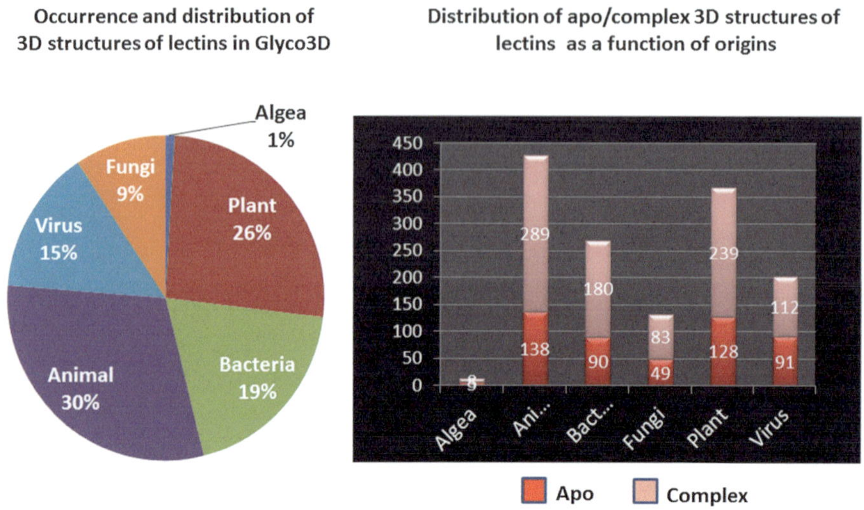

Fig. 7.11 Origin of lectin structures present in Lectin-3D database

reference to the original publication are given. The amount of information provided allows the user to make an informed choice prior to going to "Lectin Information."

Advanced Search Under the name "Select Criteria," a search box offers to select among the following items: (1) species, (2) family, (3) sugars, and (4) PDB. A search box is provided in which appears a drop-down button enlisting all the entries corresponding to the selected item. For other items, the menu guides the user in selecting the "hits" found in the database, by a simple search engine. More complex searches can be made by combining criteria from up to four search boxes.

Results The detailed results are available under two tabs: "Molecule Information" and "Display and Download."

Molecule Information Under the button, "Molecule Information" is given: origin, class, family, species, PDB code, resolution, comment, and reference. The comment section indicates whether the lectin has been solved in the form of a protein-carbohydrate complex. In that case, the nature of the sugar is indicated along with its sequence. Provision is also given to view an image of the source of the protein, along with a still three-dimensional ribbon-type representation of the three-dimensional structure, together with access to original 3D information at the Protein Database. Links to NIH sites for references and taxonomy are also provided. If the lectin has been submitted to the Consortium for Functional Glycomics for analysis of specificity through glycan arrays (Agravat et al. 2014), a link to the data page is also provided.

Display and Download On this page, is represented one or more graphical representations of the lectin or of the binding site with carbohydrate ligands that have

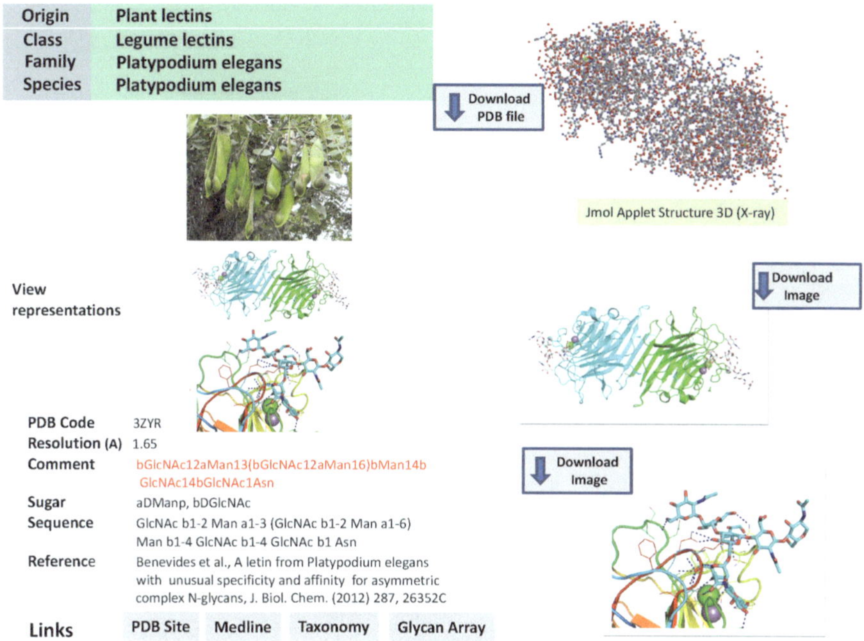

Fig. 7.12 The lectin page

been constructed from the reported atomic coordinates. A particular emphasis is given to indicate the location and conformation of the bound carbohydrate (Fig. 7.12 provides an overall presentation of the "Lectin page").

7.4 Sequence Search in the Whole Glyco-3D Portal

Glyco3D encompasses a family of databases covering the 3D features of monosaccharides, disaccharides, oligosaccharides, polysaccharides, glycosyltransferases, lectins, monoclonal antibodies, and glycosaminoglycan-binding proteins. Each of these databases has been set to account for the specific features of the class of molecules covered. Nevertheless, a logical network has been established that links all these databases together (Fig. 7.13). A search engine has been developed that scans the full content of all the databases for queries related to sequential information of the carbohydrates or other related descriptors. This is performed under the "search sequence" command.

For example, when looking for all information related to the H-type 1 trisaccharide, one needs to insert the request "Fuc a1-2 Gal b1-3 GlcNAc" in the "search sequence." This results in 22 hits in the "BiOligo" Database, 23 NMR spectra, and

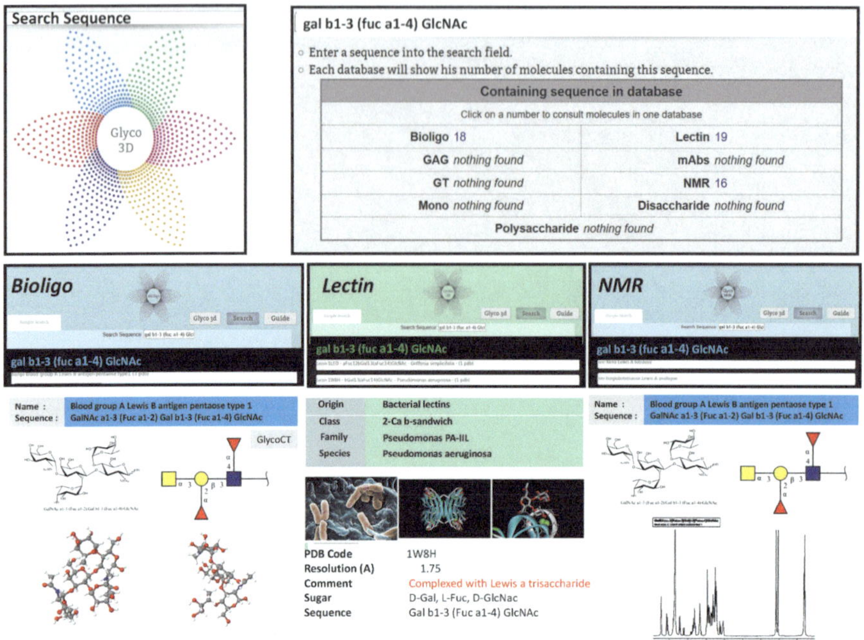

Fig. 7.13 Example of oligosaccharide search in the Glyco-3D portal

11 crystal structures of lectins. The large number of hits is due to the fact that the H-type 1 epitope is embedded not only in type 1 ABO oligosaccharides but also in several Lewis epitopes such as Lewis a, sialyl Lewis a, and Lewis b. The resulting structures for which NMR data are available are shown, whereas those lectin structures which have been co-crystallized with this glycan are given. They include lectins from virus, plants, and animals, as well as the bacterial lectin BambL from *Burkholderia ambifaria* complexed with H-type 1 tetrasaccharide depicted in Fig. 7.13 (Audfray et al. 2012).

7.5 Conclusions

The present Glyco3D portal offers a single entry to access three-dimensional features of glycans, polysaccharides, and glycan-recognizing proteins. A particular emphasis was given to the use of a common nomenclature for the structural encoding of the carbohydrates. Yet every glycan molecule is described by four different types of representations in order to cope with the different usages in chemistry and biology. As the information content of each representation may vary or highlight a particular aspect compared to others, this offers the most complete description of the many features that characterize glycans. If the use of GlycoCT code is meant to facilitate future exchanges with other databases, this may not be sufficient, and other coding such as InChi and SMILES may be required.

As with any other databases, the value of the repository lies very much on the quality of the annotation and the curation of data. Now that the overall architecture of the databases and their interconnection are established, there remains the daily intervention to maintain the content updated and validated by expert scientists. To this end, an administration interface has been designed to facilitate greatly this task. Access to this interface can be open to extend the size of the curators.

The scientific value of the Glyco3D portal and its constituting databases has proven to be of great help at the level of structural biology. Obviously, connections with other 3D structural databases would increase the number of the glycans covered, as well as the scope of the applications toward rational design of complex glycan-containing architectures. Another expected enhanced value would result from mutual online access between other databases dealing with the functional aspects of complex carbohydrates. Glyco3D should be an asset to the community for probing further into the behavior of the very important class of glycomolecules and would open the way to establish a closer collaboration with bioinformatics groups in proteomics and genomics.

Acknowledgments The research leading to this publication has received funding from the European Commission's Seventh Framework Programme FP7/2007–2013 under grant agreement n° 215536 and from the Agence Nationale de la Recherche under the "Genomic and Plant Biotechnology" Action throughout the "Wall-Array" project. We are grateful to the Marie Curie Initial Training Network as part of the FP7 People Programme for training and funding.

References

Agravat SB, Saltz JH, Cummings RD, Smith DF (2014) GlycoPattern: a web platform for glycan array mining. Bioinformatics 30:3417–3418

Allcorn LC, Martin AC (2002) SACS–self-maintaining database of antibody crystal structure information. Bioinformatics 18:175–181

Allen FH (2002) The Cambridge Structural Database: a quarter of a million crystal structures and rising. Acta Crystallogr Sect B Struct Sci 58:380–388

Ambrosi M, Cameron NR, Davis BG (2005) Lectins: tools for the molecular understanding of the glycocode. Org Biomol Chem 3:1593–1608

Arnaud J, Audfray A, Imberty A (2013) Binding sugars: from natural lectins to synthetic receptors and engineered neolectins. Chem Soc Rev 42:4798–4813

Audfray A, Claudinon J, Abounit S, Ruvoën-Clouet N, Larson G, Smith DF, Wimmerová M, Le Pendu J, Römer W, Varrot A, Imberty A (2012) The fucose-binding lectin from opportunistic pathogen *Burkholderia ambifaria* binds to both plant and human oligosaccharidic epitopes. J Biol Chem 287:4335–4347

Berman H, Henrick K, Nakamura H (2003) Announcing the worldwide protein data bank. Nat Struct Biol 10:980

Breton C, Fournel-Gigleux S, Palcic MM (2012) Recent structures, evolution and mechanisms of glycosyltransferases. Curr Opin Struct Biol 22:540–549

Degtyarenko K, De Matos P, Ennis M, Hastings J, Zbinden M, Mcnaught A, Alcantara R, Darsow M, Guedj M, Ashburner M (2008) ChEBI: a database and ontology for chemical entities of biological interest. Nucleic Acids Res 36:D344–D350

Demarco ML, Woods RJ (2008) Structural glycobiology: a game of snakes and ladders. Glycobiology 18:426–440

Engelsen SB, Hansen P, Perez S (2013) POLYS: an open source software package for building three-dimensional structures of polysaccharides. Biopolymers 101:733–743

Frank M, Schloissnig S (2010) Bioinformatics and molecular modeling in glycobiology. Cell Mol Life Sci 67:2749–2772

Gandhi NS, Mancera RL (2008) The structure of glycosaminoglycans and their interactions with proteins. Chem Biol Drug Des 72:455–482

Imberty A, Varrot A (2008) Microbial recognition of human cell surface glycoconjugates. Curr Opin Struct Biol 18:567–576

Imberty A, Lortat-Jacob H, Perez S (2007) Structural view of glycosaminoglycan-protein interaction. Carbohydr Res 342:430–439

Jo S, Im W (2013) Glycan fragment database: a database of PDB-based glycan 3D structures. Nucleic Acids Res 41:D470–D474

Loris R (2002) Principles of structures of animal and plant lectins. Biochim Biophys Acta 1572:198–208

Lutteke T, Bohne-Lang A, Loss A, Goetz T, Frank M, Von Der Lieth CW (2006) GLYCOSCIENCES.de: an internet portal to support glycomics and glycobiology research. Glycobiology 16:71R–81R

Marchessault RH, Perez S (1979) Conformations of the hydroxymethyl group in crystalline aldohexopyranoses. Biopolymers 18:2369–2374

Mcnaught AD (1997) Nomenclature of carbohydrates (Recommendations 1996). Adv Carbohydr Chem Biochem 52:43–177

Nakahara T, Hashimoto R, Nakagawa H, Monde K, Miura N, Nishimura S (2008) Glycoconjugate data bank: structures–an annotated glycan structure database and N-glycan primary structure verification service. Nucleic Acids Res 36:D368–D371

Pazur JH (1998) Anti-carbohydrate antibodies with specificity for monosaccharide and oligosaccharide units of antigens. Adv Carbohydr Chem Biochem 53:201–261

Perez S (2007) Oligosaccharide and polysaccharide conformations by diffraction methods. In: Kamerling JP (ed) Comprehensive glycosciences: analysis of glycans. Elsevier, Oxford, pp 193–220

Perez S, Mulloy B (2005) Prospects for glycoinformatics. Curr Opin Struct Biol 15:517–524

Perez S, Tvaroška I (2014) Carbohydrate-protein interactions: molecular modeling insights. Adv Carbohydr Chem Biochem 71:9–136

Perez S, Gautier C, Imberty A (2000) Oligosaccharide conformations by diffraction methods. In: Ernst B, Hart G, Sinay P (eds) Oligosaccharides in chemistry and biology: a comprehensive handbook. Wiley/VCH, Weinheim, pp 969–1001

Perez S, Tubiana T, Imberty A, Baaden M (2015) Three-dimensional representations of complex carbohydrates and polysaccharides: a video game based computer graphic software. Glycobiology 25:483–491

Pettersen EF, Goddard TD, Huang CC, Couch GS, Greenblatt DM, Meng EC, Ferrin TE (2004) UCSF Chimera–a visualization system for exploratory research and analysis. J Comput Chem 25:1605–1612

Priem B, Gilbert M, Wakarchuk WW, Heyraud A, Samain E (2002) A new fermentation process allows large-scale production of human milk oligosaccharides by metabolically engineered bacteria. Glycobiology 12:235–240

Raman R, Sasisekharan V, Sasisekharan R (2005) Structural insights into biological roles of protein-glycosaminoglycan interactions. Chem Biol 12:267–277

Rosen J, Miguet L, Perez S (2009) Shape: automatic conformation prediction of carbohydrates using a genetic algorithm. J Cheminf 1:16

Sarkar A, Perez S (2012) PolySac3DB: an annotated data base of 3 dimensional structures of polysaccharides. BMC Bioinf 13:302

Sarkar A, Drouillard S, Rivet A, Perez S (2015) Databases of conformations and NMR structures of glycan determinants. Glycobiology 25:1480–1490

Sharon N (2007) Lectins: carbohydrate-specific reagents and biological recognition molecules. J Biol Chem 282:2753–2764

Varki A, Cummings RD, Aebi M, Packer NH, Seeberger PH, Esko JD, Stanley P, Hart G, Darvill A, Kinoshita T, Prestegard JJ, Schnaar RL, Freeze HH, Marth JD, Bertozzi CR, Etzler ME, Frank M, Vliegenthart FG, Lütteke T, Perez S, Bolton E, Rudd P, Paulson J, Kanehisa M, Toukach P, Aoki-Kinoshita KF, Dell A, Narimatsu H, York W, Taniguchi N, Kornfeld S (2015) Symbol nomenclature for graphical representations of glycans. Glycobiology 25:1323–1324

Varrot A, Basheer SM, Imberty A (2013) Fungal lectins: structure, function and potential applications. Curr Opin Struct Biol 23:678–685

Wang Y, Bolton E, Dracheva S, Karapetyan K, Shoemaker BA, Suzek TO, Wang J, Xiao J, Zhang J, Bryant SH (2010) An overview of the PubChem BioAssay resource. Nucleic Acids Res 38:D255–D266

Weininger D (1988) SMILES, a chemical language and information system. 1. Introduction to methodology and encoding rules. J Chem Inf Comput Sci 28:31–36

Woods RJ (2014) GLYCAM Web (http://www.glycam.com). Complex Carbohydrate Research Center, University of Georgia, Athens

Ziolkowska NE, O'keefe BR, Mori T, Zhu C, Giomarelli B, Vojdani F, Palmer KE, Mcmahon JB, Wlodawer A (2006) Domain-swapped structure of the potent antiviral protein griffithsin and its mode of carbohydrate binding. Structure 14:1127–1135

Chapter 8
GlycoGene Database (GGDB) on the Semantic Web

Hisashi Narimatsu, Yoshinori Suzuki, Kiyoko F. Aoki-Kinoshita, Noriaki Fujita, Hiromichi Sawaki, Toshihide Shikanai, Takashi Sato, Akira Togayachi, Takehiko Yoko-o, Kiyohiko Angata, Tomomi Kubota, and Erika Noro

Abstract Glycogenes are genes that are related to glycan synthesis, such as glycosyltransferases, sugar nucleotide synthases, sugar nucleotide transporters, sulfotransferases, etc. We have accumulated and curated glycogene data into the GGDB database to allow for glycogene analysis. This chapter introduces the GlycoGene Database (GGDB) which consists of all known glycogenes in human and mouse (Togayachi et al., A database system for glycogenes (GGDB). In: Taniguchi N, Suzuki A, Ito Y, Narimatsu H, Kawasaki T, Hase S (eds) Experimental Glycoscience. Springer, Tokyo, pp 423–425, 2008). GGDB was developed at AIST and has been updated with the latest data and a new user interface utilizing Semantic Web technologies.

Keywords Glycogenes • Database • Semantic Web

8.1 System Overview

Glycogenes are genes that are related to glycan synthesis, such as glycosyltransferases, sugar nucleotide synthases, sugar nucleotide transporters, sulfotransferases, etc. We have accumulated and curated glycogene data into the GGDB database to allow for glycogene analysis.

H. Narimatsu (✉)
Research Center for Medical Glycoscience (RCMG), National Institute of Advanced Industrial Science and Technology (AIST), Ibaraki, 305-8568, Japan
e-mail: h.narimatsu@aist.go.jp

Y. Suzuki • N. Fujita • H. Sawaki • T. Shikanai • T. Sato • A. Togayachi • T. Yoko-o • K. Angata • T. Kubota • E. Noro
National Institute of Advanced Industrial Science and Technology (AIST), Ibaraki, 305-8566, Japan

K.F. Aoki-Kinoshita
Soka University, Tokyo, 192-8577, Japan

K.F. Aoki-Kinoshita (ed.), *A Practical Guide to Using Glycomics Databases*,
DOI 10.1007/978-4-431-56454-6_8

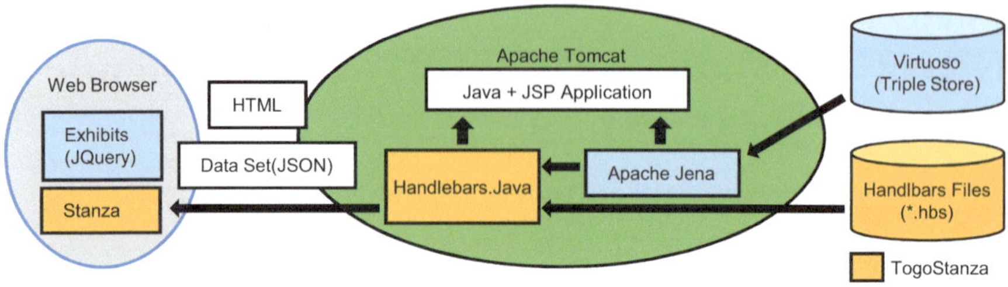

Fig. 8.1 The overall GGDB Web system, consisting of Java and JSP, supplemented by Exhibit and Handlebars

The GlycoGene Database (GGDB; http://jcggdb.jp/rcmg/ggdb) was originally developed under the "Construction of Glycogene Library" project from April 2001 to March 2004 (Narimatsu 2004). At the time, over 183 genes of human glycosyltransferases and sulfotransferases were identified, cloned, and expressed in various expression systems, in order to analyze the carbohydrate synthesis activity and related biological function. These were all accumulated into GGDB to allow users to easily access the data and relevant metadata information. The new GGDB has been revamped, with the development of a new ontology for GGDB data and a new user interface to allow users to more easily find their data of interest. The new URL is http://acgg.asia/db/ggdb.

The Web system of GGDB mainly consists of Java and JavaServer Pages (JSP). We have also used Handlebars in some parts of the user interface, based on the TogoStanza framework, to enable user-friendly Web pages that can be reused. Because GGDB has been RDFized (RDF: resource description framework, a format to specify data using triples such that semantics can be incorporated into the data on the Internet), the GGDB data can be accessed using SPARQL (https://www.w3.org/TR/sparql11-overview/) queries; we thus used the Java API of Apache Jena and Handlebars.java to handle Handlebars processing. SPARQL is the query language for RDF data. Moreover, to allow users to intuitively extract data of their target, we developed a facet query interface using the Exhibit API. Figure 8.1 illustrates the Web system of GGDB, and Table 8.1 describes each of these components in detail.

8.2 User Interface

This section describes the user interface for interacting with GGDB via the Web.

8.2.1 Top Page

The top page of GGDB can be accessed from http://acgg.asia/db/ggdb. Figure 8.2 is a snapshot of the top page, which lists the glycogenes on the right, with filters

Table 8.1 GGDB System description

Component and URL	Description
TogoStanza http://www.togostanza.org/	Stanzas are reusable Web components; they are developed so that other Web sites can display the data formatted within the stanza simply by referencing the URL of the stanza from within a <div> tag. TogoStanza is a framework for developing Semantic Web components as reusable Web components such that they can be embedded into other Web applications. Handlebars are used for the HTML templates, and Ruby or Java can be used for developing the server-side components
Handlebars http://handlebarsjs.com/	A template engine for creating dynamic Web pages. In addition to the original Javascript version named Hanlebars.js, modules have also been developed in other languages such as Java, Ruby, and PHP
Exhibit http://simile-widgets.org/exhibit/	A JavaScript toolkit that allows faceted searches of data from within a Web page

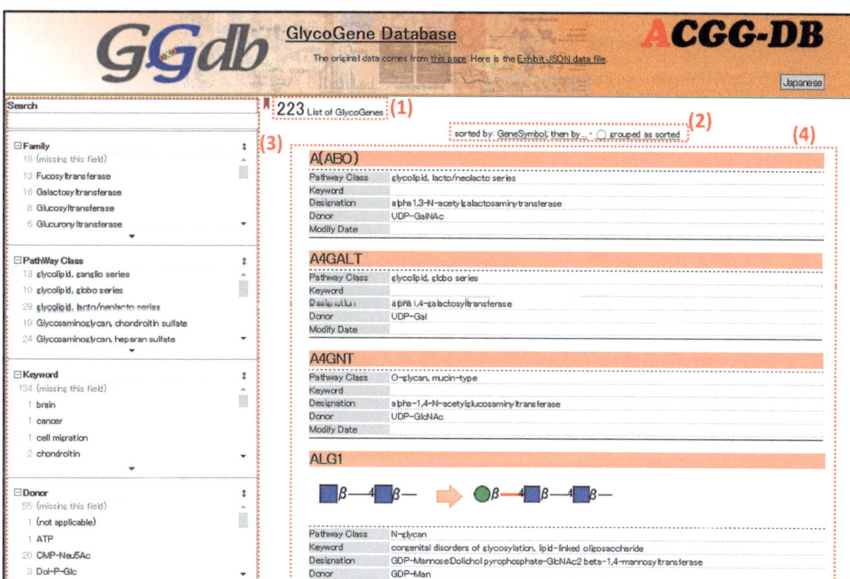

Fig. 8.2 The top page of GGDB, listing the glycogenes on the right and filters on the left

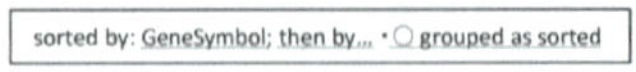

Fig. 8.3 The menu by which to change the sorting and display of the glycogenes list

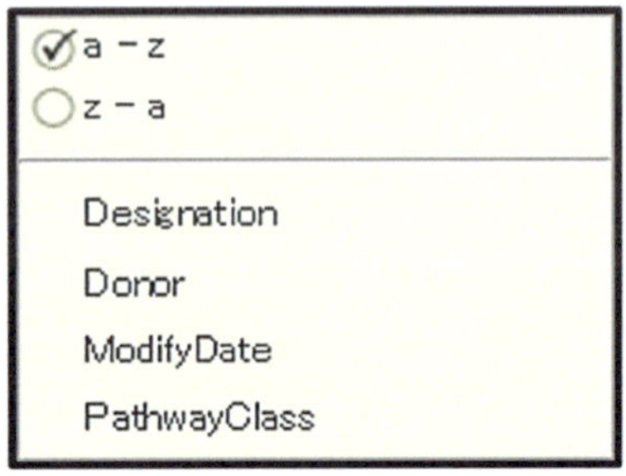

Fig. 8.4 The sorting options for each component selected in Fig. 8.3, where a-z refers to ascending order, and z-a refers to descending order. The other options are additional components by which sorting can be applied

(facets) available on the left. These filters allow the user to intuitively find their glycogene of interest. The glycogene list is initially listed in order of gene symbol. Each component of this page will be described next, following the numbers in Fig. 8.2.

1. **Number of glycogenes**
   The number of glycogenes listed in (4) is shown. As the filters are used and number of glycogenes decrease, this number will be updated to show the number of genes currently being displayed.
2. **Glycogene display menu**
   Figure 8.3 shows how the display menu can be changed, by clicking on the items following “sorted by:”. The order by which the glycogenes are listed can be modified using this menu option. Multiple options can be selected to sort in order. Each option can be right-clicked to display detailed options, as in Fig. 8.4. The sorting can be in ascending (a-z) or descending (z-a) order. In this figure, “grouped as sorted” is selected, meaning that the list of glycogenes will be grouped according to the items selected by which to sort. An example is shown in Fig. 8.6, where the glycogenes are grouped by donor. By unchecking this option, the grouping will be removed and each gene will be shown individually.
3. **Facet and text search for filtering**
   The list of glycogenes can be filtered by using the facet and/or text search. The Search field at the top can be used to specify any keyword; the list of glycogenes will then be filtered to only those that contain the given keyword. Facets allow users to filter based on the given options. Multiple options will be combined using logical AND such that the list of glycogenes only satisfying all of the selected options will be displayed (Fig. 8.5). The five options listed in Table 8.2 are currently available. The up and down arrow at the top-right of each facet can

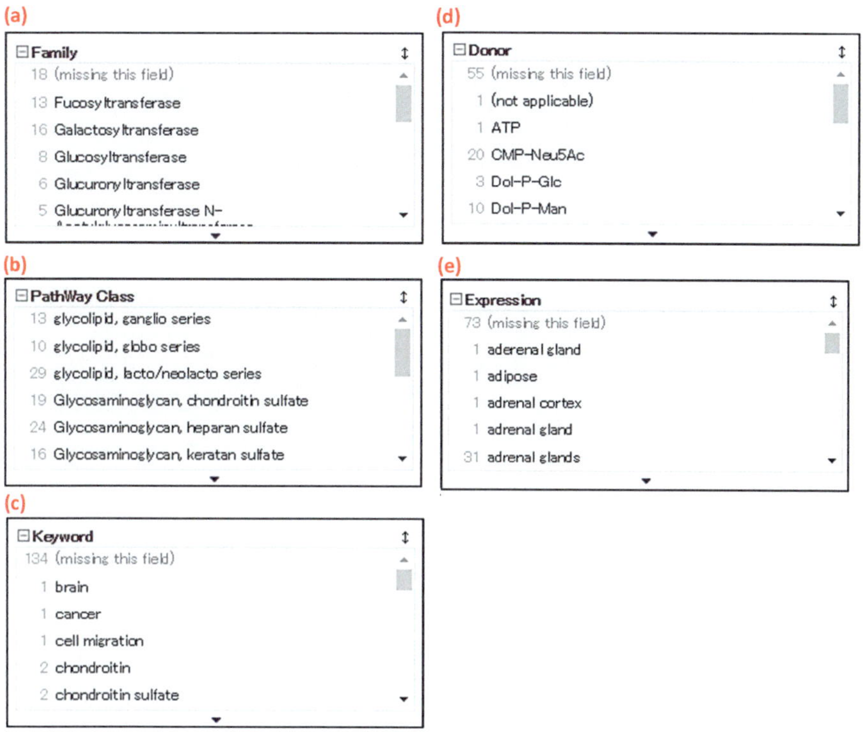

Fig. 8.5 Snapshots of the facets available in GGDB. Each of these serve as filters for the list of genes, and multiple selections are combined using logical AND operations

Table 8.2 Facet options: five facets are currently available to allow users to filter the list of glycogenes displayed

Family	The glycogene family name
Pathway class	The class name of the glycogene, such as *N*-glycans, *O*-glycans, and sphingolipids
Keyword	Related keywords, mainly extracted from the Gene Ontology (The Gene Ontology Consortium 2014) and PhenomeNet (Hoehndorf et al. 2011)
Donor	The sugar nucleotide donor of each glycogene
Expression	The tissue or cell line in which the glycogene is found to be expressed

be used to maximize the window to display the full list of options. The triangle at the bottom of each facet can be used to manually extend the window.

4. **Facet and text search for filtering**

The list of (selected) glycogenes. This list changes according to the selected options in the facets (3) on the left. This list is also affected by the sort and grouping options (2). Figure 8.6 shows how the display changes when grouping is (un)selected. When grouping is selected, the label of each group is displayed,

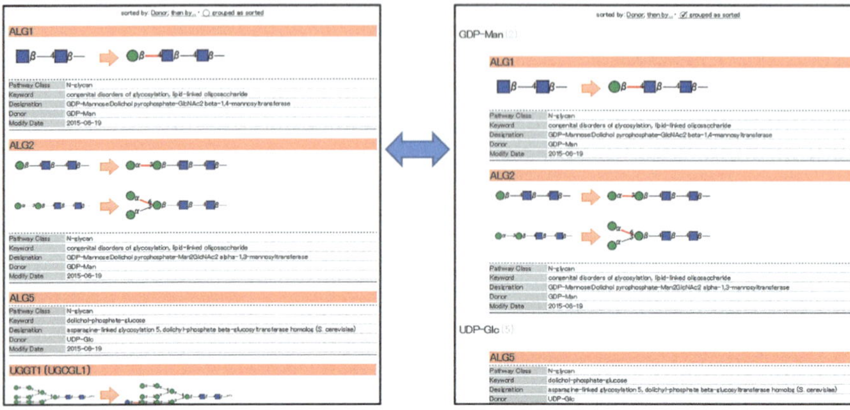

Fig. 8.6 Difference between grouped (*right*) and ungrouped (*left*) views of the glycogene list. In the grouped view, in this case *Donor* is selected, so the genes are grouped by donor

followed by the group contents. When grouping is not selected, all of the genes are listed at once.

The variety of functionality available in this new version of GGDB allows users to easily find their glycogenes of choice. Once found, the details of the selected glycogene can be viewed by clicking on the gene name. The next subsection describes the detail view of each gene.

8.2.2 *Detail View*

This view displays details regarding the selected glycogene, such as its reaction information. The display uses stanzas, such that by referencing the ID or GeneID of the glycogene, the same view can be displayed in other Web sites. This section will first describe the view, followed by an explanation of how stanzas can be used.

8.2.2.1 Detail View Overview

From the glycogene list described in Sect. 8.2.1, by selecting a particular glycogene, a Web page as shown in Fig. 8.7 will be displayed, containing detailed information regarding the glycogene. This figure describes the components of the detailed page, including orthologous genes, substrate information, and expression information, if available. Initially, the detailed sections are hidden, but they can be displayed at once by clicking the "Open All" button. Individual sections can be opened by clicking on the name of the section. Moreover, they can all be hidden again by clicking the "Close All" button.

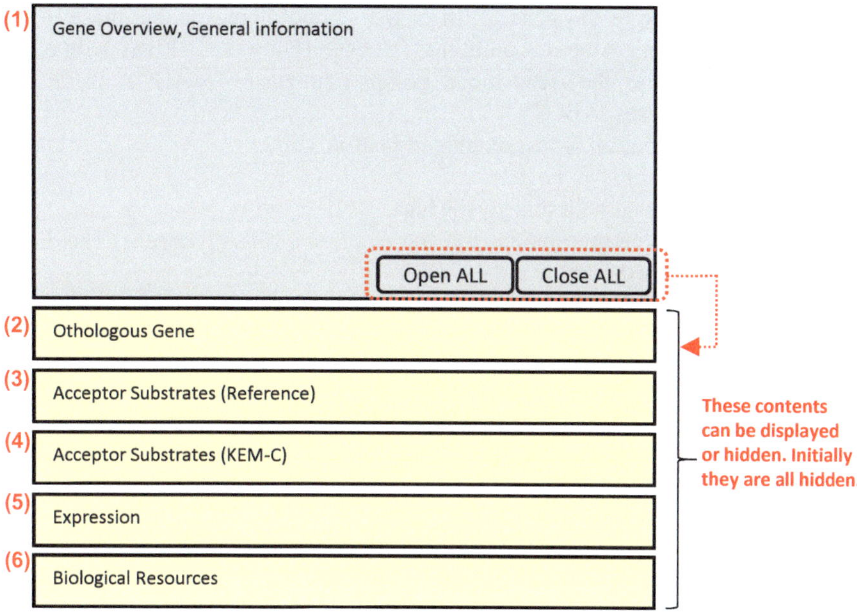

Fig. 8.7 The components of the detail view page of a selected glycogene

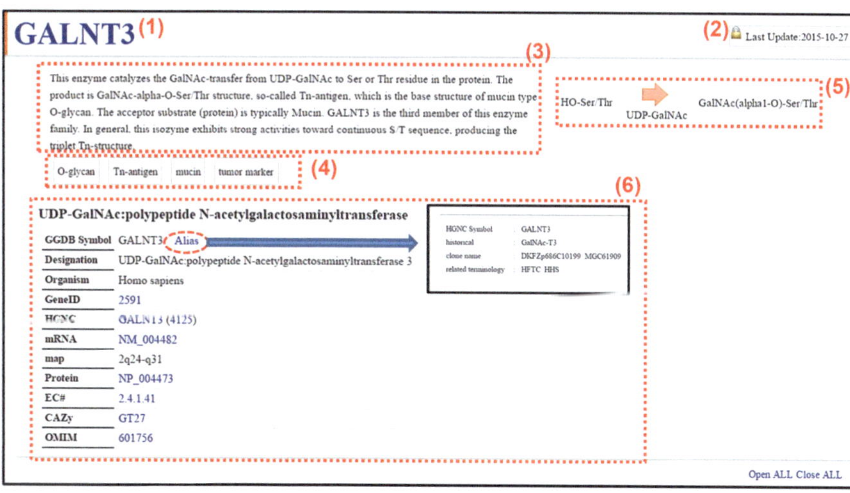

Fig. 8.8 An example of the general information section of a particular glycogene (in this case, GALNT3)

8.2.2.2 Overview, General Information

An example of part (1) in Fig. 8.7 is shown in Fig. 8.8. The components numbered in this figure are described below.

1. The symbol name of the gene. Most of the symbol names match those in the HUGO Gene Nomenclature Committee (HGNC) (Gray et al. 2016), with a few exceptions, such as the ABO blood groups (shown as "A(ABO)" in GGDB whereas HGNC uses "ABO").
2. The date of registration or last modification of this entry.
3. A summary about this glycogene.
4. Keywords associated with this glycogene.
5. An example of the representative transfer reaction of this glycogene. The donor is displayed below the arrow.
6. Other names and IDs of this glycogene as used elsewhere, summarized as follows:

   GGDB Symbol: the symbol name used in GGDB.
   Alias: when clicked, a table of other names of this glycogene will be displayed, along with their types. Types include HGNC symbol, historical names, clones, related terminology, and alleles.
   Designation: the formal name of this gene.
   Organism: organisms in which this gene is expressed.
   GeneID: the NCBI Gene ID (linked to NCBI).
   HGNC: the HGNC symbol name and ID (linked to HGNC).
   mRNA: the RefSeq ID of this gene's mRNA (linked to NCBI).
   Map: information regarding the genomic location of this gene's mRNA.
   Protein: the NCBI ID of this gene's protein (linked to NCBI).
   EC#: the (Enzyme Commission) EC number (linked to the integrated relational enzyme database).
   CAZy: the enzyme family in the CAZy database (Lombard et al. 2014) (linked to CAZy).
   OMIM: the Online Mendelian Inheritance in Man (OMIM) database ID (linked to OMIM at http://omim.org).
   GDGDB: the Glyco-Disease Genes Database (GDGDB) database ID (linked to GDGDB at http://acgg.asia/db/diseases/gdgdb).

8.2.2.3 Orthologous Genes

Part (2) of Figure 8.7 is shown in Fig. 8.9. This section lists the orthologous gene information for this glycogene, along with links to the respective Web pages. In particular, it displays the NCBI IDs for the mRNA and proteins of each orthologous gene, grouped by species. As emphasized in Fig. 8.9, the Mouse Genome Informatics (MGI) ID (Eppig et al. 2015) is additionally listed for mouse genes, if available.

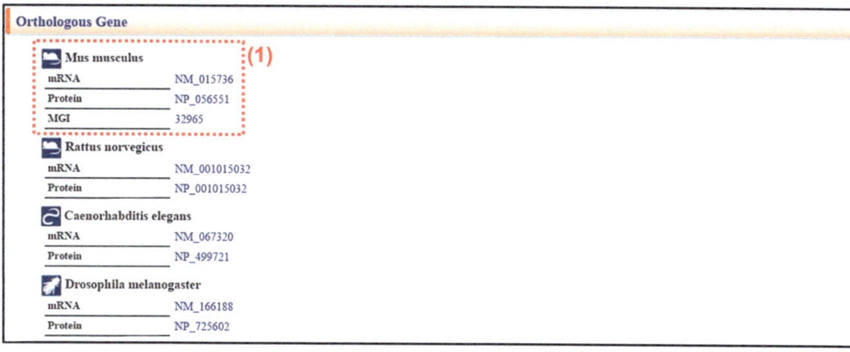

Fig. 8.9 The Orthologous Gene section of the glycogene detail page

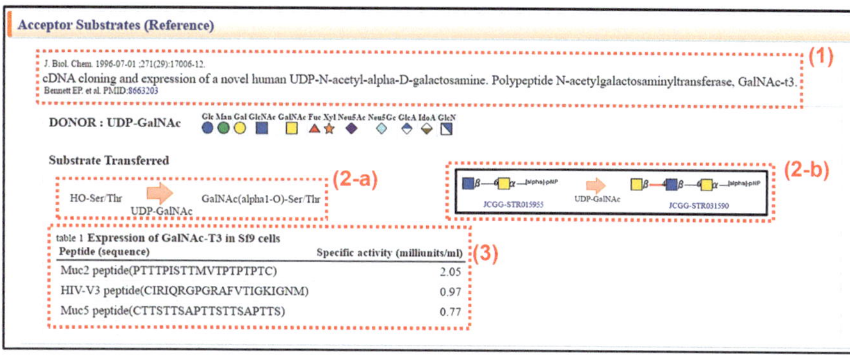

Fig. 8.10 The main components of the Acceptor Substrate section of the glycogene detail page

8.2.2.4 Acceptor Substrates

Part (3) of Fig. 8.7 is shown in Fig. 8.10. This section displays the reaction information for the glycogene. "Acceptor Substrates (Reference)" is information extracted from the literature. On the other hand, "Acceptor Substrates (KEM-C)" reflects experimentally derived data by the Research Center for Medical Glycoscience at AIST.

Each component of Fig. 8.10 is described below. Note that similar information for substrates that are not transferred are also listed, if such information is known.

1. The citation related to the reaction listed underneath. If a particular figure or table within the cited reference contains reaction details such as activity, this information is displayed after the text "Refer to:".
2. The actual glycan structures involved in the reaction is displayed here. If the structures are registered in JCGGDB, then the reaction is displayed as in (2-b) using the symbol nomenclature of the Essentials of Glycobiology (Varki

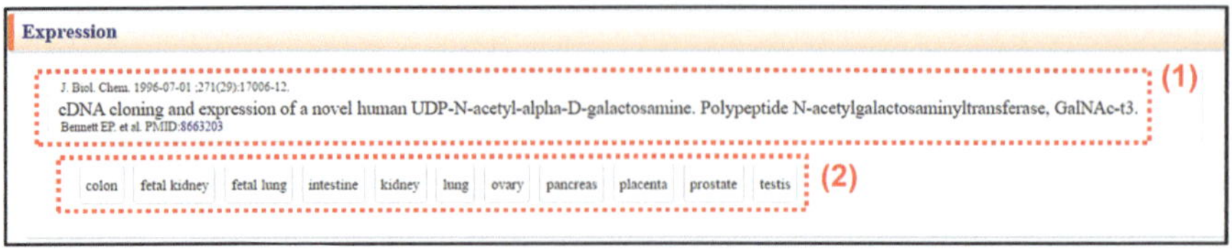

Fig. 8.11 The main components of the Expression section of the glycogene detail page

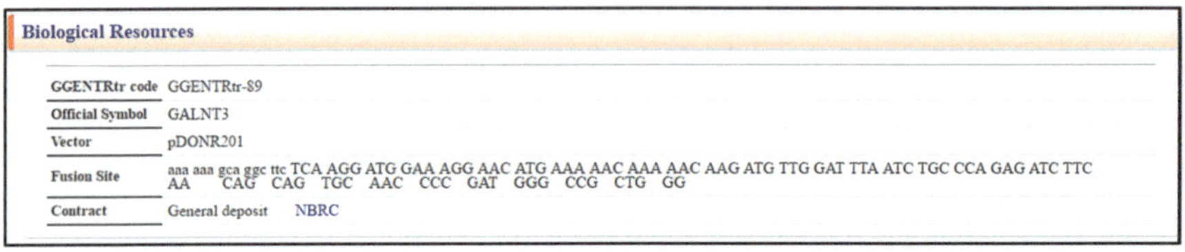

Fig. 8.12 The main components of the Human Glycogene Gateway™ Entry Clone section of the glycogene detail page

et al. 2009) along with links to the structure information. Moreover, if multiple reactions exist, they are listed in order of activity.

3. Activity information in the form of tables.

8.2.2.5 Expression Information

Part (4) of Fig. 8.7 is shown in Fig. 8.11. This section displays the citation (1) for expression information regarding this glycogene and related terminology (2) regarding the expression location.

8.2.2.6 Human Glycogene Gateway™ Entry Clone Information

Part (5) of Fig. 8.7 is shown in Fig. 8.12. This section displays the information of the Human Glycogene Gateway™ entry clone for this glycogene. These clones contain the predicted ORF region, minus the N-terminal transmembrane domain, and are available from the Biological Resource Center, NITE (NBRC). The link to NBRC is available in this section.

8.3 Using GGDB Entries as Stanzas

In addition to using the Web page to access GGDB data, programmers may be interested in accessing the data via Web protocols. An example of an HTML file follows, which contains a stanza that displays GGDB gene ID 2591.

```
<!DOCTYPE html>
<html>
<head>
<!-- js, css for stanza -->
<link rel="stylesheet" href="//cdnjs.cloudflare.com/ajax/libs/
     twitter-bootstrap/2.2.2/css/bootstrap.min.css" />
<style type="text/css">
P  text-indent: 1em;
</style>
<script src="//cdnjs.cloudflare.com/ajax/libs/jquery/1.8.3/
     jquery.min.js"></script>
<script src="//cdnjs.cloudflare.com/ajax/libs/twitter-bootstrap/
     2.2.2/bootstrap.min.js"></script>
<script src="http://togostanza.org/stanza/assets/stanza.js"
     type="text/javascript"></script>
<script>
$(function()
height = this.body.offsetHeight + 30;
parent.postMessage(JSON.stringify(height: height, id: name), "*");
);
</script>

<!-- styles from ggdb2 -->
<link rel="stylesheet" href="http://acgg.asia/db/ggdb/css/stanza.css" />
<link rel="stylesheet" href="http://acgg.asia/ggdb2/css/generic.css"
     type="text/css" />
<link rel="stylesheet" href="http://acgg.asia/ggdb2/css/ggdb.css"
     type="text/css" />
 </head>
<body>

<div data-stanza="http://acgg.asia/db/ggdb"
     data-stanza-gene_id="2591"> </div>
 </body>
</html>
```

The section in red is the actual statement that specifies the GGDB stanza. The statement "data-stanza-gene_id="2591"" indicates the gene ID to display. This code can be used in other Web pages by simply modifying the gene ID appropriately. The rest need not be modified. The section in blue is provided by TogoStanza and allows the stanza to be processed. The green section is the stylesheet used to format the display of GGDB stanzas.

Because the stanza section is displayed within an inner frame (div), care will need to be taken in that it may be necessary to apply customized styles, especially in complex Web pages. When the HTML file above is opened in a Web browser, it will be displayed as in Fig. 8.13. The top section of the GGDB Web page containing the database name is not a part of the stanza; the section below is all contained within the GGDB stanza.

Fig. 8.13 A snapshot of the GGDB stanza

8.4 Summary

In summary, the updated version of GGDB utilizes facets and stanzas to allow users to more easily find their target glycogene data. Detailed views also provide links to literature references and other database, including clone information, which is not provided elsewhere, to our knowledge. Although details were not described, all of the GGDB data has been translated to Resource Description Framework (RDF) format such that it may eventually be accessible via a SPARQL Protocol and RDF Query Language (SPARQL) end point on the Semantic Web (Aoki-Kinoshita et al. 2013). Thus, this new version of GGDB has been developed such that the data and interface can be reutilized in the future.

Acknowledgements This work has been supported by the Life Science Database Integration Project conducted by the National Bioscience Database Center (NBDC) of the Japan Science and Technology Agency (JST).

References

Aoki-Kinoshita K, Bolleman J, Campbell M, Kawano S, Kim J, Luetteke T, Matsubara M, Okuda S, Ranzinger R, Sawaki H, Shikanai T, Shinmachi D, Suzuki Y, Toukach P, Yamada I, Packer N, Narimatsu H (2013) Introducing glycomics data into the Semantic Web. J Biomed Semant 4(1):39

Eppig J, Blake J, Bult C, Kadin J, Richardson J, Group TMGD (2015) The Mouse Genome Database (MGD): facilitating mouse as a model for human biology and disease. Nucleic Acids Res 43:D726–D736

Gray KA, Seal RL, Tweedie S, Wright MW, Bruford EA (2016) A review of the new HGNC gene family resource. Hum Genomics 10(1):6

Hoehndorf R, Schofield P, Gkoutos G (2011) PhenomeNET: a whole-phenome approach to disease gene discovery. Nucleic Acids Res 39(18):e119

Lombard V, Ramulu HG, Drula E, Coutinho PM, Henrissat B (2014) The carbohydrate-active enzymes database (CAZy) in 2013. Nucleic Acids Res 42:D490–D495

Narimatsu, H (2004) Construction of a human glycogene library and comprehensive functional analysis. Glycoconjugate J 21:17–24

The Gene Ontology Consortium (2014) Gene Ontology Consortium: going forward. Nucleic Acids Res 43(D1):D1049–D1056

Varki A et al (eds) Essentials of glycobiology, vol 2. Cold Spring Harbor Laboratory Press, Cold Spring Harbor (2009)

Chapter 9
KEGG GLYCAN

Minoru Kanehisa

Abstract KEGG (Kyoto Encyclopedia of Genes and Genomes) is an integrated database resource for biological interpretation of genome sequences and other molecular data including glycomics data. It contains, among others, reference knowledge on cellular- and organism-level functions represented in terms of the KEGG molecular networks in the PATHWAY, BRITE, and MODULE databases and reference sequence datasets in the GENES and GENOME databases. The KO (KEGG Orthology) database plays the role of linking genes in the genome to KEGG molecular networks, enabling interpretation of high-level functions. The GLYCAN structure database was released in 2003 to supplement the COMPOUND database for small molecules and to integrate with the KEGG pathway maps for glycan biosynthesis and metabolism. KEGG GLYCAN originally represented this particular database, but it is now used in a broader sense to represent a collection of glycan-related datasets and software tools in the KEGG database, which is also the subject of this chapter.

Keywords Genome annotation • KEGG mapping • KEGG pathway map • KEGG glycan structure map • Glycan structure database • Glycome informatics

9.1 KEGG Databases

9.1.1 KEGG Identifiers

KEGG (Kanehisa et al. 2016a; Kanehisa and Goto 2000) consists of fifteen main databases categorized into systems, genomic, chemical, and health information as shown in Table 9.1. All the databases, except GENOME, GENES, and ENZYME, are manually developed based on published literature. GENOME and GENES are derived from RefSeq (O'Leary et al. 2016) and GenBank (Clark et al. 2016) databases, and ENZYME is taken from the enzyme nomenclature database ExplorEnz (McDonald and Tipton 2014), but they are given KEGG

M. Kanehisa (✉)
Institute for Chemical Research, Kyoto University, Uji, Kyoto, 611-0011, Japan
e-mail: kanehisa@kuicr.kyoto-u.ac.jp

K.F. Aoki-Kinoshita (ed.), *A Practical Guide to Using Glycomics Databases*,
DOI 10.1007/978-4-431-56454-6_9

Table 9.1 The KEGG databases

Category	Database name	Content	KEGG identifier	Example
Systems information	PATHWAY	KEGG pathway maps	Map number	map00510
	BRITE	BRITE functional hierarchies and BRITE tables	br/ko number	ko01003
	MODULE	KEGG modules	M number	M00055
Genomic information	KO	KEGG Orthology (KO) groups	K number	K01001
	GENOME	KEGG organisms (complete genomes) and selected viral genomes	T number (org code)	T01001 (hsa)
	GENES	Gene catalogs of KEGG organisms, viruses, plasmids, and addendum category	org:gene	hsa:1798
Chemical information (KEGG LIGAND)	COMPOUND	Metabolites and other small molecules	C number	C00110
	GLYCAN	Glycans	G number	G00001
	REACTION	Biochemical reactions	R number	R05969
	RCLASS	Reaction class	RC number	RC00002
	ENZYME	Enzyme nomenclature	EC number	2.7.8.15
Health information (KEGG MEDICUS)	DISEASE	Human diseases	H number	H00126
	DRUG	Drugs	D number	D09894
	DGROUP	Drug groups	DG number	DG00148
	ENVIRON	Crude drugs and health-related substances	E number	E00080

original annotations as well. The database in KEGG is a collection of entries, each of which is identified by the unique identifier called the KEGG identifier (see Table 9.1). Most of the KEGG identifiers take the form of a prefix followed by a five-digit number, such as map00510 (N-glycan biosynthesis) for a PATHWAY entry and G00001 (N-acetyl-D-glucosaminyldiphosphodolichol) for a GLYCAN entry. Exceptions are the EC (Enzyme Commission) number for an ENZYME entry and the org:gene identifier for a GENES entry. The latter is a combination of the two- to four-letter organism code and the gene accession given in the original database, mostly Locus_tag in GenBank and GeneID in RefSeq. Actual data of the examples in this chapter including those in Table 9.1 can be retrieved from the KEGG website by specifying KEGG identifiers in the KEGG WebLinks as shown in Table 9.2.

Table 9.2 KEGG WebLinks

Database	URL form	Example
PATHWAY	www.kegg.jp/pathway/<map number>	www.kegg.jp/pathway/hsa00510
BRITE (hierarchies only)	www.kegg.jp/brite/<br/ko number>	www.kegg.jp/brite/ko01003
		www.kegg.jp/brite/br08303_ndc
MODULE	www.kegg.jp/module/<M number>	www.kegg.jp/module/M00055
All databases except BRITE	www.kegg.jp/entry/<KEGG identifier>	www.kegg.jp/entry/K01001
		www.kegg.jp/entry/hsa:1798
		www.kegg.jp/entry/2.7.8.15

9.1.2 KEGG Pathway Maps

The databases in the systems information category, PATHWAY, BRITE, and MODULE, are the most unique databases containing reference knowledge on high-level functions of the cell, the organism, and the ecosystem, represented as KEGG pathway maps, BRITE hierarchies and tables, and KEGG modules, respectively. Basically, they represent molecular networks of reactions, interactions, and other types of relationships. The network nodes of pathway maps and modules, as well as BRITE hierarchies for protein families, are all represented in terms of the KO (KEGG Orthology) system in order to extend experimental knowledge in specific organisms to other organisms.

Figure 9.1a shows the KEGG pathway map for N-glycan biosynthesis, where proteins (mostly glycosyltransferases) are denoted by rectangular nodes and glycans (and chemical compounds) are denoted by circular nodes. In the reference pathway map (map00510) of N-glycan biosynthesis, which is manually drawn based on experimental evidence, rectangular nodes are linked to ortholog groups called KOs (and reactions when KOs are enzymes). For example, the node denoted by ALG7 is linked to the KO entry K01001 representing an ortholog group for DPAGT1/ALG7 genes in eukaryotes and their homologs in archaea. Figure 9.2a shows this KO entry, which also includes a reference link to functionally characterized protein sequence data.

Figure 9.1a is the human pathway map (hsa00510) for N-glycan biosynthesis, which is computationally generated from the reference pathway map by highlighting (green coloring) rectangular nodes when corresponding genes are present in the human genome (organism code hsa). The rectangular node ALG7 is now linked to the GENES entry hsa:1798 shown in Fig. 9.2b. Such organism-specific pathways are maintained for thousands of genomes in KEGG together with constantly improved annotations (KO assignments). Both reference pathways and organism-specific pathways can then be used for KEGG pathway mapping, which is to map, for example, genomic and transcriptomic data to rectangular nodes, as well as metabolomic data to circular nodes, enabling data interpretation and functional characterization.

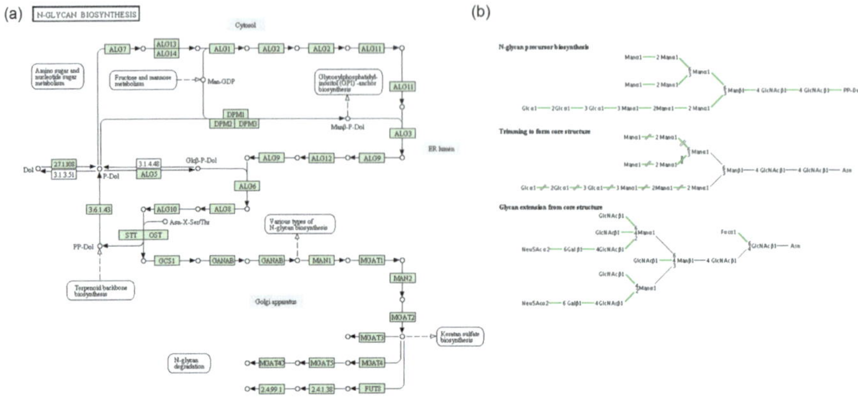

Fig. 9.1 (**a**) KEGG pathway map for N-glycan biosynthesis in human. (**b**) KEGG glycan structure map for N-glycans in human

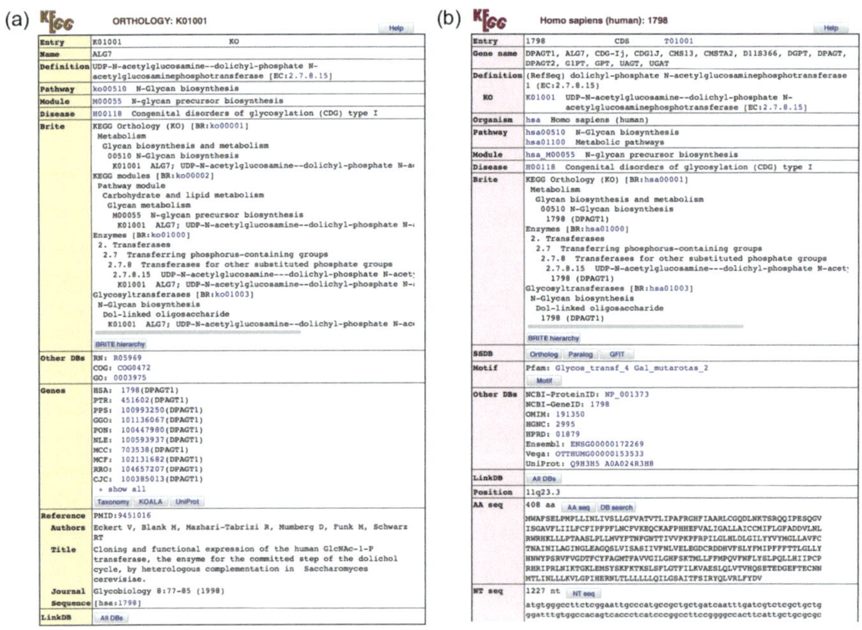

Fig. 9.2 (**a**) KO (KEGG Orthology) entry and (**b**) human gene entry for ALG7

9.1.3 Glycan Structure Maps

KEGG pathway maps for glycan biosynthesis and metabolism are sometimes associated with glycan structure maps, such as shown in Fig. 9.1b for N-glycan biosynthesis. Based on the presence or absence of specific enzymes in the genome, possible N-glycan structures are shown by highlighted edges. This is another type of KEGG mapping, where the repertoire of enzyme genes in the genome can be used to infer possible glycan structures synthesized. This approach was used for comparative analysis of glycosyltransferases in eukaryotes revealing variations of N-glycan precursor structures and GPI-anchor core structures (Hashimoto et al. 2009).

9.1.4 KEGG Modules

While KEGG pathway maps are drawn to capture overall aspects of biological processes, KEGG modules are tighter functional units for use in establishing more precise links between assigned KOs and functional meanings. In metabolic pathways, KEGG modules often correspond to subpathways and complexes in KEGG pathway maps. For example, the pathway module M00055 (N-glycan precursor biosynthesis) corresponds to the reaction steps from ALG7 to ALG10 in Fig. 9.1a, and the following reaction step by STT/OST is represented by the structural complex M00072 (N-glycosylation by oligosaccharyltransferase) shown in Fig. 9.3a. In the current implementation of organism-specific pathways, the rectangular node OST, which is linked to multiple KOs, may be highlighted when at least one KO is present in the genome. In contrast, the KEGG module is represented by a logical expression of multiple KOs, enabling to check completeness of the complex. The ortholog table (see Sect. 9.2.3 Ortholog table and module table) linked from the module page allows examination of complete modules (Fig. 9.3b) and incomplete modules indicating which KOs are missing.

Fig. 9.3 (**a**) KEGG module for oligosaccharyltransferase in human and (**b**) the ortholog table for this module

9.1.5 *BRITE Hierarchies and Tables*

KEGG BRITE is a collection of hierarchical classifications (ontologies) for genes and proteins (represented by KOs or K numbers), chemical substances (C numbers), reactions (R numbers), drugs (D numbers), diseases (H numbers), organisms (T numbers), and other biological objects. Obviously, the BRITE database captures many different types of relationships, in comparison to the PATHWAY database that is limited to molecular interactions and reactions. The BRITE file name is prefixed by ko for the hierarchy of K numbers and by br for the rest (Table 9.1). Organism-specific BRITE hierarchies, such as for protein families, are computationally generated in a way similar to organism-specific pathways by matching assigned KOs in the genome against the ko-prefixed reference hierarchy.

Figure 9.4 shows a part of the BRITE hierarchy file for glycosyltransferase (ko01003), which contains a tab-delimited column for linkage information as an attribute of glycosyltransferase reactions. Some BRITE hierarchy files are computationally expanded by adding a tab-delimited column or a new hierarchy level, which is accomplished by combining a hierarchy file and a binary relation file. For example, the BRITE hierarchy file for ATC drug classification (br08303) is combined with a binary relation file containing D number to target relationships to generate a new file with the target column (br08303_target). It is also combined with another binary relation file containing D number to NDC (National Drug Code)

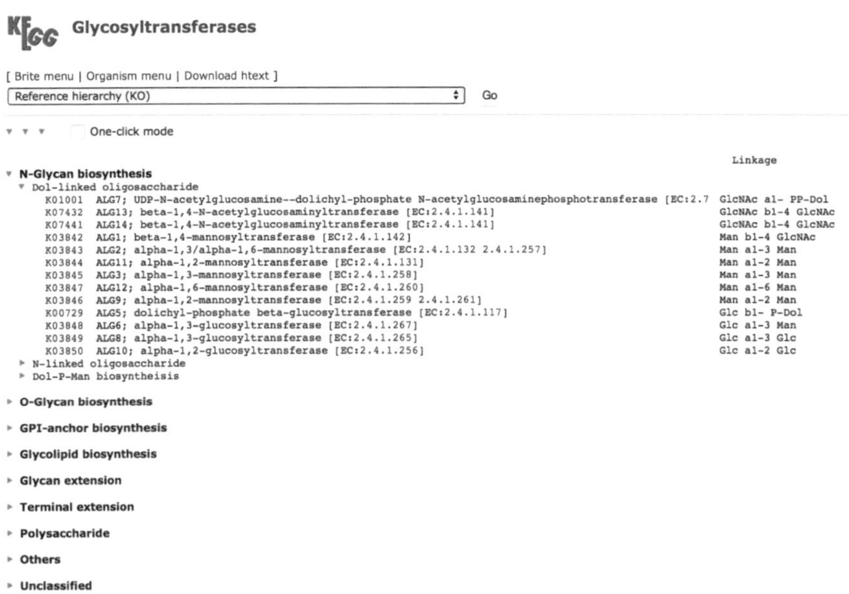

Fig. 9.4 BRITE hierarchy file for glycosyltransferases

relationships to generate a file with all marketed drug products in the USA included at the lowest level (br08303_ndc).

Although any number of columns may be added to a BRITE hierarchy file, the current htext (hierarchical text) browser in KEGG is not well suited for handling such a file. The recently introduced BRITE table file is a simple html file, which makes it much easier to understand multiple attributes and multiple relationships. The table representation is now used mostly for disease and drug information, where some BRITE hierarchy files have been converted to BRITE table files with the first few columns representing hierarchy levels.

9.1.6 KOs, GENES, and GENOME

The databases in the genomic information category are KO, GENOME, and GENES (Table 9.1). Genome annotation in KEGG is an ortholog annotation, assigning KOs to GENES database entries. The KO database is an ortholog database, where each KO represents a functional ortholog, as well as a sequence similarity group, among multiple organisms. It is also a molecular function database, where each KO entry contains, whenever possible, experimentally characterized gene/protein functions with appropriate references and links to sequence data used in the experiments. The sequence data links are made to the GENES database, indicating core sequences of defining functional orthologs.

The GENES and GENOME databases used to be limited to complete genomes, but this restriction was removed in order to make the KO database more comprehensive. Currently, the GENOME database consists of complete genomes of eukaryotes and prokaryotes, which are called KEGG organisms, and selected (disease-causing) viral genomes. As shown in Table 9.3, the GENES database consists of KEGG organisms identified by three- or four-letter organism codes and additional categories of viruses, plasmids, and addendum with two-letter codes of vg, pg, and ag, respectively. The addendum category contains an increasing number of functionally characterized protein sequence data that were directly submitted

Table 9.3 Data source of GENES database

Category	Primary data source	Content	GENES identifier	
			org	gene
Eukaryotes	RefSeq	Complete genomes	three- or four-letter code	GeneID
Prokaryotes	RefSeq			Locus_tag
	GenBank			Locus_tag
Viruses	RefSeq	RefSeq viral section	vg	GeneID
Plasmids	RefSeq	RefSeq plasmid section	pg	GeneID
Addendum	PubMed	Functionally characterized proteins	ag	ProteinID

by the authors to GenBank/ENA/DDBJ databases. These individual (non-genome) protein sequence data entries play the role of core sequences and have significantly expanded the repertoire of KOs.

The GENES database is a reference sequence database with KOs assigned to protein-coding and RNA-coding genes in thousands of genomes. Since all the KEGG molecular networks are represented by networks of KOs, once genes in a genome are assigned KOs by sequence similarity search against GENES (see Sect. 9.2.1 BlastKOALA), organism-specific pathway maps, BRITE hierarchies, and KEGG modules are automatically reconstructed (see Sect. 9.2.2 KEGG Mapper).

9.1.7 *Chemical Compounds, Glycans, and Reactions*

The databases in the chemical information category, COMPOUND, GLYCAN, REACTION, RCLASS, and ENZYME, are collectively called KEGG LIGAND (Table 9.1). The COMPOUND database contains chemical structures of metabolites and other chemical substances that affect molecular and higher-level functions of genes and proteins, including environmental substances. The GLYCAN database is a collection of glycan structures, originally derived from CarbBank and then expanded by KEGG (Hashimoto et al. 2006). The product of the reaction catalyzed by ALG7 in Fig. 9.1a is N-acetyl-D-glucosaminyldiphosphodolichol, which is shown by both the compound representation (C04500) and the glycan representation (G00001) in Fig. 9.5. The glycan representation is preferentially used in the KEGG pathway maps for glycan biosynthesis and metabolism both for circular nodes (glycans) and rectangular nodes (reactions).

The ENZYME database contains the enzyme nomenclature list (EC number list) obtained from the ExplorEnz database (McDonald and Tipton 2014). The

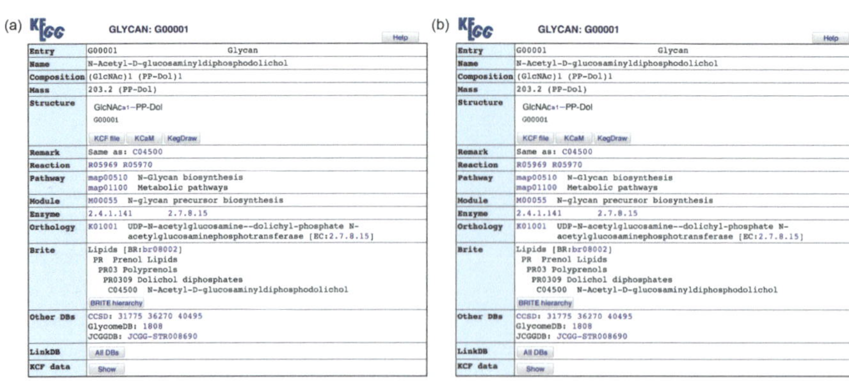

Fig. 9.5 (**a**) COMPOUND entry and (**b**) GLYCAN entry for N-acetyl-D-glucosaminyldiphosphodolichol

KEGG **Tumor markers**

[Brite table menu]

Commonly used tumor markers

Marker	Cancer Type	Gene	Glycan
ALK gene rearrangements	Non-small cell lung cancer [DS:H00014] Anaplastic large cell lymphoma [DS:H01601]	EML4-ALK fusion gene TFG-ALK fusion gene KIF5B-ALK fusion gene NPM-ALK fusion gene	
Alpha-fetoprotein (AFP)	Liver cancer [DS:H00048] Germ cell tumors [DS:H00023]	AFP	
Beta-2-microglobulin (B2M)	Multiple myeloma [DS:H00010] Chronic lymphocytic leukemia [DS:H00005] Some lymphomas [DS:H00011]	B2M	
Beta-human chorionic gonadotropin (Beta-hCG)	Choriocarcinoma [DS:H00028] Germ cell tumors [DS:H00023]	CGB	
BRCA1 and BRCA2 gene mutations	Ovarian cancer [DS:H00027]	BRCA1 BRCA2	
BCR-ABL fusion gene	Chronic myeloid leukemia [DS:H00005] Acute lymphoblastic leukemia [DS:H00001 H00002] Acute myelogenous leukemia [DS:H00003]	ABL1 BCR	
BRAF mutation V600E	Cutaneous melanoma [DS:H00038] Colorectal cancer [DS:H00020]	BRAF mutation V600E	
CA15-3/CA27.29	Breast cancer [DS:H00031]	MUC1	G00242
CA19-9	Pancreatic cancer [DS:H00019] Gallbladder cancer [DS:H00047] Bile duct cancer [DS:H00046] Gastric cancer [DS:H00018]		G12794
CA125	Ovarian cancer [DS:H00027]	MUC16 (CA125)	

Fig. 9.6 BRITE table file for tumor markers

REACTION database contains chemical reactions in the enzyme nomenclature list and in the KEGG metabolic pathways. In a similar way as genes are grouped into KOs, reactions are grouped into reaction classes based on the local structure transformation patterns of substrate-product pairs and stored in the RCLASS database (Muto et al. 2013; Kanehisa et al. 2014).

9.1.8 Diseases and Drugs

The health information category consists of the DISEASE, DRUG, DGROUP, and ENVIRON databases for disease and drug information (Table 9.1). The DISEASE database is a collection of human disease entries, each containing a list of known genetic factors (disease genes), environmental factors, diagnostic markers, and therapeutic drugs. There are a number of congenital disorders of glycans, glycoproteins, and glycolipids, whose disease genes are linked to the KEGG metabolic pathway maps. Glycans also appear as diagnostic markers in certain cancers as shown in Fig. 9.6, which is a part of the BRITE table file for tumor markers (br08442).

The DRUG database is a comprehensive collection of approved drugs in Japan, the USA, and Europe unified based on the chemical structures of active ingredients. Each drug entry is associated with target, metabolizing enzyme, and other molecular interaction network information, which are often represented in the KEGG pathway maps. The DRUG database is supplemented by the DGROUP database defining drug groups, which is like KO for GENES and RCLASS for REACTION, and by the ENVIRON database for crude drugs and other health-related substances.

These internally developed databases are integrated with drug labels (package inserts) of all marketed drugs in Japan and the USA. KEGG MEDICUS is an interface for this integrated resource, aiming to meet the needs of society. The Japanese version of KEGG MEDICUS is especially advanced in this integration and heavily accessed mostly through web search engines.

9.2 KEGG and GenomeNet Tools

9.2.1 BlastKOALA

Since 1995 the KEGG databases and various analysis tools had been developed at the GenomeNet (genome.jp) website, as part of the Japanese GenomeNet service (Kanehisa 1977). At the end of 2011, the KEGG database development was moved to the KEGG (kegg.jp) website, and since then, the GenomeNet site has become a mirror site. Some features, notably direct queries against the relational databases, are available only at the KEGG website, while most of the computational tools, including KAAS (Moriya et al. 2007), SIMCOMP (Hattori et al. 2010), and KCaM (Aoki et al. 2004), are available only at the GenomeNet website (Table 9.4). These differences, however, may not be noticeable by the users because mutual links are made as if they are a single site.

BlastKOALA (Kanehisa et al. 2016a, b) is a new service available at the KEGG website (Table 9.4) for automatic annotation (KO assignment) of genome and metagenome sequences using the GENES database as a reference sequence database. Two features of the GENES database should be mentioned as advantages. First, the GENES database can be considered as partitioned into KO groups, which facilitates the processing of sequence similarity search results. It is simply to assign the most appropriate K numbers, as implemented in both KAAS and BlastKOALA. Second, the GENES database is created from a collection of complete genomes, which allows a novel way to define a nonredundant database in order to cope with the increasing number of closely related sequences. A nonredundant GENES database is a collection of nonredundant pangenomes at the species, genus, or family level, which is created by removing redundant sequences but retaining the set of assigned K numbers (KO content) for each taxonomic rank (Kanehisa et al. 2016a, b). Different nonredundant database files are created to meet the conflicting needs of speed and accuracy.

Figure 9.7a shows a sample result page of BlastKOALA, where the genome of *Entamoeba nuttalli* P19 (RefSeq assembly: GCF_000257125.1) was analyzed. A set of 6187 amino acid sequences was compared by BLAST against the nonredundant

Table 9.4 URLs for KEGG databases and tools

Content		KEGG website	GenomeNet website
KEGG		www.kegg.jp	www.genome.jp/kegg/ (mirror site)
KEGG MEDICUS		www.kegg.jp/kegg/medicus.html	
KEGG annotation	Ortholog table and module table	www.kegg.jp/kegg/annotation/	
KEGG SSDB	Precomputed sequence similarity	www.kegg.jp/kegg/ssdb/	
KEGG Mapper	KEGG mapping tools	www.kegg.jp/kegg/mapper.html	
BlastKOALA	Automatic annotation tool	www.kegg.jp/blastkoala/	
KAAS	Automatic annotation tool		www.genome.jp/tools/kaas/
BLAST	Sequence similarity search tool		www.genome.jp/tools/blast/
SIMCOMP	Chemical structure similarity search tool		www.genome.jp/tools/simcomp/
KCaM	Glycan structure similarity search tool		www.genome.jp/tools/kcam/

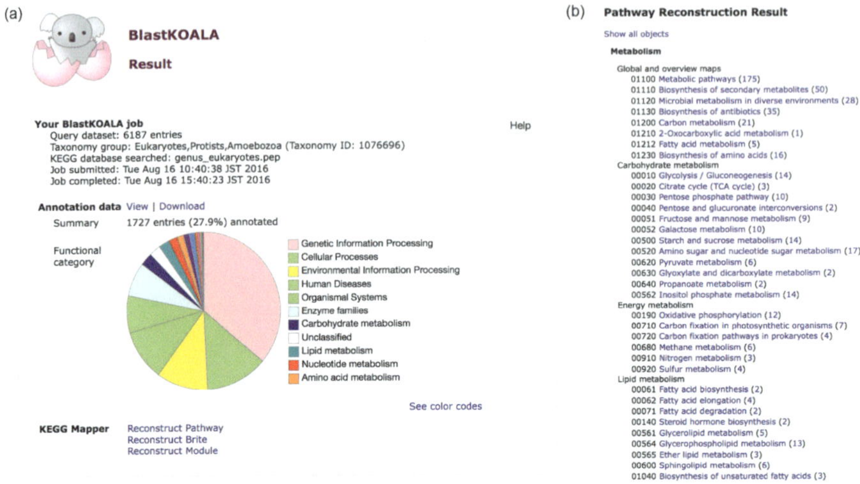

Fig. 9.7 (**a**) BlastKOALA result page and (**b**) the list of reconstructed pathways linked from Reconstruct Pathway of KEGG Mapper

file genus_eukaryotes.pep, which represents pangenome sequences at the genus level for eukaryotes. Overall the annotation rate was not high (about 28 %), which is common to a protist genome. The pie chart shows a summary of assigned gene functions based on the categorization of KEGG pathway maps and BRITE hierarchies. The View or Download link allows the list of query genes and assigned KOs to be examined in detail or to be downloaded. The result page contains links to KEGG Mapper tools described next, enabling interpretation of high-level functions encoded in the genome (Fig. 9.7b).

9.2.2 KEGG Mapper

KEGG Mapper is a collection of KEGG mapping tools shown in Table 9.5 including PATHWAY mapping, BRITE mapping, MODULE mapping, DISEASE mapping, and taxonomy mapping. Two pathway mapping tools, Search Pathway and Search&Color Pathway, were made available from the beginning of the KEGG project, and they are still the most popular tools. Except for taxonomy mapping, the mapping can be made either in the reference mode using KOs or in the organism-specific mode using gene identifiers, such as human gene identifiers used to search against human pathway maps.

The three tools, Reconstruct Pathway, Reconstruct Brite, and Reconstruct Module, are particularly useful to process and interpret the result of KO assignments by the automatic annotation servers of KAAS and BlastKOALA. The direct links to these tools in the BlastKOALA result page (Fig. 9.7b) allow this interpretation process as an integral part of the BlastKOALA server. However, it is sometimes

Table 9.5 KEGG Mapper tools

Tool	Query data	Database
Search Pathway	KOs, gene identifiers, C numbers, etc.	PATHWAY
Search&Color Pathway	KOs, gene identifiers, C numbers, etc.	PATHWAY
Color Pathway	KOs, gene identifiers	Single KEGG pathway map
Color Pathway WebGL	KOs, gene identifiers	Single KEGG pathway map
Search Brite	KOs, gene identifiers, C numbers, etc.	BRITE
Search&Color Brite	KOs, gene identifiers, C numbers, etc.	BRITE
Join Brite	KOs, D numbers, etc.	Single BRITE hierarchy
Join Brite Table	KOs, D numbers, etc.	Single BRITE table
Search Module	KOs, gene identifiers, C numbers, etc.	MODULE
Search&Color Module	KOs, gene identifiers, C numbers, etc.	MODULE
Search Disease	KOs, gene identifiers	DISEASE
Reconstruct Pathway	KOs	PATHWAY
Reconstruct Brite	KOs	BRITE
Reconstruct Module	KOs	MODULE
Map Taxonomy	Organism codes, NCBI taxonomy IDs	Taxonomy

necessary to run multiple BlastKOALA queries, download individual KO assignment files, concatenate them, and use these KEGG Mapper tools. This happens when the query dataset is too large and needs to split into subsets or when reconstructed pathways from multiple genomes need to be compared. The latter can be done by adding comment lines with optional highlight coloring specifications in the concatenated KO assignment file.

9.2.3 Ortholog Table and Module Table

The quality of KO assignments in the GENES database is critical for successful mapping of KEGG pathways and other molecular networks. One way to assess this quality is the ortholog table (Fig. 9.3b), which displays current KO assignments in KEGG organisms (complete genomes). The KEGG Annotation page (Table 9.4) contains an interface for the ortholog table. The table displays, for a given set of K numbers, each KEGG organism with assigned genes in a row. There is an inherent ordering of KEGG organisms, starting with *Homo sapiens* (hsa) and following the KEGG taxonomy (br08601), which is consistent with the NCBI taxonomy (br08610). The coloring of cells indicates adjacent genes on the chromosome.

Another tool in the KEGG Annotation page is the module table, which may be used to capture a more global picture of KO assignments in KEGG organisms. The interface accepts both K numbers for KOs and M numbers for modules, and the module table displays only the presence of KOs or complete modules by pink coloring of cells. Multiple KEGG organisms may be combined into a pangenome at the species or genus level, thus reducing the number of rows displayed.

9.2.4 SSDB and BLAST

The GENES database is associated with the SSDB database, which contains amino acid sequence similarity scores for all GENES entry pairs. It also contains information about the best hit relations for all genome pairs. The SSDB database is used internally for KO assignments, but it is also made available as part of the GENES database (Table 9.4). Ortholog and Paralog links in each GENES entry (Fig. 9.2b) allow the examination of the best hit genes in other genomes and similar genes in the same genome, respectively. Furthermore, Gene cluster link enables, whenever available, examination of conserved gene clusters along the chromosome. The GENES entry also contains DB search link, which can be used to perform on-the-fly sequence similarity search by BLAST against many different databases.

9.2.5 SIMCOMP and KCaM

The search for similar chemical structures or glycan structures can be performed from DB search or KCaM link for a given COMPOUND or GLYCAN entry (Fig. 9.5). The search programs, SIMCOMP (Hattori et al. 2010) for chemical structure search and KCaM (Aoki et al. 2004) for glycan structure search, are made available as web tools at the GenomeNet website (Table 9.5). These search tools are based on the KEGG original representation of structures, called KCF (KEGG Chemical Function) representation. It is a graph representation where monosaccharides (nodes) and specific linkages (edges) are precisely described for a glycan structure. For a chemical compound structure, atoms (nodes) are labeled by 68 KEGG atom types distinguishing functional groups and microenvironments of atoms. These representations enable graph-based structure comparisons by SIMCOMP and KCaM. In the web service, SIMCOMP accepts widely used SMILES and MOL file representations of chemical structures, but KCaM accepts only the KCF representation of glycan structures.

9.3 KEGG GLYCAN Resource

9.3.1 Glycan-Specific Datasets

KEGG is a generic database covering all organisms and all known pathways. As such it may sometimes be difficult to readily find specific contents meeting individual needs and interests. KEGG GLYCAN is an attempt to provide an entry point to glycan-related resources in KEGG. Table 9.6 is a summary of glycan-specific pathway maps and BRITE files, but glycoconjugates are ubiquitously found in other pathway maps and BRITE files, reflecting their important roles in many cellular processes. One approach to search them would be to start from a specific gene or protein of interest. KEGG provides ways to link different organisms through the KO and taxonomy systems and related genes and proteins through the KEGG molecular networks. Understanding the overall KEGG architecture described in this chapter would enable navigating through different databases more easily.

9.3.2 Linking Genes to Glycan Structures

In addition to linking genes to pathways, linking genes to chemical structures of glycans has been an objective in developing the KEGG GLYCAN resource. For example, glycosyltransferases in the genome (Hashimoto et al. 2009) or the transcriptome (Kawano et al. 2005) were used to predict possible glycan structures. Composite Structure Map (CSM) was developed as a reference for mapping genes to structures, containing all possible variations of carbohydrate structures in a tree form, where nodes and edges are linked to corresponding GLYCAN and KO (or GENES) entries, respectively. CSM is computationally generated, while glycan structure maps embedded in KEGG pathway maps are manually drawn. KEGG Mapper supports only the manually drawn datasets.

9.3.3 Linking Glycans to Diseases

The KEGG resource has been developed mostly from published research results. The health information category of KEGG MEDICUS is somewhat different in that it incorporates drug labels and other information used in medical practice and in society. Glycoconjugates play major roles in cellular communications and interactions. Research findings with relevance to cancers and infectious diseases are accumulated in the BRITE files br08441 (Cancer-associated carbohydrates) and br08431 (Carbohydrates in viral and bacterial infections). Some of the cancer-associated carbohydrates are actually used as tumor markers in practice, which are summarized in the BRITE file br08442 (Fig. 9.6) based on the compilation of the National Cancer Institute. Thus, KEGG MEDICUS is being developed for the

Table 9.6 Glycan-specific contents in KEGG

Category	Content	URL
Pathway map	N-Glycan biosynthesis	www.kegg.jp/pathway/map00510
	Various types of N-glycan biosynthesis	www.kegg.jp/pathway/map00513
	Mucin-type O-glycan biosynthesis	www.kegg.jp/pathway/map00512
	Other types of O-glycan biosynthesis	www.kegg.jp/pathway/map00514
	Glycosaminoglycan biosynthesis – chondroitin sulfate/dermatan sulfate	www.kegg.jp/pathway/map00532
	Glycosaminoglycan biosynthesis – heparan sulfate/heparin	www.kegg.jp/pathway/map00534
	Glycosaminoglycan biosynthesis – keratan sulfate	www.kegg.jp/pathway/map00533
	Glycosaminoglycan degradation	www.kegg.jp/pathway/map00531
	Glycosylphosphatidylinositol (GPI)-anchored biosynthesis	www.kegg.jp/pathway/map00563
	Glycosphingolipid biosynthesis – lacto and neolacto series	www.kegg.jp/pathway/map00601
	Glycosphingolipid biosynthesis – globo series	www.kegg.jp/pathway/map00603
	Glycosphingolipid biosynthesis – ganglio series	www.kegg.jp/pathway/map00604
	Lipopolysaccharide biosynthesis	www.kegg.jp/pathway/map00540
	Peptidoglycan biosynthesis	www.kegg.jp/pathway/map00550
	Other glycan degradation	www.kegg.jp/pathway/map00511
	Proteoglycans in cancer	www.kegg.jp/pathway/hsa05205
Brite hierarchy	Glycosyltransferases	www.kegg.jp/brite/ko01003
	Proteoglycans	www.kegg.jp/brite/ko00535
	Glycosaminoglycan-binding proteins	www.kegg.jp/brite/ko00536
	Glycosylphosphatidylinositol (GPI)-anchored proteins	www.kegg.jp/brite/ko00537
	Lectins	www.kegg.jp/brite/ko04091
Brite table	Cancer-associated carbohydrates	www.kegg.jp/kegg/disease/br08441.html
	Carbohydrates in infections	www.kegg.jp/kegg/disease/br08431.html

research community as a translational bioinformatics resource and for the wider society as a resource for understanding scientific basis of diseases and drugs.

References

Aoki KF, Yamaguchi A, Ueda N, Akutsu T, Mamitsuka H, Goto S, Kanehisa M (2004) KCaM (KEGG Carbohydrate Matcher): a software tool for analyzing the structures of carbohydrate sugar chains. Nucleic Acids Res 32:W267–W272

Clark K, Karsch-Mizrachi I, Lipman DJ, Ostell J, Sayers EW (2016) GenBank. Nucleic Acids Res 44:D67–D72

Hashimoto K, Goto S, Kawano S, Aoki-Kinoshita KF, Ueda N, Hamajima M, Kawasaki T, Kanehisa M (2006) KEGG as a glycome informatics resource. Glycobiology 16:63R–70R

Hashimoto K, Tokimatsu T, Kawano S, Yoshizawa AC, Okuda S, Goto S, Kanehisa M (2009) Comprehensive analysis of glycosyltransferases in eukaryotic genomes for structural and functional characterization of glycans. Carbohydr Res 344:881–887

Hattori M, Tanaka N, Kanehisa M, Goto S (2010) SIMCOMP/SUBCOMP: chemical structure search servers for network analyses. Nucleic Acids Res 38:W652–W656

Kanehisa M (1977) Linking databases and organisms – GenomeNet resources in Japan. Trends Biochem Sci 22:442–444

Kanehisa M, Goto S (2000) KEGG: Kyoto Encyclopedia of Genes and Genomes. Nucleic Acids Res 28:27–30

Kanehisa M, Goto S, Sato Y, Kawashima M, Furumichi M, Tanabe M (2014) Data, information, knowledge and principle: back to metabolism in KEGG. Nucleic Acids Res 42:D199–D205

Kanehisa M, Sato Y, Kawashima M, Furumichi M, Tanabe M (2016a) KEGG as a reference resource for gene and protein annotation. Nucleic Acids Res 44:D457–D462

Kanehisa M, Sato Y, Morishima K (2016b) BlastKOALA and GhostKOALA: KEGG tools for functional characterization of genome and metagenome sequences. J Mol Biol 428:726–731

Kawano S, Hashimoto K, Miyama T, Goto S, Kanehisa M (2005) Prediction of glycan structures from gene expression data based on glycosyltransferase reactions. Bioinformatics 21:3976–3982

McDonald AG, Tipton KF (2014) Fifty-five years of enzyme classification: advances and difficulties. FEBS J 281:583–592

Moriya Y, Itoh M, Okuda S, Yoshizawa A, Kanehisa M (2007) KAAS: an automatic genome annotation and pathway reconstruction server. Nucleic Acids Res 35:W182–W185

Muto A, Kotera M, Tokimatsu T, Nakagawa Z, Goto S, Kanehisa M (2013) Modular architecture of metabolic pathways revealed by conserved sequences of reactions. J Chem Inf Model 53:613–622

O'Leary NA, Wright MW, Brister JR, Ciufo S, Haddad D, McVeigh R, Rajput B, Robbertse B, Smith-White B, Ako-Adjei D, Astashyn A, Badretdin A, Bao Y, Blinkova O, Brover V, Chetvernin V, Choi J, Cox E, Ermolaeva O, Farrell CM, Goldfarb T, Gupta T, Haft D, Hatcher E, Hlavina W, Joardar VS, Kodali VK, Li W, Maglott D, Masterson P, McGarvey KM, Murphy MR, O'Neill K, Pujar S, Rangwala SH, Rausch D, Riddick LD, Schoch C, Shkeda A, Storz SS, Sun H, Thibaud-Nissen F, Tolstoy I, Tully RE, Vatsan AR, Wallin C, Webb D, Wu W, Landrum MJ, Kimchi A, Tatusova T, DiCuccio M, Kitts P, Murphy TD, Pruitt KD (2016) Reference sequence (RefSeq) database at NCBI: current status, taxonomic expansion, and functional annotation. Nucleic Acids Res 44:D733–D745

Part IV
Glycoproteomics Data

Chapter 10
Exploring the UniCarbKB Database

Matthew P. Campbell, Robyn A. Peterson, Elisabeth Gasteiger, Frederique Lisacek, and Nicolle H. Packer

Abstract UniCarbKB (http://unicarbkb.org) is an international collaboration to develop an open-source glycobioinformatics database to support glycomics and the emerging technology of glycoproteomics. It is a peer-reviewed, curated collection of information on well-characterised published glycan structures derived from glycoproteins, biological fluids and tissues. UniCarbKB provides contextual information for glycan structures attached to proteins, and where known maintains the commonly lost connection between a glycan structure and the attached protein sites as annotated in UniProtKB. This information is further supplemented with descriptions of their biological source, supporting reference(s) and the experimental methods that were employed to determine the glycan structure. In this chapter, an overview of UniCarbKB is described, as well as interfaces and workflows available for browsing curated information on glycan structures and accompanying glycoproteins, inclusive of any evidence pertinent to global and site-specific glycosylation. In addition, searches of exact structure, substructure and composition are described. This database is useful for not only glycoscientists but also for a wide range of life science researchers in that it provides an important link between glycans and their attached proteins.

Keywords Glycomics • Glycoproteomics • Database • Glycobioinformatics • Glycans

10.1 Introduction

Understanding the molecular and functional complexity of glycoproteins is challenging and requires sustainable bioinformatic resources to advance our basic understanding of glycosylation and its wider biological implications. Unfortunately,

---

M.P. Campbell • R.A. Peterson • N.H. Packer (✉)
Biomolecular Frontiers Research Centre, Macquarie University, 2109 North Ryde, NSW, Australia
e-mail: nicki.packer@mq.edu.au

E. Gasteiger • F. Lisacek
SIB Swiss Institute of Bioinformatics, CH-1211 Geneva 4, Switzerland

K.F. Aoki-Kinoshita (ed.), *A Practical Guide to Using Glycomics Databases*,
DOI 10.1007/978-4-431-56454-6_10

despite the success of several international initiatives, the glycosciences still lack a managed infrastructure that contributes to the advancement of research through the provision of comprehensive structural and experimental glycan data collections.

UniCarbKB (Campbell et al. 2011a, b) represents a continued effort to build upon the success of EUROCarbDB (von der Lieth et al. 2011) and GlycoSuiteDB (Cooper et al. 2001) by providing the glycoscience community with sustainable informatics architecture. An important characteristic of UniCarbKB is our efforts to make glycomics data more accessible and discoverable.

The UniCarbKB knowledge base is seeded with curated GlycoSuiteDB data collections and a limited subset from the EUROCarbDB-GlycoBase database (Campbell et al. 2008). Since its launch, the coverage and content of UniCarbKB has been extended by our efforts to curate over 200 publications that contain characterised glycan structures, often with their sites of attachment to proteins, with supporting experimental data. Our biocuration aspirations are to build a central hub for glycan structures with an emphasis on quality glycan and glycoprotein information. At this stage, UniCarbKB provides contextual information for glycan structures attached to proteins whether from a purified glycoprotein or from a mixture of glycoproteins derived from a biological source such as secreted fluids, cells or tissues. Such an approach has the advantage of providing a gold standard set of structural glycan data, to be extended in the future to other glycoconjugates in the knowledge base.

10.1.1 What You Can Currently Find in UniCarbKB

UniCarbKB is a manually annotated and nonredundant database of glycan structures that are found on glycoproteins. Although UniCarbKB provides annotated entries for all species, its primary focus is the annotation of protein-linked glycans from mammalian systems of distinct taxonomic groups – entries are continuously reviewed to keep up with current scientific literature.

To date, the knowledge base provides access to experimentally determined glycan structures and sites of glycosylation on a protein(s) where known; protein and gene names use standardised accepted nomenclature and synonyms from the literature and other databases, with associated glycosyltransferase enzyme-specific information, tissue localization and species of origin. In the case where the attached protein has been defined, two levels of annotation are provided where known: (1) global data denotes glycan structures characterised on a single purified glycoprotein and (2) site-specific data for individual glycan structures that have been experimentally verified to be attached at a given amino acid sequence position. Other glycan structures are included that have been characterised in a global glycomic study of a mixture of cell-derived glycoproteins.

Users can choose to search UniCarbKB by (sub)structure, monosaccharide composition, glycan mass, taxonomy, tissue, glycoprotein (accession number or

Swiss-Prot name) or literature publication. The search results are displayed either as simple lists or as mini-summaries, with links to associated data.

10.1.2 Content-Driven Design

As part of our effort to provide simple yet functional interfaces, we have put the user experience at the centre of our design. Recently, we have launched a new website documented by Campbell et al. 2014. In brief, the interface is more visual, encapsulating a simpler content layout with an emphasis placed on displaying information that researchers want to access in information-oriented design. Many of these changes are in line with the presentation of information previously available from EUROCarbDB and GlycoSuiteDB databases with a choice of notation formats.

Note This information is based on UniCarbKB September 2014 release; therefore, some of the web pages may have changed. Any modern web browser such as Firefox, Safari, Chrome and Internet Explorer 10+ will work to display UniCarbKB web pages.

10.2 Browsing UniCarbKB

The protocols in this chapter illustrate how to use the suite of tools and navigation links in UniCarbKB to find information on individual glycan structures, attached glycoproteins (at the global and site-specific level) where known, and what metadata has been sourced for each curated publication.

Note This information is based on the September 2014 release of UniCarbKB; some of the web pages may have changed since the chapter was written. An online user knowledge base is under development and will provide access to detailed examples and information about user interface improvements.

10.2.1 The UniCarbKB Front Page

This is the normal entry point for most users of UniCarbKB. To view this page, type http://www.unicarbkb.org into your browser:

- The navigation bar has drop-down menus and links giving access to the data and functionality available on the UniCarbKB website.
- The main text section includes a description of UniCarbKB including database statistics, the participating expert research institutions and our funding sources.

- The home page also lists all the web pages and collaborations accessible throughout the UniCarbKB website.

As part of our continuing effort to provide robust querying/search tools, a series of user-friendly interfaces and features are available. For example, we have (1) enhanced native selects by including a multi-select interface, (2) made greater use of modern frameworks and libraries to improve data loading and (3) extended pagination for improved presentation and navigation of large data sets.

The navigation bar at the top of the page provides links to the tools and resources of UniCarbKB including ‘Query’, ‘References’, ‘Glycoproteins’, ‘GlycanBuilder’, ‘GlycoDigest’, ‘About’ and ‘Contact’. The ‘About’ is a description of the project as a whole and “Contact” provides details on how to get assistance with UniCarbKB, to ask questions or when troubleshooting a problem. In this chapter, we will review the tools and querying options offered except for the included GlycoDigest tool, which is described elsewhere (Gotz et al. 2014).

Note Results are displayed in a context-dependent manner depending on your query. Some results will have descriptive text details, whereas others might link directly to core glycan structure or glycoprotein content. Your search terms will be highlighted and will appear in the title or descriptive text of each results page.

10.2.2 Finding Taxonomic, Tissue and Glycoprotein Content

In this section, we will describe the basic query options available and how to use them:

- The ‘Query’ page accessible from the navigation bar provides users with five search items: ‘Taxonomy’, ‘Tissue’, ‘Protein’, ‘Protein Accession’ and ‘Composition’.
- To start querying UniCarbKB, select a preferred query option, which will automatically load a search text box or multiple search text boxes in the case of ‘Composition’.
- After selecting the appropriate search option, start typing a name or keywords into the search box at the top of the query screen.
- When the user enters text, the quick navigation aid automatically provides a drop-down list of matching items. The items are matched by NCBI taxonomy, MeSH (Medical Subject Headings) or protein name.
- The search box supports the selection of multiple items and supporting expanded search criteria.
- Pressing the search button will load data available for the selected item(s).

When users perform a taxonomy, glycoprotein or tissue query, UniCarbKB will search all content in the database. The results will appear on a new screen in the main section of which you will see a text block for each item that matched the search criteria, with the following information:

- An icon representing the content type (taxonomy, tissue or glycoprotein).
- The title or name of the content search that is linked to the item's main viewing template.
- The most relevant few lines of content related to the item. In brief, taxonomy searches list a limited number of glycoproteins; for the tissue and glycoprotein results, only the number of experimentally determined and published structures is shown.
- An icon representing the number of structures is shown to the right of each text block.

10.2.2.1 Taxonomy

The taxonomy schema is based on the NCBI taxonomy database, which is supplemented with data specific to UniCarbKB. Species with manually annotated and reviewed glycoprotein and glycan structure data are named according to NCBI convention. You can query the UniCarbKB taxonomy by taxon names. Searches by names are case insensitive, and to assist, we have provided an autocomplete search feature – so that when you search for a taxon name, the feature will automatically list matching taxonomies with entries in the database – this implementation ensures that taxonomy searches use the correct names and form as defined in the database. In addition, users can search for multiple taxonomies simply by searching and selecting a name from the autocomplete list.

If searching for viruses, the species field describes the organism from which the DNA of the protein originates.

Subsequently, each taxonomy entry is annotated with a corresponding NCBI taxonomy identifier that allows for interoperability between glyco-focused databases that support NCBI terms including EUROCarbDB, GlycoBase and GlycomeDB (Ranzinger et al. 2008).

10.2.2.2 Tissue

The actions for searching tissue data are similar to the taxonomy interface. Here, the tissue database is based on the NCBI MeSH (Medical Subject Headings) controlled vocabulary. As such, each tissue or biological source entry is linked with a NCBI MeSH identifier. Again, (single or multiple) searches are case insensitive, and the search bar provides suggestions after entering three characters.

10.2.2.3 Protein and Protein Accession

UniCarbKB protein descriptions are based on the preferred name and primary accession identifier denoted by UniProtKB (UniProt 2015). However, in cases where UniProtKB does not describe the protein, we use the protein name described in the cited publication – often this occurs for data curated from older publications – although we have attempted to update these records where possible. Similar to searching by tissue and taxonomy, we have utilised the autocomplete feature for efficiently searching (single or multiple) records with corresponding protein names and accession numbers.

10.2.2.4 Why Is the Autocomplete Useful?

As you type in the search box, you can find information quickly by seeing terms and keywords that match your search criteria. For example, as shown in Fig. 10.1, it is possible to search by multiple glycoproteins, and when you start to type 'major', protein names partially matching the entered text will be listed. This feature serves two major advantages notably:

- Save time searching – Choose from keywords to find information faster while typing less.
- Spelling corrections – When searching for keyword, autocomplete will show matching or similar terms; this is particularly useful in the case of abbreviated classifications.

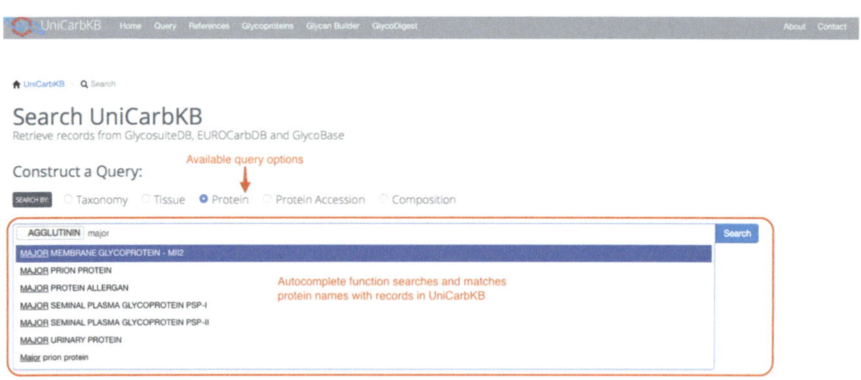

Fig. 10.1 Users can search the database content by monosaccharide composition, attached protein, taxonomy or tissue – to start searching UniCarbKB, click the Query tab in the navigation bar at the top of all pages, and select your search preference. As you type in the search box, the autocomplete feature will start to show terms that match data entries in UniCarbKB. Autocomplete possesses search terms that are supported by the database; for example, as you start typing 'major' in the protein search bar, a list of proteins with 'major' in their name will be listed. To find information about a category, check the search button

10.2.3 Searching by Composition

The search fields correspond to the monosaccharide (and modification, e.g. sulphate, phosphate) compositions of glycan structures listed in UniCarbKB. The composition search form comprises 14 boxes corresponding to individual monosaccharide residues and modifications. By default, each composition is set to zero and treated as underivatised. To perform a composition, search simply:

- On the home page, in the navigation bar at the top of the page, click the Query tab. After a few seconds, you will be presented with the Query page. Select the Composition option.
- Select the N-linked or O-linked button to limit searches to the specified family of glycans.
- Enter whole integer values into any of the monosaccharide composition boxes and click the Search button.

If the composition search is successful, the results page will list all structures matching the exact composition. The results page consists of a table listing the structures with corresponding glycoprotein(s) and species information if known. Notable features include a filter text box (positioned above the structure table) that can be used to refine results based on matching glycoprotein or taxon name; the 'Query' text is displayed in line with the page header, and akin to many pages, the 'Structure Format' button enables users to convert between supported nomenclature formats:

- The composition results page lists structures matching the exact composition only.
- Each listed glycan structure is linked to its corresponding summary page, which can be accessed by clicking the image.

The simple searches shown above will suffice for many situations. However, the default search options may cast a wide net of hits. If this is the case, one may wish to use the GlycanBuilder Search, which gives finer control on finding content relevant to particular detailed structures and substructures/epitopes.

10.2.4 GlycanBuilder and Searching for (Sub)Structures

UniCarbKB glycan structural queries can also be made by using the versatile tool GlycanBuilder. We have integrated the Vaadin release of GlycanBuilder (Ceroni et al. 2007; Damerell et al. 2012), which is an upgraded build of the original tool developed by members of the EUROCarbDB project (Ceroni et al. 2007), for searching and retrieving information. As part of the UniCarbKB initiative, we have further modified and updated GlycanBuilder to support the latest software packages. Aligned to our open-source commitment, the source code for GlycanBuilder is available at https://github.com/alternativeTime/GlycanBuilderVaadin7Version including instructions for developers interested in deploying and contributing to the tool.

GlycanBuilder, accessed at http://unicarbkb.org/builder, allows users to (1) build a new glycan structure, (2) build a substructure or epitope or (3) extend a structure by using the library of predefined structures. Users are encouraged to refer to guides and documentation provided by the developers (https://bitbucket.org/daviddamerell/glycanbuilderv/wiki/Home). In brief, the tool comprises three main sections: (1) drawing canvas, (2) monosaccharide and linkage panel and (3) menu options for exporting images, switching between symbol/notation formats and selecting monosaccharide residues.

10.2.4.1 Query by Structure

By default, only database entries that have information exactly matching the topology, linkage and anomeric configuration of the submitted structure will be retrieved. In the case of partial structure searching, a level of fuzziness is introduced, whereby unknown information is handled as wildcards by the search algorithm. For substructure searching, only structures matching the epitope or extended motif will be listed in the results page.

10.2.4.2 Glycan Structure Summary Page

In UniCarbKB, the central data content is the glycan structure and its supporting evidence. The glycan structure page is the place where all information connected with a structure can be easily accessed. Each page includes a standard set of data types and provides users with an overview of content available. Finding content or supporting evidence for the glycan structure is quick and easy. This section leads users through the basics of navigating content linked to each glycan entry in UniCarbKB that is most relevant; an example screenshot is shown in Fig. 10.2.

There are several routes for finding a structure of interest: (1) by using GlycanBuilder or the composition query engine or (2) by using the contextual navigation options presented in the viewing templates (for more information, refer to individual page descriptions below). Keep in mind that UniCarbKB considers structures to be unique based on the topology and sequence annotation, e.g. monosaccharide anomericity and glycosidic linkage.

The template is divided into four main sections:

- Structure image – the left-side column displays the structure that by default is displayed using the Essentials/CFG nomenclature.
- Supporting publications – a table of published literature references, available in UniCarbKB, which reports the experimental basis on which the glycan structure

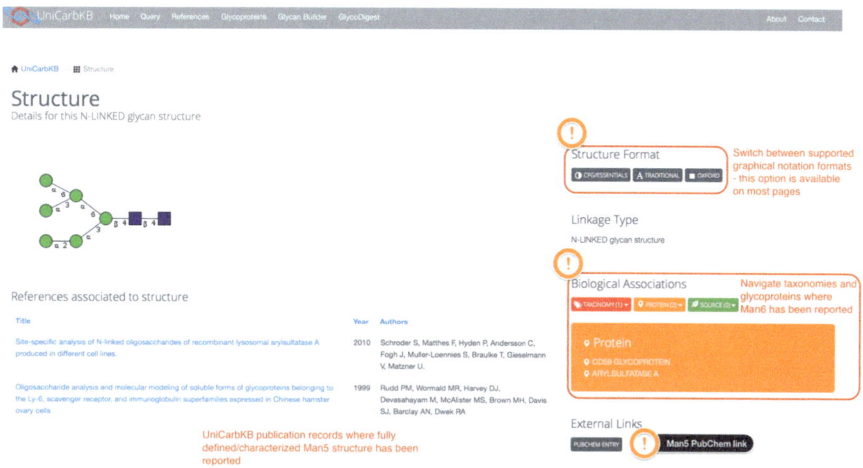

Fig. 10.2 Every structure result has three parts: (*i*) structure listing (*left panel*) that displays the structure (in a defined nomenclature format) with a description of supporting references; (*ii*) the 'Biological Associations' subsection, which summarises the glycan distribution in the context of taxonomies, tissues and glycoproteins; and (*iii*) links to external database including PubChem

was determined, is clearly summarised. To view any of the publications in detail, click the title to open the Reference page.

- Glycosyltransferases (limited to N-glycans) – if the structure is fully defined, i.e. no sequence fuzziness or linkage ambiguity, an enzyme tab will be displayed. This feature, called GlycanSynth, documents the enzyme glycosyltransferases (GTs) involved in the biosynthesis of each disaccharide unit in the N-glycan structure. A list of these gene names, enzymes and reactions is available at http://unicarbkb.org/enzymes.
- Sidebar – displays biological association metadata, links to external databases, calculated mass and structure classification. The biological association submenu concisely groups taxonomic, glycoprotein and tissue source information pertinent to the displayed structure (refer to Sect. 10.2.7.1).

10.2.4.3 Switching Between Symbol Nomenclatures

Symbol notation provides a compact, simpler-to-visualise approach for describing complex glycans and contextually provides an efficient, easier-to-understand means for annotating complex datasets. Unfortunately, until recently, no standard notation has been proposed for the representation of complex glycan structures, and many different types of symbols and conventions can be found in the literature. Examples of cartoons are the representation scheme developed by the *Essentials of Glycobiology* textbook editors (and adopted by the CFG, Raman et al. 2006; Varki 2009) and the scheme developed by the Oxford Glycobiology Institute (Harvey et al. 2009,

2011). A combination of both these formats, in which the linkages are depicted as angles as in the Oxford scheme, on the coloured residue symbols of the *Essentials in Glycobiology* scheme, is gaining acceptance (Varki et al. 2015) and is supported by the UniCarbKB suite of tools and databases, as well as planned to provide a linked option to UniCarbKB in the next (3rd) edition of the *Essentials in Glycobiology* textbook.

By default, the hybrid CFG/Essentials with linkage format is set for displaying structures in UniCarbKB. To convert between the 'CFG/Essentials', 'Oxford' and 'Traditional (IUPAC)' formats, users can select an option in the 'Structure Format' sidebar panel. The viewing symbol preference will be retained for the session – the duration of the time spent on the website.

10.2.5 Exploring and Finding Curated References

The UniCarb KnowledgeBase offers the community access to a growing, curated database of information on the glycan structures on glycoproteins. Primarily, at this stage, UniCarbKB is a eukaryotic glycoprotein-centric resource built upon a corpus of curated information originating from GlycoSuiteDB and a select number of data sets from EUROCarbDB. We are actively curating more recent publications that contain (partially or completely characterised) glycan structures with supporting experimental data, predominantly from glycoprotein(s), including site-specific data when available.

A complete overview of publications that have been curated for entry into UniCarbKB can be viewed by clicking the 'References' tab in the navigation bar. The References page tabulates all publication details inclusive of title, year of publication, authors and journal name, in addition to the number of described glycan structures. By default, the number of publications visible is limited to ten entries, and pagination is used to improve content layout. At the bottom of the page, users can use the page counter to scan through lists of results to find the publication of interest.

To help users quickly find a publication of interest, we have provided an intuitive filter box, which can be used to filter publications by matching an author name or publication title. For example, entering mucin will retrieve publication records with the word 'mucin' in the title; alternatively typing an author name will list all appropriate papers:

- On the home page, select the References tab from the navigation bar at the top of the page.
- Search for a title or author by entering text into the search bar.
- To view individual publication content including glycan structures, biological context and experimental information, users can click the publication title.

10.2.5.1 Viewing Individual Publications

The publication template is divided into these sections:

- Bibliographic header displays information about the publication with an external link to PubMed.
- Abstract and Structure section displays the abstract provided by PubMed and a graphical listing of glycan structures curated from the publication that are linked to the 'Glycan Structure' page.
- Contextual navigation options accessible from the sidebar include 'Biological Associations', 'Validation Method' and 'Connections'.

An example publication listing is shown in Fig. 10.3.

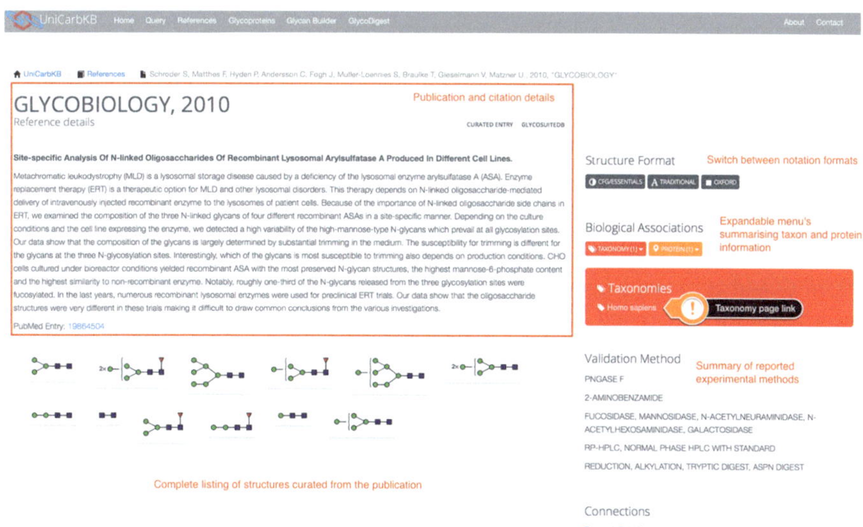

Fig. 10.3 Our biocuration aspirations are to provide the community with an accurate and long-term resource. For each publication, we carefully review the publication and record the following methodology details: (*i*) sample preparation procedures and glycan release techniques and/or methods that alter glycan structure, including exoglycosidase treatment and derivatization, (*ii*) the analytical (mass, sequencing and linkage) approach, and (*iii*) complementary validation methods. Such information is listed under the 'Validation Methods' subsection. It is our intention that by providing such metadata, users can assess the reliability, specifics and accuracy of the contained information. In addition, we capture the biological content reported, i.e. species, tissue and glycoprotein(s), that are grouped into drop-down lists in the 'Biological Associations' subsection. Finally, all structures reported are clearly summarised in the left panel under the abstract which are linked to the structure summary page

10.2.5.2 More About the 'Biological Associations'

'Biological Associations' refers to the reported (individual or multiple) taxonomic, glycoprotein and tissue sources. Each classification is grouped into expandable drop-down lists, as show in Fig. 10.2. To find related biological content, you can use the embedded links in the quick navigation bar to access individual taxonomy, glycoprotein or tissue source pages.

10.2.5.3 Inference and Confidence 'Validation Methods'

The 'Validation Methods' subsection is the primary means for obtaining information on the experimental methods reported in the publication, including (1) the described sample preparation and glycan release techniques and/or methods that alter glycan structure, including exoglycosidase treatment and derivatization, (2) the analytical (mass, sequencing and linkage) approach used and (3) any complementary methods described for validation of the reported structure.

- An expandable list of terms/keywords is used to describe instrumentation, sample preparation/clean-up and orthogonal techniques during the manual curation process.

The provision of this experimental detail allows users to assess the reliability, specifics and accuracy of the contained information, and such provenance metadata is provided for all curated publications.

10.2.6 Using the Glycoprotein Navigation Page to Find Global and Site-Specific Content

UniCarbKB stores information that provides useful context on both global and site-specific glycosylation of individual and mixtures of glycoproteins. The glycoprotein page, accessed from the navigation bar, is a quick navigation aid to find content when the glycosylation of an individual glycoprotein has been characterised:

- Point your browser at the UniCarbKB glycoprotein page at http://www.unicarbkb.org/proteins.
- Start typing your glycoprotein of interest into the filter box, and UniCarbKB will start to match names in the database as you type, by showing a quickly adjusting drop-down list of glycoproteins.
- To view more information available for a glycoprotein of interest, select the item from the drop-down list.
- Press the Filter button.
- Click an item's name or accession number to open the Glycoprotein summary page.

Immediately after pressing the filter button, the table will be updated listing entries that match your selected glycoprotein name(s). The table lists the UniProtKB accession number, taxonomy, reported number of glycan structures and a label indicating if site-specific glycosylation information is available.

It is also possible to search by UniProt Accession number by using the 'Protein Accession' option listed at the main query page http://www.unicarbkb.org/query. Similar to above, an autocomplete feature is enabled that will match stored accession numbers with the text entered.

10.2.6.1 Viewing Glycoprotein Information

Every glycoprotein entry is linked to a comprehensive summary page. The glycoprotein templates make it quick and easy to navigate and find information about the specified glycoprotein. It takes information curated from the literature and when possible maps known site-specific data with data available in UniProtKB. Here are some of the ways you can use the data:

- The main section summarises information that is relevant to the glycoprotein.
  - A summary of experimentally verified site-specific Glycosylation Sites clearly denotes the amino acid position and number of structures associated with that site.
  - Users can view glycan structures reported at a specific site by clicking the 'Associated Structures' label.
  - Below the Glycosylation Sites subsection, all glycan structures associated with the specified glycoprotein are listed. For more information on the glycan structures, including supporting publications and biological associations, users are prompted to click the structure of interest.
- The sidebar shows three sections:
  - The primary sequence of the protein, which is dynamically loaded from UniProtKB.
  - Information regarding Biological Associations (see below) including taxonomy and tissue source.
  - Peer-reviewed publications that were sourced to compile the global and site-specific glycosylation information. For further details on the publication cited, users can navigate to the Reference page by clicking the publications title.

Figure 10.4 shows the summary page for the protein alpha-2-HS-glycoprotein, which provides a description of the attached glycan structures and any knowledge of site-specific glycosylation that has been curated from the literature. In addition, for each protein summary, we provide a comprehensive summary of associated metadata including biological source and publications citing the data. Information about the glycoprotein can also be displayed in UniProtKB by clicking the listed protein accession identifier.

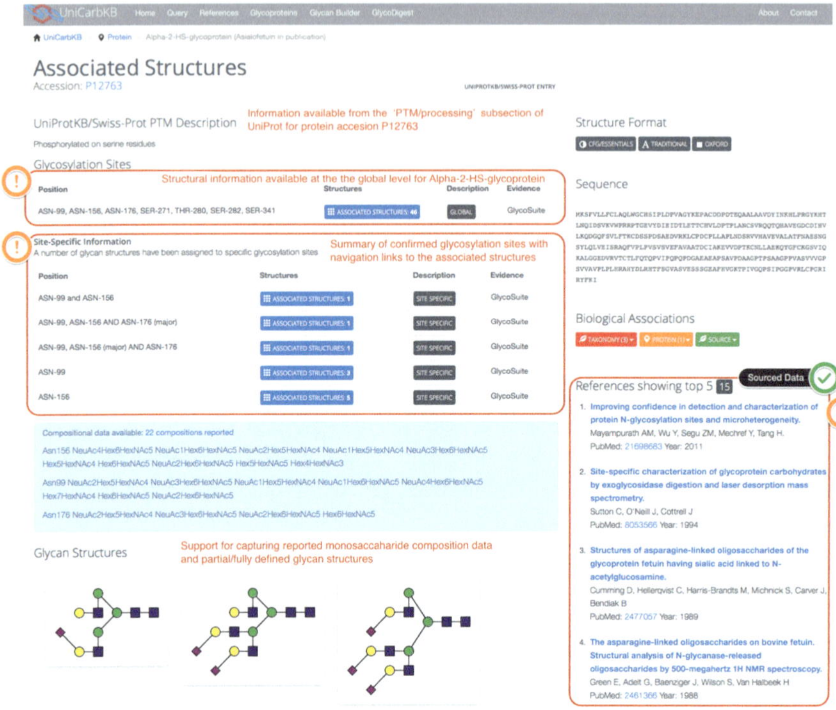

Fig. 10.4 Screenshot for the glycoprotein entry alpha-2-HS-glycoprotein. For each summary page, a description of the glycoprotein is obtained from the UniProtKB 'PTM/processing' subsection in addition to the protein primary sequence. The level of information displayed can vary; for example, not all entries will catalogue supporting site-specific or compositional data as shown for alpha-2-HS-glycoprotein as glycosylation site information is often not known. However, in general, all protein summary pages will list glycan structures characterised on the glycoprotein(s) at the global level. Users can click the 'Associated Structures' label(s) to retrieve further information pertinent to glycan structures determined at the global and/or site-specific level where known. Finally, details on the relevant biological associations can be displayed by clicking the 'Taxonomy', 'Protein' or 'Source' boxes. All information presented has been sourced from the publications listed in the sidebar, which can be accessed by clicking the manuscript title

10.2.7 Information Available for Taxonomy and Tissue Searches

UniCarbKB organises content in a number of ways. So far we have mentioned the layout for the glycoprotein and glycan structure entries. Similar layouts are used to display information about individual taxons and tissues. Both templates apply the common theme; however, the content can be very rich and complex:

- Go to the *Homo sapiens* taxonomy by the following steps: (1) select the Query tab from the navigation bar; (2) select taxonomy and enter Homo sapiens and press the search button; and (3) click the blue structures label.

- The left column lists all glycan structures and compositions in UniCarbKB associated with the taxon or tissue, e.g. *Homo sapiens*. By default, only ten glycan structures are displayed – to view the complete catalogue of structures, click the 'Show all structures' button. Note this option will only appear when the size of the structure data set exceeds ten.
- You will notice that the right sidebar looks different. Since we have restricted the database search to a tissue or taxonomy, the sidebar will only list relevant 'Biological Associations', for example, all glycoproteins characterised from the species *Homo sapiens*.
- To view all glycoproteins, click the 'Biological Associations' Proteins label.
- You can navigate to the glycoprotein page by scrolling the list to find any protein of interest.

10.2.7.1 More About the Sidebar

In order to improve navigation and the logical organisation of UniCarbKB, we placed page-specific links in the sidebar. The sidebar appears on a majority of pages, except for the References and Glycoprotein summaries. By using the sidebar you can:

- Choose your preferred 'Structure Format' for the remainder of your visit.
- Perform various operations via the links on the sidebar.
- Use the contextual navigation options that appear under each subheading, based on the type of content you are viewing, for example, viewing relevant Biological Associations.
- Access external databases and supporting tools.

The links in this area change, depending on the section of database you are viewing. The sidebar is a good place to display important links to internal and external pages.

10.2.7.2 'Connections' – What You Can Find Through UniCarbKB

Specialised information within the scope of UniCarbKB is made available via cross-references to relevant resources, such as PubChem (NCBI Chemical database), SugarBind (Shakhsheer et al. 2013), UniCarb-DB (a MS/MS experimental database; Hayes et al. 2011), GlycoMob (an Ion Mobility Collision Cross Section database, Struwe et al. 2015) and GlycoDigest (a predictive glycosidase database, Gotz et al. 2014), and additional links are created for research publications included in the Research Data Australia service. For each glycan structure entry and publication summary page, implicit links to these affiliated databases are listed under 'Connections' (refer to Fig. 10.2). A list of cross-referenced databases and current state of development is available at http://www.unicarbkb.org/crossreferences including the pending release of UniCorn a theoretical database of N-glycan structures.

In addition, as part of the GlycoRDF (Ranzinger et al. 2015) project, a series of new user interfaces are under development to support connections with affiliated databases including GlyTouCan (Aoki-Kinoshita et al. 2016), GlycomeDB (Ranzinger et al. 2008) and the Carbohydrate Structure Database (Egorova and Toukach 2014).

A key feature of the central glycan structure feature of UniCarbKB is our effort to forge connections with UniProtKB, albeit many described structures have been determined from global cellular or tissue analysis and are not necessarily linked to known, specific glycoproteins. To this end, links to UniProtKB are only provided for each annotated known glycoprotein. Users can navigate from UniCarbKB to UniProtKB when the attached glycoprotein is known, via the stable and unique primary protein accession number links; an example is shown in Fig. 10.4.

10.3 Summary

The UniCarbKB database is far from complete; in fact it is just beginning! At this time, although glycan structures can be characterised in detail on a global glycomic scale (and are included in UniCarbKB), there is only limited data available on the detail of glycan structures assigned to specific sites on individual glycoproteins. In addition, site glycosylation is known to change both spatially and temporally in response to a cell's environment. The emergence of new analytical techniques is only slowly unravelling this complexity of the glycome and glycoproteome, let alone the knowledge accumulating on other glycoconjugates. As such, the content of UniCarbKB needs to continue to grow, user interfaces to mature and new tools developed to exploit the data being increasingly acquired.

UniCarbKB is an open access and open-source project. All the software developed for use is available for distribution subject to Creative Commons licencing, and the data itself is available in a variety of formats. The relational database is compatible with PostgreSQL 9.1, and in addition, the data is available in the GlycoRDF format and is also accessible via a dedicated SPARQL endpoint (http://bit.ly/24ejAsD).

Submission of Updates, New Data and Troubleshooting

The initiative is driven as a community endeavour; as such, the team encourages end-user feedback. To submit updates and/or corrections to UniCarbKB and for any enquiries, use the e-mail address matthew.campbell@mq.edu.au.

References

Aoki-Kinoshita K, Agravat S, Aoki NP, Arpinar S, Cummings RD, Fujita A, Fujita N, Hart GM, Haslam SM, Kawasaki T et al (2016) GlyTouCan 1.0–the international glycan structure repository. Nucleic Acids Res 44(D1):D1237–D1242

Campbell M, Royle L, Radcliffe C, Dwek R, Rudd P (2008) GlycoBase and autoGU: tools for HPLC-based glycan analysis. Bioinformatics 24(9):1214–1216

Campbell M, Hayes C, Struwe W, Wilkins M, Aoki-Kinoshita K, Harvey D, Rudd P, Kolarich D, Lisacek F, Karlsson N, Packer N (2011a) UniCarbKB: putting the pieces together for glycomics research. Proteomics 11(21):4117–4121

Campbell MP, Hayes CA, Struwe WB, Wilkins MR, Aoki-Kinoshita KF, Harvey DJ, Rudd PM, Kolarich D, Lisacek F, Karlsson NG, Packer NH (2011b) UniCarbKB: putting the pieces together for glycomics research. Proteomics 11(21):4117–4121

Campbell MP, Peterson R, Mariethoz J, Gasteiger E, Akune Y, Aoki-Kinoshita KF, Lisacek F, Packer NH (2014) UniCarbKB: building a knowledge platform for glycoproteomics. Nucleic Acids Res 42(1):D215–D221

Ceroni A, Dell A, Haslam S (2007) The GlycanBuilder: a fast, intuitive and flexible software tool for building and displaying glycan structures. Source Code Biol Med 2:3

Cooper C, Harrison M, Wilkins M, Packer N (2001) GlycoSuiteDB: a new curated relational database of glycoprotein glycan structures and their biological sources. Nucleic Acids Res 29(1):332–335

Damerell D, Ceroni A, Maass K, Ranzinger R, Dell A, Haslam S (2012) The GlycanBuilder and GlycoWorkbench glycoinformatics tools: updates and new developments. Biol Chem 393(11):1357–1362

Egorova KS, Toukach PV (2014) Expansion of coverage of Carbohydrate Structure Database (CSDB). Carbohydr Res 389:112–114

Gotz L, Abrahams JL, Mariethoz J, Rudd PM, Karlsson NG, Packer NH, Campbell MP, Lisacek F (2014) GlycoDigest: a tool for the targeted use of exoglycosidase digestions in glycan structure determination. Bioinformatics 30(21):3131–3133

Harvey DJ, Merry AH, Royle L, Campbell MP, Dwek RA, Rudd PM (2009) Proposal for a standard system for drawing structural diagrams of N- and O-linked carbohydrates and related compounds. Proteomics 9(15):3796–3801

Harvey DJ, Merry AH, Royle L, Campbell MP, Rudd PM (2011) Symbol nomenclature for representing glycan structures: extension to cover different carbohydrate types. Proteomics 11(22):4291–4295

Hayes C, Karlsson N, Struwe W, Lisacek F, Rudd P, Packer N, Campbell M (2011) UniCarb-DB: a database resource for glycomic discovery. Bioinformatics 27(9):1343–1344

Raman R, Venkataraman M, Ramakrishnan S, Lang W, Raguram S, Sasisekharan R (2006) Advancing glycomics: implementation strategies at the consortium for functional glycomics. Glycobiology 16(5):82R–90R

Ranzinger R, Herget S, Wetter T, von der Lieth C (2008) GlycomeDB – integration of open-access carbohydrate structure databases. BMC bioinf 9:384

Ranzinger R, Aoki-Kinoshita KF, Campbell MP, Kawano S, Lutteke T, Okuda S, Shinmachi D, Shikanai T, Sawaki H, Toukach P, Matsubara M, Yamada I, Narimatsu H (2015) GlycoRDF: an ontology to standardize glycomics data in RDF. Bioinformatics 31(6):919–925

Shakhsheer B, Anderson M, Khatib K, Tadoori L, Joshi L, Lisacek F, Hirschman L, Mullen E (2013) SugarBind database (SugarBindDB): a resource of pathogen lectins and corresponding glycan targets. J Mol Recognit 26(9):426–431. doi:10.1002/Jmr.2285

Struwe WB, Pagel K, Benesch JL, Harvey DJ, Campbell MP (2015) GlycoMob: an ion mobility-mass spectrometry collision cross section database for glycomics. Glycoconj J [Epub ahead of print]

UniProt C (2015) UniProt: a hub for protein information. Nucleic Acids Res 43(Database issue):D204–D212

Varki A (2009) Essentials of glycobiology, 2nd edn. Cold Spring Harbor Laboratory Press, Cold Spring Harbor

Varki A, Cummings RD, Aebi M, Packer NH, Seeberger PH, Esko JD, Stanley P, Hart G, Darvill A, Kinoshita T, Prestegard JJ, Schnaar RL, Freeze HH, Marth JD, Bertozzi CR, Etzler ME, Frank M, Vliegenthart JF, Lutteke T, Perez S, Bolton E, Rudd P, Paulson J, Kanehisa M, Toukach P, Aoki-Kinoshita KF, Dell A, Narimatsu H, York W, Taniguchi N, Kornfeld S (2015) Glycobiology 25:1323–1324

von der Lieth CW, Freire AA, Blank D, Campbell MP, Ceroni A, Damerell DR, Dell A, Dwek RA, Ernst B, Fogh R, Frank M, Geyer H, Geyer R, Harrison MJ, Henrick K, Herget S, Hull WE, Ionides J, Joshi HJ, Kamerling JP, Leeflang BR, Lutteke T, Lundborg M, Maass K, Merry A, Ranzinger R, Rosen J, Royle L, Rudd PM, Schloissnig S, Stenutz R, Vranken WF, Widmalm G, Haslam SM (2011) EUROCarbDB: an open-access platform for glycoinformatics. Glycobiology 21(4):493–502

Chapter 11
GlycoProtDB: A Database of Glycoproteins Mapped with Actual Glycosylation Sites Identified by Mass Spectrometry

Hiroyuki Kaji, Toshihide Shikanai, Yoshinori Suzuki, and Hisashi Narimatsu

Abstract Asn(N)-linked glycans are attached to the consensus sequence of *N*-glycosylation, Asn-Xaa-[Ser/Thr], where Xaa $\neq$ Pro, of a nascent protein by the oligosaccharyltransferase (*OST*) complex. Because the modification is carried out in the endoplasmic reticulum (*ER*), all predictable potential sequences are not necessarily glycosylated. Therefore, the information on the "actual" glycosylated site is important for glycobiology research. GlycoProtDB was constructed to provide such information collected by *MS*-based glycoproteomics technology. Now, the database contains the data of 3122 glycoproteins obtained from *C. elegans*, mouse, and human samples and is available at the URL: http://jcggdb.jp/rcmg/gpdb/index.

Keywords Glycoprotein • Mass spectrometry (*MS*) • *N*-glycosylation site • Stable isotope labeling • Lectin capture • Glycoproteomics • Peptide-*N*-glycanase (PNGase)

11.1 Introduction

Asn(N)-linked glycans are attached co-translationally to a nascent protein by the oligosaccharyltransferase (*OST*) complex in the luminal space of the endoplasmic reticulum (*ER*). Therefore, potential sites of N-glycosylation can be predicted based on a combination of the specificity of the enzyme *OST* and localization of the protein of interest. *OST* has an acceptor (protein) sequence specificity, i.e., Asn-Xaa-[Ser/Thr], where Xaa is any amino acid residue except Pro; the sequence is called a "sequon." *OST* is localized on the *ER* membrane at the luminal side near the protein translocation channel, translocon. Thus, the acceptor protein must enter into the *ER* to be glycosylated and so must have translocation signal to the *ER*. Two types of this

H. Kaji (✉) • T. Shikanai • Y. Suzuki • H. Narimatsu
Biotechnology Research Institute for Drug Discovery, National Institute of Advanced Industrial Science and Technology (AIST), Tsukuba, Japan
e-mail: kaji-rcmg@aist.go.jp

K.F. Aoki-Kinoshita (ed.), *A Practical Guide to Using Glycomics Databases*,
DOI 10.1007/978-4-431-56454-6_11

signal are known, one is a cleavable signal peptide and another is a transmembrane segment. The uncleaved signal peptide is called a signal anchor and acts as a transmembrane segment. Therefore, the potential N-glycosylation site is present in the sequon on the segment translocated into the *ER*. However, because there are some other factors limiting glycosylation and unknown translocation mechanisms independent of these signals, the information on the actual glycosylated site is important for glycobiology research. GlycoProtDB was constructed to provide such information collected by *MS*-based glycoproteomics technology, which was named *IGOT* method which we developed (Kaji et al. 2003, 2006, 2012; Kaji and Isobe 2013; Shinkawa et al. 2005).

11.2 The Contents of GlycoProtDB

Organisms: *C. elegans*, mouse (C57BL/6 J, male), and human

Information: Glycosylation sites, glycosylated peptide sequences identified, and lectins used to collect glycopeptides identified, which suggest structure motif present on the glycan. The data to construct this database are secondary, based on published papers (Kaji et al. 2003, 2007, 2012, 2013; Sogabe et al. 2014; Hirao et al. 2014; Sugahara et al. 2012, 2015).

11.3 The Method to Collect Data on the Actual Glycosylation Sites

The method to collect data for the database construction was reported previously (Kaji et al. 2003, 2006, 2012; Kaji and Isobe 2013). The outline of the method is shown in Fig. 11.1. The sample is a preparation of glycopeptides, which is enriched by lectin affinity chromatography or hydrophilic interaction chromatography (*HILIC*) from a protease digest of a sample protein mixture. To identify the core peptides at high throughput and sensitively, the glycan moiety was removed before

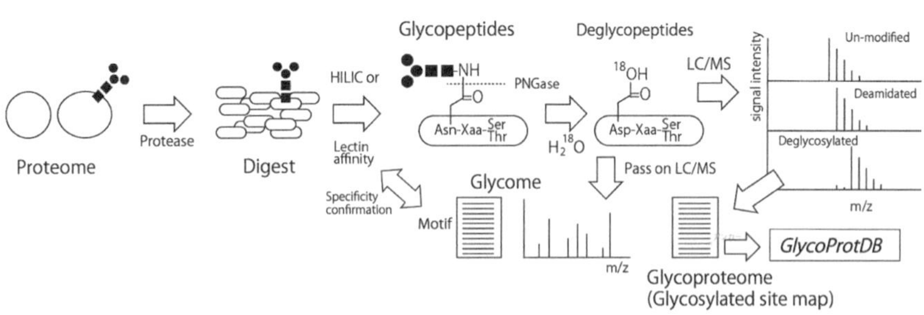

Fig. 11.1 Outline of a method to collect large-scale data on protein glycosylation sites, which was used for the construction of glycoprotein database, GlycoProtDB

LC/MS analysis with peptide-N-glycanase (PNGase). By the enzyme reaction, glycosylated Asn is converted to Asp in the deglycosylated peptides, and the mass of the peptide shifts by 1 Da compared to that of the original sequence. This mass shift serves as a tag of the formerly glycosylated position; however, the shift also occurs by deamidation of the Asn side chain in vivo and in vitro and may cause false identification of the deamidated peptides as formerly glycosylated peptides. Therefore, the deglycosylation with PNGase is carried out in solvent water labeled with stable isotope ^{18}O ($H_2{}^{18}O$). The ^{18}O is incorporated into the Asn side chain to form Asp in the peptide (Fig. 11.1). As a result, the formerly glycosylated peptide labeled with ^{18}O shows a +3 Da mass shift. Fortunately, because the deamidation reaction proceeds slowly, little ^{18}O incorporation occurs in the reaction time of deglycosylation with PNGase, ca 16 h. The labeled peptides are identified by *LC/MS* analysis followed by database search using an appropriate software and sequence database, e.g., Mascot and *NCBI* Refseq protein database, respectively. If sample glycopeptides are captured with a lectin, the specificity can be confirmed by *MS* profiling of glycans released from the obtained glycopeptides (Sugahara et al. 2015). In a general *LC*-coupled *MS* analysis system, a reversed phase chromatograph using an octadecylsilica (*ODS*) column is used, and the sample peptides are trapped at once on a pre-column (*ODS*) for desalting. Thus, the released glycans are able to recover as a flow-through fraction of the trap column (Fig. 11.1).

11.4 Construction of Data Resource from *MS*-Based Identification List of Glycopeptides

The glycopeptides for the database construction are identified by using the Mascot search engine and the *NCBI* Refseq protein sequence database. By the search, multiple proteins (Gene IDs (GI number) such as isoforms or family proteins) are often assigned from a single *MS/MS* spectrum. In the study of mouse tissues, about 70 % peptides identified were assigned to a single protein sequence; however, the remaining 30 % were shared by multiple proteins. Because the actual origins of the peptide cannot be determined, all assigned proteins are included in the database, where these proteins (gi) are summarized under the higher gene symbol. For example, two gi numbers of asialoglycoprotein receptor 1 were identified, gi| 308387363 and 4502251, from a single peptide. Then these are summarized under their common gene symbol, *ASGR1*.

11.5 How to Use the Database, GlycoProtDB

URL: http://jcggdb.jp/rcmg/gpdb/index

The outline of this database is available by clicking the links in "Documents" on the left side of the page.

11.6 Search

The database is searchable by specifying a keyword in “Gene name” or “Protein description” (e.g., receptor or transferase, not case sensitive) or in “Gene symbol” (e.g., afp or egfr), as well as by “ID search,” a digit in “Gene ID (GI)/Accession” (Fig. 11.2a). If a search is started without any input (by a blank), all entries will be listed. A search by multiple keywords is available under “Advanced search” (Fig. 11.2b). To perform an “*AND*” search, input multiple words in the text box separating them by blanks or comma, e.g., “amino acid transporter” which means “amino” *AND* “acid” *AND* “transporter.” To perform an “*OR*” search, add conditions and input a word in each box, e.g., “transporter” in box 1 and “channel” in box 2.

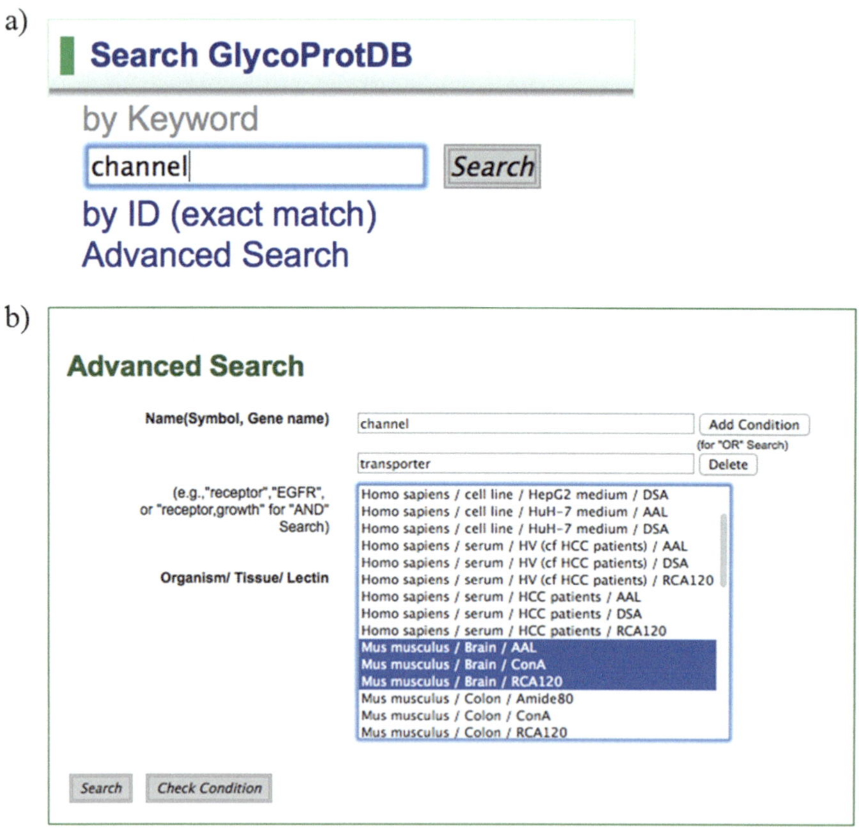

Fig. 11.2 Start point of a search in GlycoProtDB. (**a**) To start a search by keyword, input a word or number into the text box and click Search button. (**b**) A search by multiple keywords is available under “Advanced Search.” See text for detail

Sample preparation(s) including organism, tissue, and lectin (e.g., *Mus musculus*, liver, conA) can be selected in the Advanced Search.

11.7 Search Results

Search results are shown in tabular format with information about the assigned protein including Gene Symbol, Gene Name, GlycoProtDB_ID, GI/Accession, (Protein) Description, Length, and Molecular weight calculated (*MW*) and of the sample (*IGOT* analysis) including organism, tissue, and lectin (Fig. 11.3) with search conditions. The words matching the search conditions are highlighted in yellow. The terms linked to other information are shown in blue. By clicking GlycoProtDB_ID or GI/Accession of an interested protein, a series of detailed information of the protein is shown as tables or graphics, i.e., glycosylation site(s), protein sequence with sequon(s) of both potential and identified, identified glycopeptide(s), and quotation from external databases.

Search Conditions

Taxonomy / Tissue / Lectin: Mus musculus / Kidney / AAL, Mus musculus / Kidney / ConA, Mus musculus / Kidney / RCA120

Search Results

451 to 500 of 574 [1] [2] ... [7] [8] [9] 10 [11] [12] Next>

Assigned Protein							IGOT Analysis	
Gene Symbol	Gene Name	Protein Variant(s)					Organism	Tissue/Lectin
		GlycoProtDB_ID	GI/Accession	Description	Length	MW		
Serpina7	serine (or cysteine) peptidase inhibitor, clade A (alpha-1 antiproteinase, antitrypsin), member 7	GPMMU00001313	46559749	thyroxine-binding globulin precursor [Mus musculus].	418	47051.9	Mus musculus	Kidney/RCA120 Liver/RCA120 Liver: b4GalT-I(+/+)/RCA120 Liver: b4GalT-I(-/-)/RCA120 Serum/Amide80
		GPMMU00006410	274321952	thyroxine-binding globulin precursor [Mus musculus].	426	48080.1	Mus musculus	Kidney/RCA120 Liver/RCA120 Liver: b4GalT-I(+/+)/RCA120 Liver: b4GalT-I(-/-)/RCA120 Serum/Amide80
Serpinc1	serine (or cysteine) peptidase inhibitor, clade C (antithrombin), member 1	GPMMU00000091	18252782	antithrombin-III precursor [Mus musculus].	465	52003.1	Mus musculus	Brain/RCA120 Colon/Amide80 Colon/RCA120 Heart/Amide80 Heart/ConA Heart/RCA120 Kidney/AAL Kidney/ConA Kidney/RCA120 Liver/AAL Liver/ConA Liver/RCA120 Liver/SSA Liver/WGA Liver: b4GalT-I(+/+)/RCA120 Liver: b4GalT-I(-/-)/RCA120 Lung/AAL

Fig. 11.3 Search results. Information on glycoprotein(s) matched with the search conditions is shown by tabular format. Gene symbol is the highest level of a protein. If multiple proteins have the same symbol, e.g., isoforms, they are shown in parallel. The words matched with search conditions are highlighted in *yellow*. The detailed information is shown by clicking a *blue text*, e.g., GI/Accession

11.8 Glycosylation Sites

At the top of the result, glycosylation sites identified by *IGOT-LC/MS* method are provided graphically as shown in Fig. 11.4a. The full sequence of the selected protein is illustrated as a horizontal line with scales. Vertical lines on the scales indicate potential sites of N-glycosylation (sequon). The identified sites are indicated with "ball and stick" symbols on the horizontal line for each sample. The order of the samples is sortable by clicking "Tissue" or "Lectin." The sequence lines can be enlarged in the horizontal direction by dragging the mouse, and the enlarged figure

a)

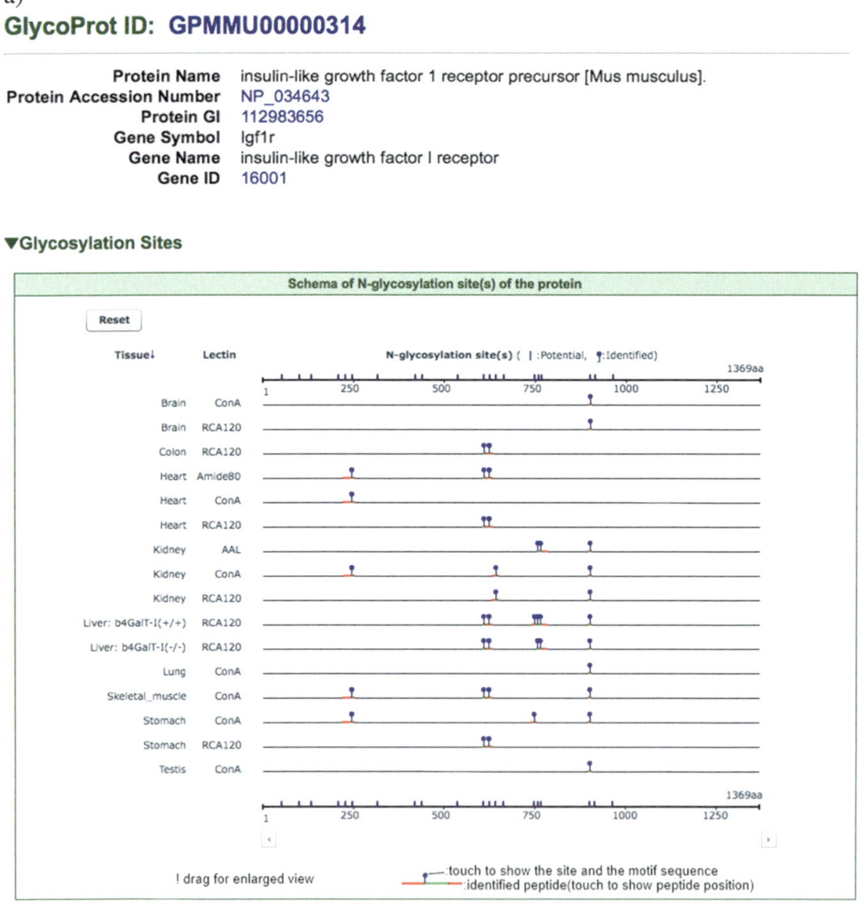

Fig. 11.4 A part of detailed search results; glycosylation sites. Potential and actually identified sites are shown by graphics at the top of the result (**a**). Whole amino acid sequence of the glycoprotein selected and both potential and identified glycosylation sites are shown under the graphic and detailed sequence of glycopeptides provided by table (**b**)

b)

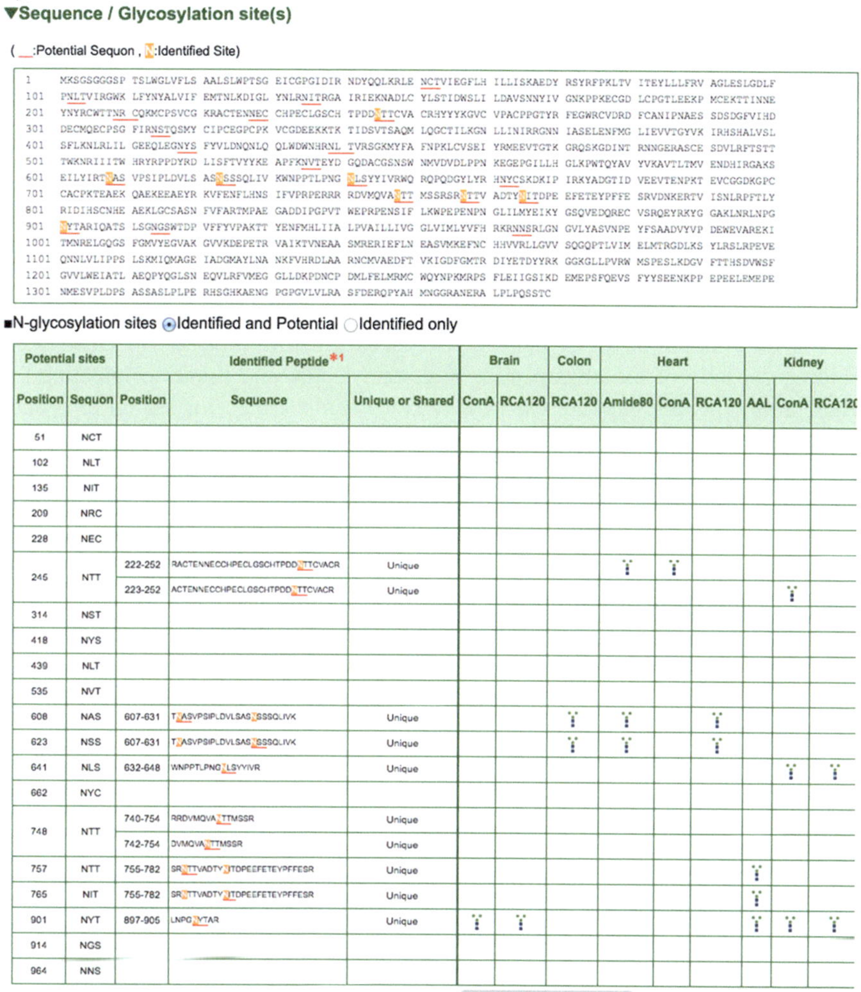

Potential sites		Identified Peptide *1			Brain		Colon	Heart			Kidney		
Position	Sequon	Position	Sequence	Unique or Shared	ConA	RCA120	RCA120	Amide80	ConA	RCA120	AAL	ConA	RCA120
51	NCT												
102	NLT												
135	NIT												
209	NRC												
228	NEC												
245	NTT	222-252	RACTENNECCHPECLGSCHTPDDNTTCVACR	Unique									
		223-252	ACTENNECCHPECLGSCHTPDDNTTCVACR	Unique									
314	NST												
418	NYS												
439	NLT												
535	NVT												
608	NAS	607-631	TNASVPSIPLDVLSASNSSSQLIVK	Unique									
623	NSS	607-631	TNASVPSIPLDVLSASNSSSQLIVK	Unique									
641	NLS	632-648	WNPPTLPNGNLSYYIVR	Unique									
662	NYC												
748	NTT	740-754	RRDVMQVANTTMSSR	Unique									
		742-754	DVMQVANTTMSSR	Unique									
757	NTT	755-782	SRNTTVADTYNITDPEEFETEYPFFESR	Unique									
765	NIT	755-782	SRNTTVADTYNITDPEEFETEYPFFESR	Unique									
901	NYT	897-905	LNPGNYTAR	Unique									
914	NGS												
964	NNS												

*1: _:Potential site ■:Asn (glycosylated) ■:Gln (deaminated:pyroGlu) ■:Met (oxidized) ■:Cys (carbamidomethylated and deaminated)

Fig. 11.4 (continued)

is scrollable. By touching the ball on the ball and stick, the site position and the sequon are shown, e.g., 608–610: NAS. The identified peptide portion is shown on the sequence line in red with the sequon in green and is indicated as, e.g., 607–631 by pointing with the mouse.

Under the graphics, the whole sequences of the selected glycoprotein and glycopeptides identified are shown as in Fig. 11.4b. In the protein sequence, potential sites are underlined in red, and identified Asn is shown in orange text. In the table below the sequence, potential sites and their sequons are shown. The position and sequence of the identified glycopeptide are also shown for each potential site. If multiple peptides were identified for a single site, all sequences are provided with modifications detected if any. If the sequence is assigned to a specific protein, the peptide is marked as "Unique." Conversely, if multiple proteins were assigned from the sequence, the peptide is marked "Shared (n)", where "n" is the number of proteins identified from the sequence. The sample information is provided for each site on the right half of the table, which is scrollable.

The information of the actually glycosylated proteins, their glycosylated sites, binding capacity of the glycopeptide to certain lectin, and tissues originating the glycoprotein/peptide may be useful in various studies, e.g., for search of tissues expressing an interested protein, the presence of site-specific glycan motif, and prediction of membrane topology. However, it is important to note that "not identified" does not mean "not present," "not glycosylated," or "not having a specific glycan motif," especially in considering the tissue distribution, unglycosylated sites, or glycan motif of intended glycoprotein(s) based on the specificity of the lectin used for the capture of identified peptide, respectively. There are various reasons for why a glycopeptide was not identified, e.g., less abundant, lower ionization efficiency due to the length or negative charge, size of peptide (too long or short to identify by *MS/MS* measurement), or unexpected modifications in the database search. Data-dependent selection of precursor ions for *MS/MS* analyses often misses the detectable ions due to insufficient speed of the spectra acquisition, especially in highly complex mixtures. Thus, care must be taken when dealing with information that appears to be missing.

11.9 Links to Other Databases

The various information of the protein is collectable from other databases. To get them easily, links to other databases are provided (Fig. 11.5).

11.10 Download of Whole Data Resources

Whole data resources of this database will be downloadable; however, it is now under construction (March, 2016).

Quotation from External Databases

▼NCBI Protein

Gene Name	Accession	Organism	Length	Molecular Weight
insulin-like growth factor I receptor	NP_034643	Mus musculus	1369	155284.3

▼NCBI Gene

Gene ID	Official Symbol	Alias	Chromosome	Location	Description	External Links
16001	Igf1r	Igf1r, A330103N21Rik, CD221, D930020L01, IGF-1R, hyf	7	7 D1|7 33.0 cM	insulin-like growth factor I receptor	MGI:96433, Ensembl:ENSMUSG00000005533

▼External Links of Protein

ENSEMBL	GPIDB	IPI	JGPI	NCBI_GI	REFSEQ	SWISSPROT	TREMBL
ENSMUSP00000005671	GPMMU00000314	IPI01008146.1	314	112983656	NP_034643	E9QNX9 Q60751	E9QNX9

▼Related Proteins

Organism (ID)	Symbol(GENE_ID)	Accession(GI)
Caenorhabditis elegans (6239)	daf-2 (175410)	NP_497650.3 (71995160)
Anopheles gambiae (7165)	INR (1280455)	XP_320130.3 (118792063)
Drosophila melanogaster (7227)	InR (42549)	NP_001138094.1 (221458230)
Danio rerio (7955)	igf1ra (245701)	NP_694500.1 (23308649)
Danio rerio (7955)	igf1rb (245702)	NP_694501.1 (23308651)
Gallus gallus (9031)	IGF1R (395889)	NP_990363.1 (45384296)
Pan troglodytes (9598)	IGF1R (453676)	XP_001136377.1 (114659080)
Homo sapiens (9606)	IGF1R (3480)	NP_000866.1 (4557665)
Canis lupus familiaris (9615)	IGF1R (442951)	XP_858440.1 (73951091)
Bos taurus (9913)	IGF1R (281848)	XP_606794.3 (119913553)
Mus musculus (10090)	Igf1r (16001)	NP_034643.2 (112983656)
Rattus norvegicus (10116)	Igf1r (25718)	NP_434694.1 (16258823)

Fig. 11.5 Links to other databases for a selected glycoprotein

References

Hirao Y, Matsuzaki H, Iwaki J, Kuno A, Kaji H, Ohkura T, Togayachi A, Abe M, Nomura M, Noguchi M, Ikehara Y, Narimatsu H (2014) Glycoproteomics approach for identifying glycobiomarker candidate molecules for tissue type classification of non-small cell lung carcinoma. J Proteome Res 13(11):4705–4716

Kaji H, Isobe T (2013) Stable isotope labeling of N-glycosylated peptides by enzymatic deglycosylation for mass spectrometry-based glycoproteomics. Methods Mol Biol 951:217–227

Kaji H, Saito H, Yamauchi Y, Shinkawa T, Taoka M, Hirabayashi J, Kasai K, Takahashi N, Isobe T (2003) Lectin affinity capture, isotope-coded tagging and mass spectrometry to identify N-linked glycoproteins. Nat Biotechnol 21(6):667–672

Kaji H, Yamauchi Y, Takahashi N, Isobe T (2006) Mass spectrometric identification of N-linked glycopeptides using lectin-mediated affinity capture and glycosylation site-specific stable isotope tagging. Nat Protoc 1(6):3019–3027

Kaji H, Kamiie J, Kawakami H, Kido K, Yamauchi Y, Shinkawa T, Taoka M, Takahashi N, Isobe T (2007) Proteomics reveals N-linked glycoprotein diversity in Caenorhabditis elegans and suggests an atypical translocation mechanism for integral membrane proteins. Mol Cell Proteomics 6(12):2100–2109

Kaji H, Shikanai T, Sasaki-Sawa A, Wen H, Fujita M, Suzuki Y, Sugahara D, Sawaki H, Yamauchi Y, Shinkawa T, Taoka M, Takahashi N, Isobe T, Narimatsu H (2012) Large-scale identification of N-glycosylated proteins of mouse tissues and construction of a glycoprotein database, GlycoProtDB. J Proteome Res 11(9):4553–4566

Kaji H, Ocho M, Togayachi A, Kuno A, Sogabe M, Ohkura T, Nozaki H, Angata T, Chiba Y, Ozaki H, Hirabayashi J, Tanaka Y, Mizokami M, Ikehara Y, Narimatsu H (2013) Glycoproteomic discovery of serological biomarker candidates for HCV/HBV infection-associated liver fibrosis and hepatocellular carcinoma. J Proteome Res 12(6):2630–2640

Shinkawa T, Taoka M, Yamauchi Y, Ichimura T, Kaji H, Takahashi N, Isobe T (2005) STEM: a software tool for large-scale proteomic data analyses. J Proteome Res 4(5):1826–1831

Sogabe M, Nozaki H, Tanaka N, Kubota T, Kaji H, Kuno A, Togayachi A, Gotoh M, Nakanishi H, Nakanishi T, Mikami M, Suzuki N, Kiguchi K, Ikehara Y, Narimatsu H (2014) Novel glycobiomarker for ovarian cancer that detects clear cell carcinoma. J Proteome Res 13(3):1624–1635. 6

Sugahara D, Kaji H, Sugihara K, Asano M, Narimatsu H (2012) Large-scale identification of target proteins of a glycosyltransferase isozyme by Lectin-*IGOT-LC/MS*, an *LC/MS*-based glycoproteomic approach. Sci Rep 2:680

Sugahara D, Tomioka A, Sato T, Narimatsu H, Kaji H (2015) Large-scale identification of secretome glycoproteins recognized by *Wisteria floribunda* agglutinin: a glycoproteomic approach to biomarker discovery. Proteomics 5(17):2921–2933

Part V
Glycan Interactions

Chapter 12
GlycoEpitope

Shujiro Okuda, Hiromi Nakao, and Toshisuke Kawasaki

Abstract Glycan research is an important field in life science. Glycans are carbohydrate sugar chains, which are ubiquitously found in a wide variety of cell types and interact with various proteins, viruses, bacteria, and antibodies. Antibodies that recognize carbohydrates have been widely used for analyzing glycan structures and functions. GlycoEpitope has been developed as a database to integrate carbohydrate antigens and carbohydrate-recognizing antibodies. It has been developed by the cooperation of researchers in the field of glycobiology. GlycoEpitope provides information of not only epitopes and antibodies but also of proteins (glycoproteins) and lipids (glycolipids) that carry carbohydrate epitopes, on enzymes that take part in synthesis and degradation of carbohydrate epitopes and on a lot of related information. This database is useful not only for glycobiologists but also for a wide range of life scientists. Here, we would like to introduce a general outline of GlycoEpitope and how to use it.

Keywords Epitope • Antibody • Glycoprotein • Glycolipid • Enzyme

12.1 Introduction

Glycan research has been an area of focus for a wide range of life science research and biotechnology. They are often found on the cell surface, serving as "switches" in toggling various cellular functions (Hakomori 1984). To better understand glycan function, polyclonal and monoclonal antibodies that detect carbohydrate expression are extremely important (Feizi 1985; Sell 1990). GlycoEpitope is a novel database that integrates a variety of useful information on carbohydrate antigens and their

S. Okuda (✉)
Graduate School of Medical and Dental Sciences, Niigata University, 1-757 Asahimachi-dori, 951-8510 Chuo-ku, Niigata, Japan
e-mail: okd@med.niigata-u.ac.jp

H. Nakao • T. Kawasaki (✉)
Research Center for Glycobiotechnology, Ritsumeikan University, 1-1-1 Nojihigashi, 525-8577 Kusatsu, Shiga, Japan
e-mail: tkawasak@fc.ritsumei.ac.jp

K.F. Aoki-Kinoshita (ed.), *A Practical Guide to Using Glycomics Databases*,
DOI 10.1007/978-4-431-56454-6_12

antibodies (Kawasaki et al. 2006). It has been developed with the cooperation of glycobiologists. GlycoEpitope provides a wealth of information including glycoproteins that express carbohydrate epitopes, carbohydrate epitopes on glycolipids, enzymes that take part in the synthesis and degradation of epitopes, the times and sites of expression of carbohydrate epitopes, diseases to which carbohydrate epitopes are related, and suppliers from which carbohydrate-recognizing antibodies can be obtained. This database is useful not only for glycobiologists but also for a wide range of life scientists.

12.2 Features of GlycoEpitope

As of March 27, 2016, GlycoEpitope (see Fig. 12.1) provides 173 epitope and 614 antibody information, organized into five categories. GlycoEpitope has several features classifying the information about carbohydrate epitopes and antigens, described below.

1. The "General" category contains general information on epitopes including name, identifier, carbohydrate structure, species or tissues expressing the epitope, subcellular locations, expression changes during development, cell lines, recep-

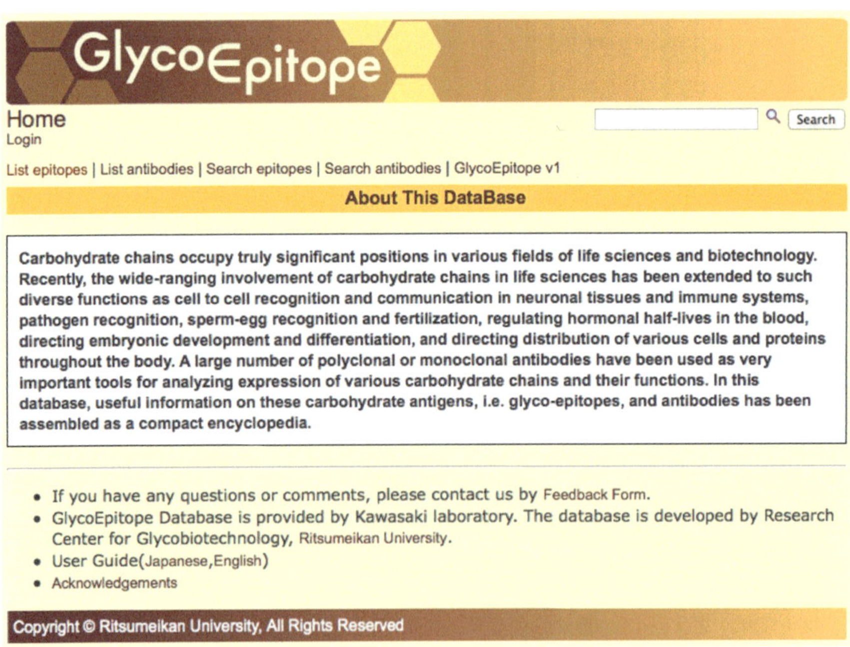

Fig. 12.1 Top page of GlycoEpitope

tor (binding proteins), basic functions, related disease, and examples of epitopes and antibodies which are actually used in applied fields.

2. The "Antibody" category contains information on the antibody recognizing an epitope. This includes the name of the antibody, the species expressing the antibody, isotypes, the recognition sequence, antibody type (monoclonal or polyclonal), methods used to detect the epitope and antigen (immunoprecipitation, immunoblot, and histochemistry), and availability information regarding how to obtain the antibody.
3. The "Glycoprotein" category contains information on the carrier protein of the epitope, such as whether it is found as an *N*-glycan or *O*-glycan, the glycosylation sites, epitope attachment sites, and other gene-sequence related information such as amino acid sequences and 3D structures.
4. The "Glycolipid" category consists of information on the lipid, on which the epitope is found. This includes the name, aliases, classification of the glycolipid, the carbohydrate sequence, molecular weight, and the external database links related to the glycolipid.
5. The "Enzyme" category provides such information as the enzymes involved in the synthesis and degradation of the epitope, name, catalytic reaction, description, EC number, external database links, and the information regarding availability of gene deficient organisms.

12.3 Web Interface

GlycoEpitope provides a web interface to display epitope and antibody entries. Users can sort the entries by epitope or antibody identifiers, names, and related information to easily access the entry of their interest. An example of the epitope entry for HNK-1 is illustrated in Fig. 12.2. As shown in this figure, each entry page describes the detailed information described above, separated by tabs for each category. In these detailed pages, GlycoEpitope provides several web links to go to the original information page (Fig. 12.3). On the General page, GlycoEpitope provides a link to JCGGDB (http://jcggdb.jp/) when the epitope structure matches a glycan structure stored in JCGGDB, which is the database collection of carbohydrate-related information including glycan structure, glycoprotein, experimental protocols, etc. Therefore, from the epitope linked to JCGGDB, users can obtain more information about the epitope and antibody. The JCGGDB

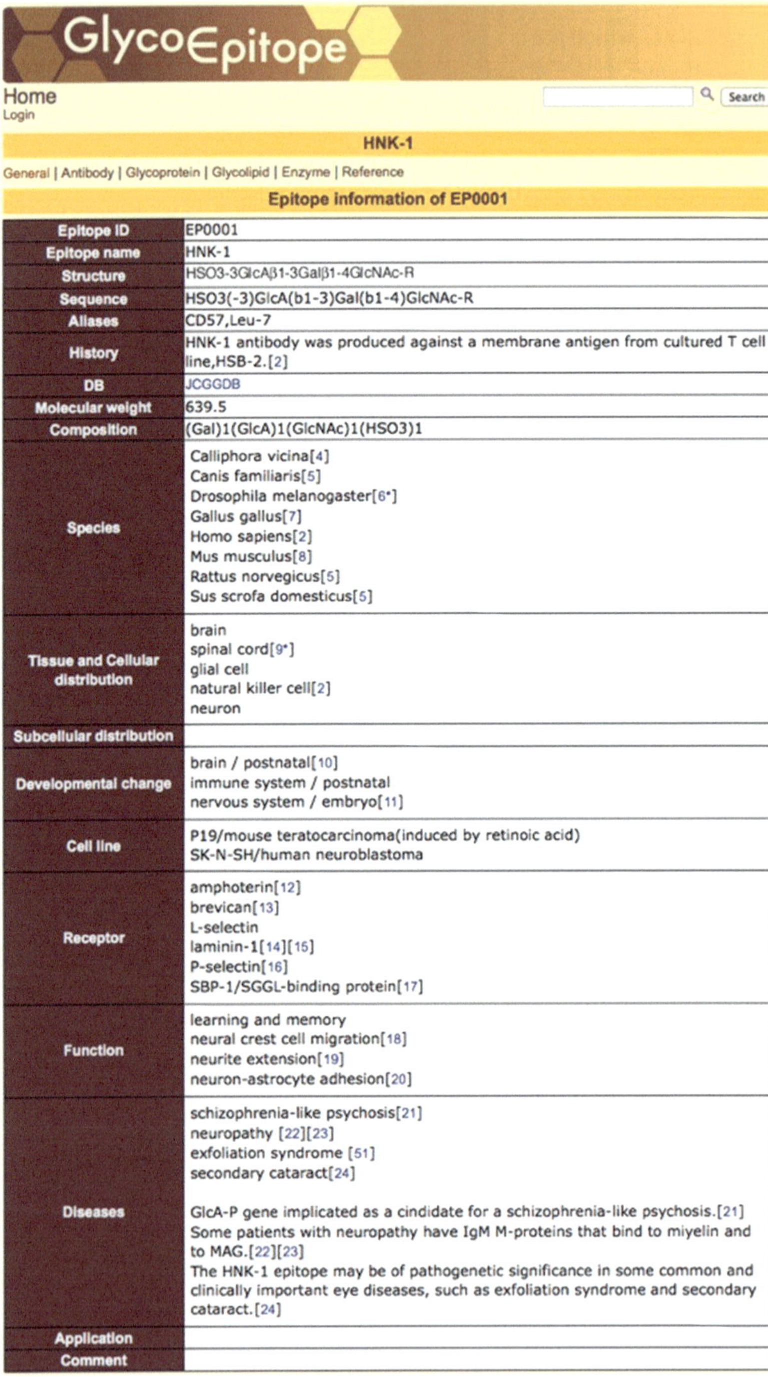

GlycoEpitope

Home
Login

Search

HNK-1

General | Antibody | Glycoprotein | Glycolipid | Enzyme | Reference

Epitope information of EP0001

Epitope ID	EP0001
Epitope name	HNK-1
Structure	HSO3-3GlcAβ1-3Galβ1-4GlcNAc-R
Sequence	HSO3(-3)GlcA(b1-3)Gal(b1-4)GlcNAc-R
Aliases	CD57,Leu-7
History	HNK-1 antibody was produced against a membrane antigen from cultured T cell line,HSB-2.[2]
DB	JCGGDB
Molecular weight	639.5
Composition	(Gal)1(GlcA)1(GlcNAc)1(HSO3)1
Species	Calliphora vicina[4] Canis familiaris[5] Drosophila melanogaster[6*] Gallus gallus[7] Homo sapiens[2] Mus musculus[8] Rattus norvegicus[5] Sus scrofa domesticus[5]
Tissue and Cellular distribution	brain spinal cord[9*] glial cell natural killer cell[2] neuron
Subcellular distribution	
Developmental change	brain / postnatal[10] immune system / postnatal nervous system / embryo[11]
Cell line	P19/mouse teratocarcinoma(induced by retinoic acid) SK-N-SH/human neuroblastoma
Receptor	amphoterin[12] brevican[13] L-selectin laminin-1[14][15] P-selectin[16] SBP-1/SGGL-binding protein[17]
Function	learning and memory neural crest cell migration[18] neurite extension[19] neuron-astrocyte adhesion[20]
Diseases	schizophrenia-like psychosis[21] neuropathy [22][23] exfoliation syndrome [51] secondary cataract[24] GlcA-P gene implicated as a cindidate for a schizophrenia-like psychosis.[21] Some patients with neuropathy have IgM M-proteins that bind to miyelin and to MAG.[22][23] The HNK-1 epitope may be of pathogenetic significance in some common and clinically important eye diseases, such as exfoliation syndrome and secondary cataract.[24]
Application	
Comment	

Fig. 12.2 The epitope entry page for HNK-1

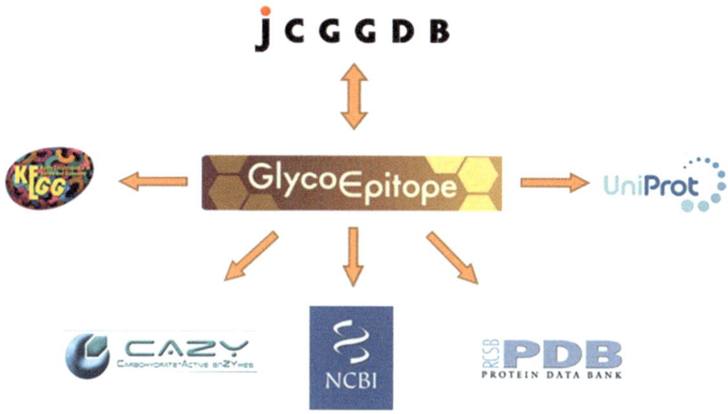

Fig. 12.3 Website links in GlycoEpitope

link is shown in the "DB" row not only on the General page but also on the Glycolipid, Glycoprotein, and Enzyme pages. On each page, other links to related databases are also provided: a link to PDB (Rose et al. 2013) (http://www.rcsb.org/) provides access to 3D protein structures, NCBI (Benson et al. 2014) (http://www.ncbi.nlm.nih.gov/) to gene information, SwissProt/TrEMBL (Consortium 2014) (http://www.uniprot.org/) to protein information, CAZy (Lombard et al. 2014) to carbohydrate-active enzyme information, and KEGG (Kanehisa et al. 2014) (http://www.genome.jp) to EC numbers for the corresponding enzyme, glycoprotein, or biological pathway map information. Some of these links are shown in a comma-delimited format. The left side of the comma indicates the database name, the right side is the identifiers, and the link is created based on these information to lead to the proper page. If the identifier is not provided, the link is created by the content of the corresponding name.

GlycoEpitope also provides a lot of literature information. References to the corresponding scientific paper are appended as a citation. The citation is shown as a number in the order in which it is cited. The number is linked to the abstract of the paper in the PubMed database (Fig. 12.4). Thus, users can easily access the related paper of the epitope and antibody of their interest. In addition, papers published in the *Journal of Biological Chemistry* (JBC) are directly linked to the PDF file, so users can also obtain the paper itself (kindly approved by American Society for Biochemistry and Molecular Biology, ASBMB). The reference information is summarized on the Reference page. Users can sort the references by published year, so as to obtain access to the latest information.

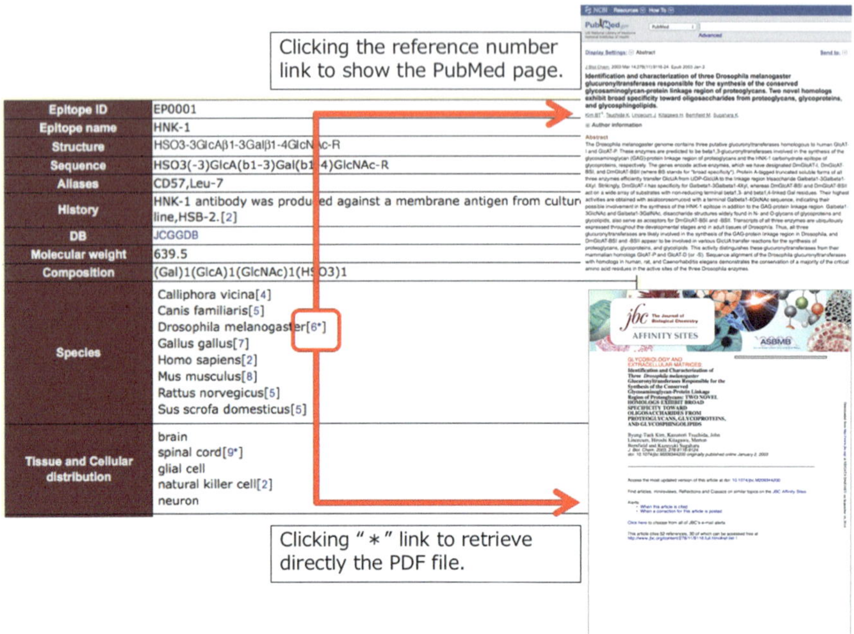

Fig. 12.4 Link to the abstract of a paper in the PubMed database

12.4 Search Tool

GlycoEpitope provides some keyword search tools. One is the epitope search tool (Fig. 12.5), which allows users to search epitope entries by a combination of keywords including identifiers, names, sequences, molecular weight, carbohydrate composition, species, tissue specificity, receptor proteins, enzymes, diseases, and references. In addition, users can directly search a keyword from all the fields. The second search function is the antibody search tool to retrieve entries by information related to antibodies (Fig. 12.6), in a similar fashion to the epitope search tool. The Antibody search tool is based on a combination of keywords including antibody identifiers and names. GlycoEpitope also provides a keyword search for the entire database from the top page. Using this keyword search tool, users can retrieve both epitope and antibody entries at the same time. Figure 12.7 is an example of using "HNK-1" as a keyword query. First, users enter the keyword into the search box located at the top right of the GlycoEpitope website. Then they can obtain the list of the matching keyword hits as a table, in which the query word is highlighted in yellow. The table includes links to the epitopes and antibodies related to the keyword. This highlight function continues to the next web page. Note that because an epitope can be related to several different antibodies, the same epitope ID may appear multiple times in the results. In this table view, users can sort the list of the search results by each column, so they can easily reach to the entry of the interest.

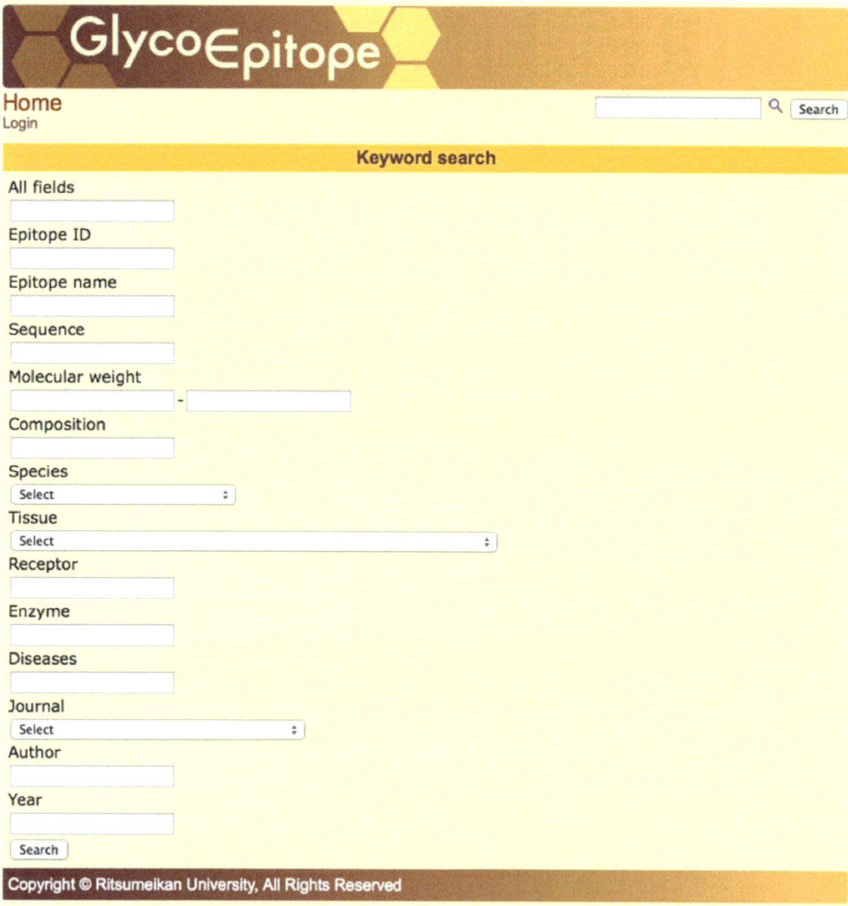

Fig. 12.5 Epitope search interface

By clicking the link of the epitope, users can get the detailed information of the epitope, and by clicking the antibody link, they can directly obtain the information about the antibody.

12.5 How to Operate the GlycoEpitope Database

In this database, records are cross-linked, i.e., each epitope record is linked to records of its corresponding antibodies. Since these data are based on published literature, there are also links to some external database and online journals, if available, so you can look into original manuscripts. You can find epitopes and antibodies you are looking for by their names or by using search functions in this database.

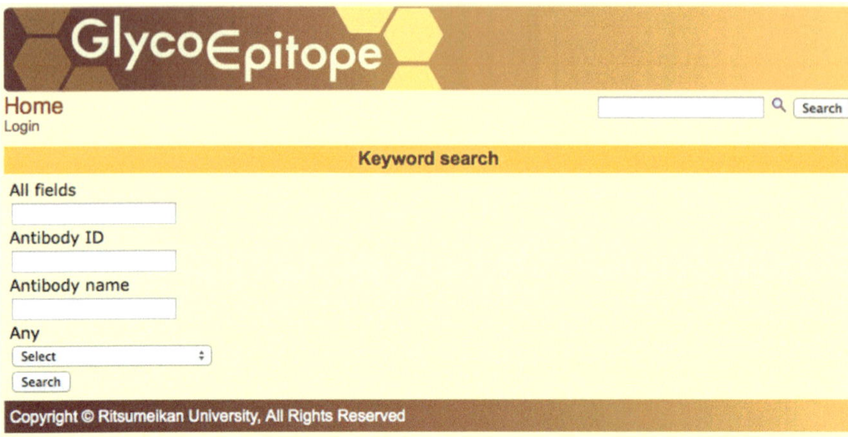

Fig. 12.6 Antibody search interface

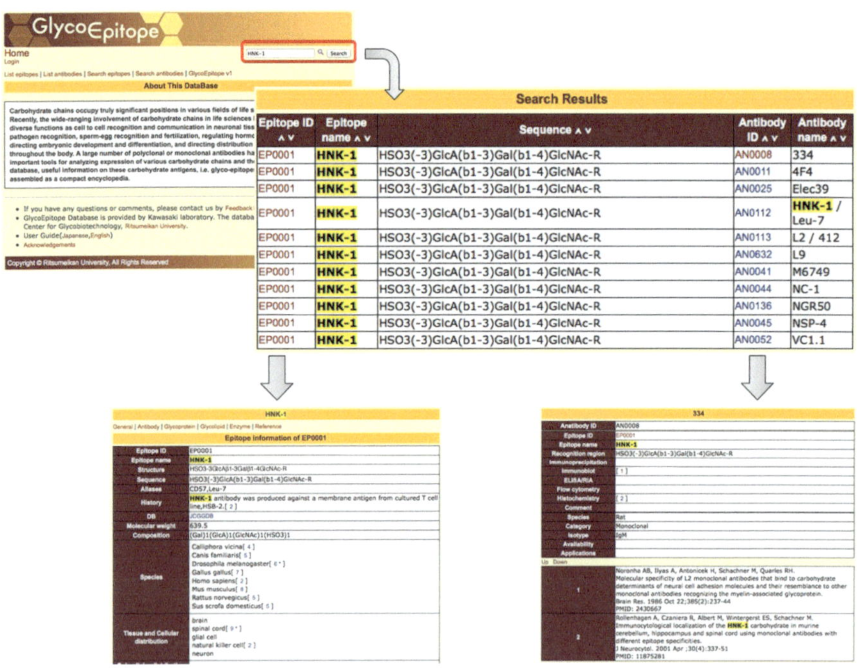

Fig. 12.7 A search example of using “HNK-1” as a keyword query

12.5.1 To Find an Epitope Record from "List Epitopes"

First, you will see the list of epitopes by clicking the "List epitopes" button at the top page (Fig. 12.8). Epitopes are listed in alphabetical order except those names that start with numbers that are listed at the beginning of the list. You can change the order by choosing "Epitope ID" in the box at the top left of the list and click "sort" and then the list will be arranged in the order of Epitope ID. Each epitope ID links to its General page, which consists of the following information wherever possible: epitope structure, sequence, aliases, history, 3D structure, molecular weight, composition, species, tissue and cellular distribution, subcellular distribution, developmental change, cell line, receptor, function, diseases, and application. You can also look at related information listed under each tab as follows: "Antibody" "Glycoprotein" "Glycolipid" "Enzyme" and "References"

12.5.2 To Find an Antibody Record from "List Antibodies"

All antibodies that are already released in this database are listed in "List antibodies" in alphabetical order except those names that start with numbers which are listed at the beginning of the list (Fig. 12.9). You can change the order by choosing "Antibody ID" in the box at the top left of the list and click "sort" and then the list will be arranged in order of Antibody ID. An antibody record consists of information about its recognizing epitopes as well as information such as immunized animals/systems (species), its category (polyclonal, monoclonal, or scFv), its isotype, and its supplier and application.

12.5.3 Using Search Functions

12.5.3.1 Epitope Search Rules

You can conduct a search by selecting an item from the pull-down menu or by entering keywords in any column (see Fig. 12.5). The keyword search is case insensitive and supports partial matching (except for Epitope ID, whose search is exact matching). No logic or Boolean operators (e.g., AND, OR, NOT, etc.) can be used. If several keywords are entered, entries containing all the keywords will be listed (i.e., "AND" search). Keyword(s) entered in "All fields" will be highlighted in the search results.

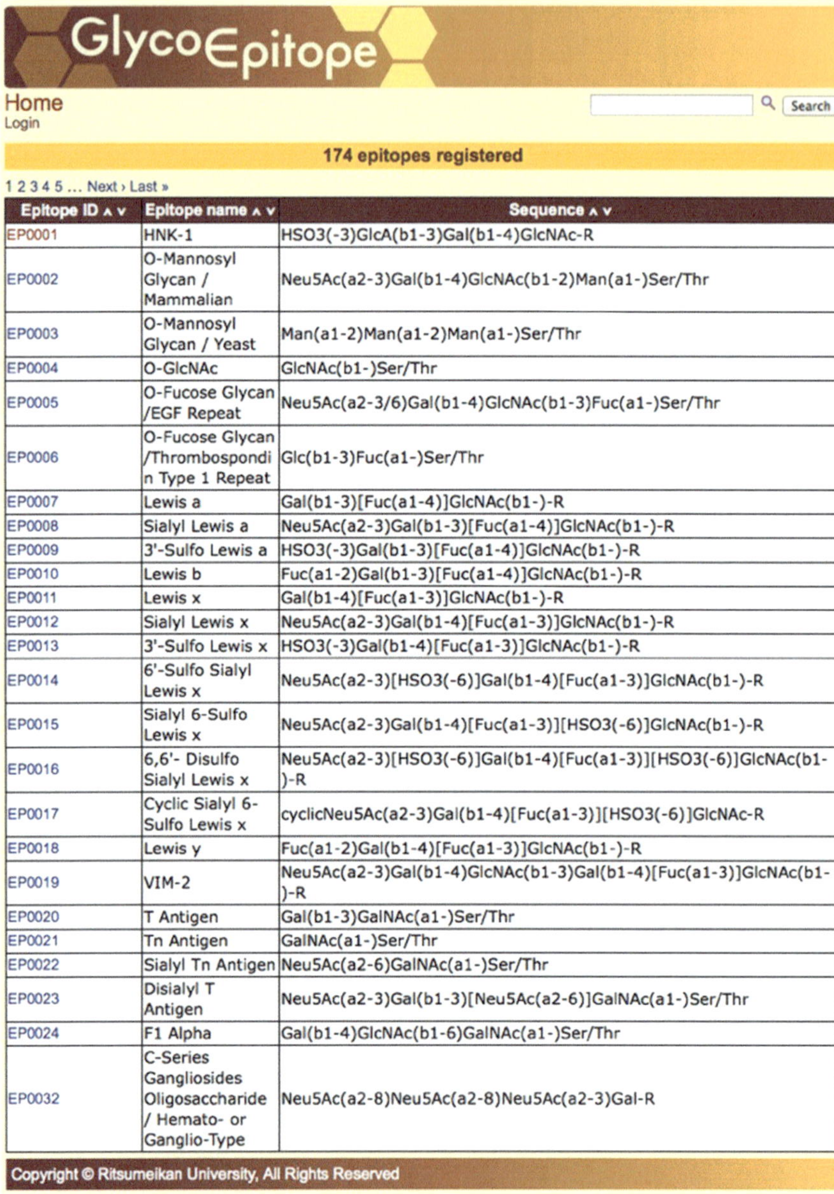

GlycoEpitope

Home

Login

Search

174 epitopes registered

1 2 3 4 5 … Next › Last »

Epitope ID	Epitope name	Sequence
EP0001	HNK-1	HSO3(-3)GlcA(b1-3)Gal(b1-4)GlcNAc-R
EP0002	O-Mannosyl Glycan / Mammalian	Neu5Ac(a2-3)Gal(b1-4)GlcNAc(b1-2)Man(a1-)Ser/Thr
EP0003	O-Mannosyl Glycan / Yeast	Man(a1-2)Man(a1-2)Man(a1-)Ser/Thr
EP0004	O-GlcNAc	GlcNAc(b1-)Ser/Thr
EP0005	O-Fucose Glycan /EGF Repeat	Neu5Ac(a2-3/6)Gal(b1-4)GlcNAc(b1-3)Fuc(a1-)Ser/Thr
EP0006	O-Fucose Glycan /Thrombospondin Type 1 Repeat	Glc(b1-3)Fuc(a1-)Ser/Thr
EP0007	Lewis a	Gal(b1-3)[Fuc(a1-4)]GlcNAc(b1-)-R
EP0008	Sialyl Lewis a	Neu5Ac(a2-3)Gal(b1-3)[Fuc(a1-4)]GlcNAc(b1-)-R
EP0009	3'-Sulfo Lewis a	HSO3(-3)Gal(b1-3)[Fuc(a1-4)]GlcNAc(b1-)-R
EP0010	Lewis b	Fuc(a1-2)Gal(b1-3)[Fuc(a1-4)]GlcNAc(b1-)-R
EP0011	Lewis x	Gal(b1-4)[Fuc(a1-3)]GlcNAc(b1-)-R
EP0012	Sialyl Lewis x	Neu5Ac(a2-3)Gal(b1-4)[Fuc(a1-3)]GlcNAc(b1-)-R
EP0013	3'-Sulfo Lewis x	HSO3(-3)Gal(b1-4)[Fuc(a1-3)]GlcNAc(b1-)-R
EP0014	6'-Sulfo Sialyl Lewis x	Neu5Ac(a2-3)[HSO3(-6)]Gal(b1-4)[Fuc(a1-3)]GlcNAc(b1-)-R
EP0015	Sialyl 6-Sulfo Lewis x	Neu5Ac(a2-3)Gal(b1-4)[Fuc(a1-3)][HSO3(-6)]GlcNAc(b1-)-R
EP0016	6,6'- Disulfo Sialyl Lewis x	Neu5Ac(a2-3)[HSO3(-6)]Gal(b1-4)[Fuc(a1-3)][HSO3(-6)]GlcNAc(b1-)-R
EP0017	Cyclic Sialyl 6-Sulfo Lewis x	cyclicNeu5Ac(a2-3)Gal(b1-4)[Fuc(a1-3)][HSO3(-6)]GlcNAc-R
EP0018	Lewis y	Fuc(a1-2)Gal(b1-4)[Fuc(a1-3)]GlcNAc(b1-)-R
EP0019	VIM-2	Neu5Ac(a2-3)Gal(b1-4)GlcNAc(b1-3)Gal(b1-4)[Fuc(a1-3)]GlcNAc(b1-)-R
EP0020	T Antigen	Gal(b1-3)GalNAc(a1-)Ser/Thr
EP0021	Tn Antigen	GalNAc(a1-)Ser/Thr
EP0022	Sialyl Tn Antigen	Neu5Ac(a2-6)GalNAc(a1-)Ser/Thr
EP0023	Disialyl T Antigen	Neu5Ac(a2-3)Gal(b1-3)[Neu5Ac(a2-6)]GalNAc(a1-)Ser/Thr
EP0024	F1 Alpha	Gal(b1-4)GlcNAc(b1-6)GalNAc(a1-)Ser/Thr
EP0032	C-Series Gangliosides Oligosaccharide / Hemato- or Ganglio-Type	Neu5Ac(a2-8)Neu5Ac(a2-8)Neu5Ac(a2-3)Gal-R

Copyright © Ritsumeikan University, All Rights Reserved

Fig. 12.8 The interface for the whole list of epitopes

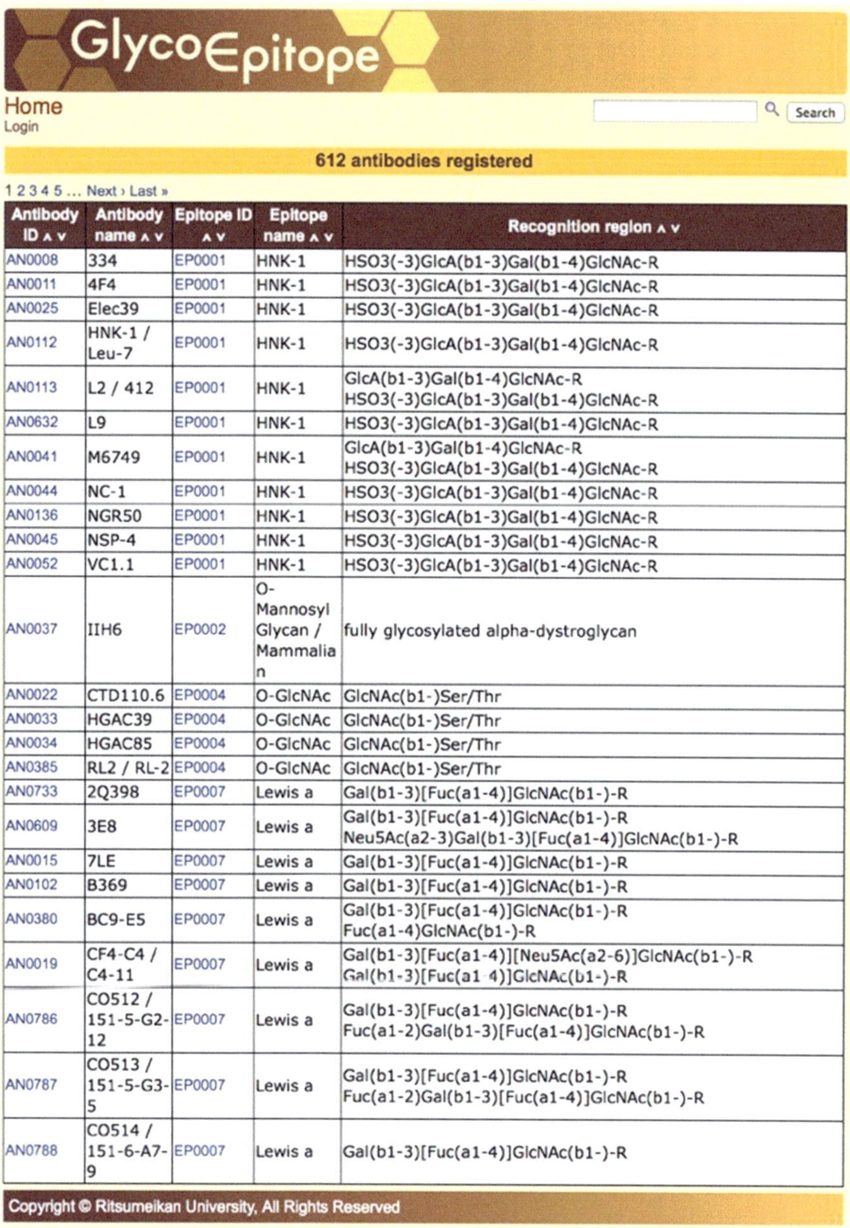

GlycoEpitope

Home

Login

Search

612 antibodies registered

1 2 3 4 5 … Next › Last »

Antibody ID ∧ ∨	Antibody name ∧ ∨	Epitope ID ∧ ∨	Epitope name ∧ ∨	Recognition region ∧ ∨
AN0008	334	EP0001	HNK-1	HSO3(-3)GlcA(b1-3)Gal(b1-4)GlcNAc-R
AN0011	4F4	EP0001	HNK-1	HSO3(-3)GlcA(b1-3)Gal(b1-4)GlcNAc-R
AN0025	Elec39	EP0001	HNK-1	HSO3(-3)GlcA(b1-3)Gal(b1-4)GlcNAc-R
AN0112	HNK-1 / Leu-7	EP0001	HNK-1	HSO3(-3)GlcA(b1-3)Gal(b1-4)GlcNAc-R
AN0113	L2 / 412	EP0001	HNK-1	GlcA(b1-3)Gal(b1-4)GlcNAc-R HSO3(-3)GlcA(b1-3)Gal(b1-4)GlcNAc-R
AN0632	L9	EP0001	HNK-1	HSO3(-3)GlcA(b1-3)Gal(b1-4)GlcNAc-R
AN0041	M6749	EP0001	HNK-1	GlcA(b1-3)Gal(b1-4)GlcNAc-R HSO3(-3)GlcA(b1-3)Gal(b1-4)GlcNAc-R
AN0044	NC-1	EP0001	HNK-1	HSO3(-3)GlcA(b1-3)Gal(b1-4)GlcNAc-R
AN0136	NGR50	EP0001	HNK-1	HSO3(-3)GlcA(b1-3)Gal(b1-4)GlcNAc-R
AN0045	NSP-4	EP0001	HNK-1	HSO3(-3)GlcA(b1-3)Gal(b1-4)GlcNAc-R
AN0052	VC1.1	EP0001	HNK-1	HSO3(-3)GlcA(b1-3)Gal(b1-4)GlcNAc-R
AN0037	IIH6	EP0002	O-Mannosyl Glycan / Mammalian	fully glycosylated alpha-dystroglycan
AN0022	CTD110.6	EP0004	O-GlcNAc	GlcNAc(b1-)Ser/Thr
AN0033	HGAC39	EP0004	O-GlcNAc	GlcNAc(b1-)Ser/Thr
AN0034	HGAC85	EP0004	O-GlcNAc	GlcNAc(b1-)Ser/Thr
AN0385	RL2 / RL-2	EP0004	O-GlcNAc	GlcNAc(b1-)Ser/Thr
AN0733	2Q398	EP0007	Lewis a	Gal(b1-3)[Fuc(a1-4)]GlcNAc(b1-)-R
AN0609	3E8	EP0007	Lewis a	Gal(b1-3)[Fuc(a1-4)]GlcNAc(b1-)-R Neu5Ac(a2-3)Gal(b1-3)[Fuc(a1-4)]GlcNAc(b1-)-R
AN0015	7LE	EP0007	Lewis a	Gal(b1-3)[Fuc(a1-4)]GlcNAc(b1-)-R
AN0102	B369	EP0007	Lewis a	Gal(b1-3)[Fuc(a1-4)]GlcNAc(b1-)-R
AN0380	BC9-E5	EP0007	Lewis a	Gal(b1-3)[Fuc(a1-4)]GlcNAc(b1-)-R Fuc(a1-4)GlcNAc(b1-)-R
AN0019	CF4-C4 / C4-11	EP0007	Lewis a	Gal(b1-3)[Fuc(a1-4)][Neu5Ac(a2-6)]GlcNAc(b1-)-R Gal(b1-3)[Fuc(a1-4)]GlcNAc(b1-)-R
AN0786	CO512 / 151-5-G2-12	EP0007	Lewis a	Gal(b1-3)[Fuc(a1-4)]GlcNAc(b1-)-R Fuc(a1-2)Gal(b1-3)[Fuc(a1-4)]GlcNAc(b1-)-R
AN0787	CO513 / 151-5-G3-5	EP0007	Lewis a	Gal(b1-3)[Fuc(a1-4)]GlcNAc(b1-)-R Fuc(a1-2)Gal(b1-3)[Fuc(a1-4)]GlcNAc(b1-)-R
AN0788	CO514 / 151-6-A7-9	EP0007	Lewis a	Gal(b1-3)[Fuc(a1-4)]GlcNAc(b1-)-R

Copyright © Ritsumeikan University, All Rights Reserved

Fig. 12.9 The interface for the whole list of antibodies

Rules Specific to Individual Fields

Epitope ID Epitope ID is the database-specific ID, and an entry for this search has to be an exact match.

Composition You can put the name of monosaccharide residues in parentheses and specify the number of monosaccharide residues after the parentheses.

Examples:

More than one GlcA: *(GlcA)*

One GlcA: *(GlcA)1*

Two GlcA and one Gal: *(GlcA)2 (Gal)1*

Molecular Weight You can set lower and/or upper limits for the molecular weight.

Examples:

Lower limit in left box only: 1234.5

Upper limit in right box only: 2345.6

Both lower and upper limits in both boxes: 1234.5–2345.6

Tissue Select the name of the tissue or a keyword from the pull-down menu.

Author Enter authors' last names.

Year You can enter the publication year of an article (e.g., 2004). Only one keyword (year) can be specified in this field. Search by publication year is available only in the Reference section.

12.5.3.2 Antibody Search Rules

The keyword search is case insensitive and uses partial matching, except for Antibody ID, whose search uses exact matching (see Fig. 12.6). No logic or Boolean operators (e.g., AND, OR, NOT, etc.) can be used. If several keywords are entered, only entries containing all the keywords will be listed (i.e., "AND" search). Keyword(s) entered in "All fields" will be highlighted in the search results.

Rules Specific to Individual Fields

Antibody ID Antibody ID is database-specific ID and the search for this field is exact matching.

Antibody Name You can input an antibody name or select an antibody from the pull-down menu. When the antibody search result is shown, click an Antibody ID, and detailed antibody record will be displayed. Clicking Epitope IDs also shows epitope records. There are hyperlinks for external database if available.

12.5.3.3 Example of How to Search

GlycoEpitope provides several search methods. Users can search the information of their interest according to their purposes. In this section, a guide of how to search is shown. Although all examples described below are about searches related to HNK-1, there are many methods to approach the information of HNK-1.

To Find Antibodies Recognizing Specific Epitope

When You Know the Epitope Name

When you go to "Search epitopes" enter an epitope name in the "Epitope name" column and press the "Search" button. The search result will be shown, and you can click the Epitope ID of the epitope that you are looking for. The general page of the epitope record will be shown first. Choose the "Antibody" tab on the top, and information of corresponding antibodies will be shown. If no entry was found, try to enter the epitope name in "All fields" and conduct the search again since there are some aliases for some epitopes.

Example

To obtain antibody information recognizing epitope CD57 (Fig. 12.10):

1. Input "CD57" in the "Epitope name" column in the epitope search page.
2. Start searching by clicking the "Search" button.
3. No entry is found.
4. Input "CD57" in the "All fields" column and start searching again by clicking the "Search" button.
5. The search result is shown.
6. When the Epitope ID is clicked, the General page about the epitope record is shown. You can find from this page that CD57 is an alias name of HNK-1.
7. When the Antibody tag in the General page is clicked, you can obtain several sources of information on antibodies recognizing CD57.

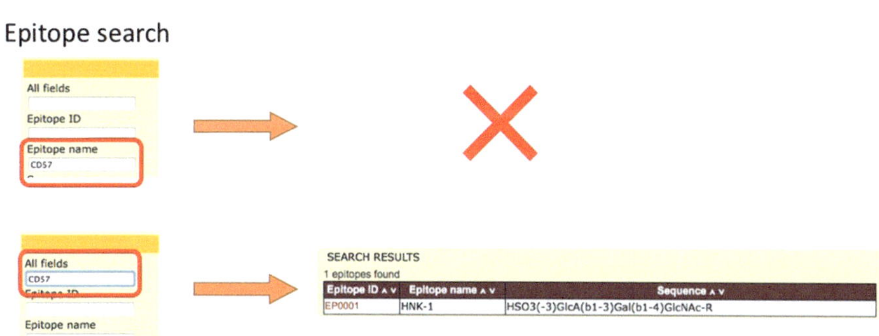

Fig. 12.10 An example of how to obtain information about antibodies that recognize epitope CD57

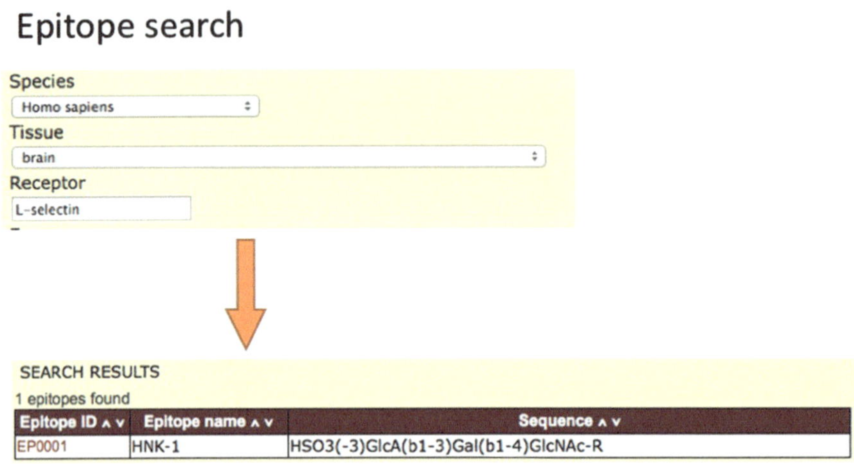

Fig. 12.11 An example of how to find the antibodies recognizing the epitope that is expressed in human brain and whose receptor is L-selectin

When You Do Not Know the Epitope Name

When you go to "Search epitopes" you can enter known information in any of these columns: "Composition" "Molecular Weight" "Species" "Tissue" "Receptor" "Diseases" "Enzyme" and/or "Reference"

Example

To find the antibodies recognizing the epitope that is expressed in human brain and whose receptor is L-selectin (Fig. 12.11):

1. Input the following keywords in the epitope search page.
    I. Choose "Homo sapiens" in the Species column.
    II. Choose "brain" in the Tissue column.
    III. Enter "L-selectin" in the Receptor column.
2. Start searching by clicking the "Search" button.
3. The search results are shown. You can see that the epitope satisfying the search condition is HNK-1.
4. When Epitope ID is clicked, the General page about the epitope record is shown.
5. Clicking the antibody tab on the General page allows you to get the information about HNK-1.

To Find Epitopes Recognized by a Specific Antibody

You can find this information from both "Search epitopes" and "Search antibodies" Enter the antibody name in "Antibody name" or choose name from the pull-down menu in "Search antibodies" Click Epitope ID, and the General page of the epitope will be shown.

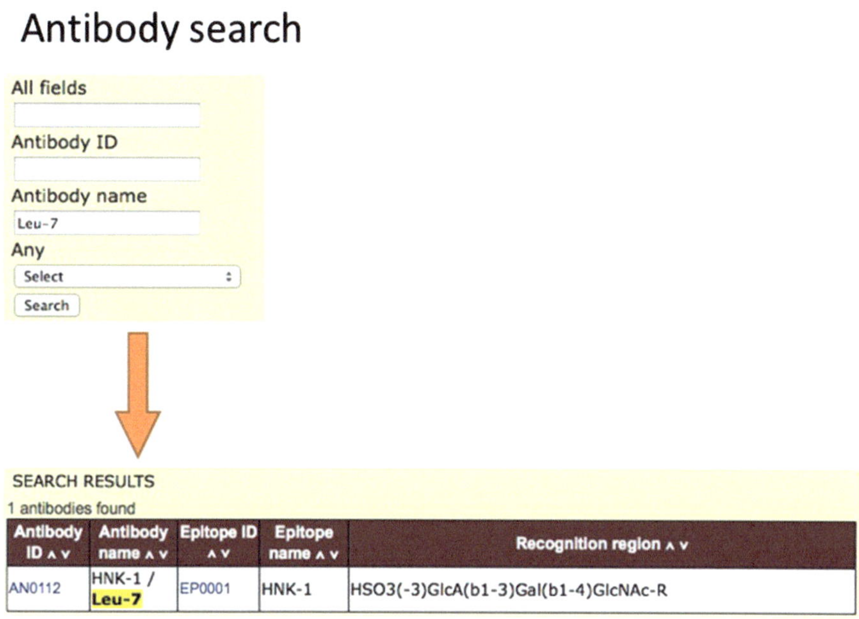

Fig. 12.12 An example of how to find epitopes recognized by a specific antibody

Example
Search from antibody (Fig. 12.12):

1. Input "Leu-7" in the "Antibody name" column in the antibody search page, or choose "Leu-7" from the pull-down menu of antibody name.
2. Start searching by clicking the "Search" button.
3. The search results are shown. You can see that the epitope recognized by "Leu-7" is HNK-1.
4. Clicking the Epitope ID provides the General page of the epitope.
5. You can get the information about HNK-1 from the General page.

You can also perform the same search from "Search epitopes" Input the antibody name, Leu-7 into the "All fields" column in the epitope search page. You can get the search results showing that Leu-7 is an alias name of HNK-1. Clicking the Epitope ID provides you the information about HNK-1 from the General page.

To Find Information About the Function of a Carbohydrate Whose Molecular Weight or Composition Is Known

Go to "Search epitopes" and enter the known information in the "Molecular weight" column or "Composition" column.

Fig. 12.13 An example of how to find information about the function of a carbohydrate whose molecular weight or composition is known

The search results will be shown. Clicking an Epitope ID, you can find information about the epitope's function in the General page.

Example

1. Input the following keywords in the epitope search page (Fig. 12.13).
   I. Enter 500–1000 in the molecular weight column.
   II. Enter (GlcA)1(GlcNAc)1(Gal)1 in the Composition column.
2. Start searching by clicking the "Search" button.
3. The search results are shown. You can see that the epitope satisfying the above conditions is HNK-1.
4. You can obtain the information about HNK-1 function by clicking the Epitope ID.

To Find Epitopes That Are Specifically Expressed in a Particular Tissue

Go to "Search epitopes" and choose the tissue name from the pull-down menu. The search result will be shown. If you cannot find the proper tissue name, try to enter the tissue name in "All fields" and conduct the search.

Example

Epitope specifically expressed in cancer tissue (Fig. 12.14):

1. Choose "cancer/colon" in the "Tissue" column in the epitope search page.
2. Start searching by clicking the "Search" button.
3. The search results are shown.
4. You can obtain the information about the epitope specifically expressed in colon cancer.

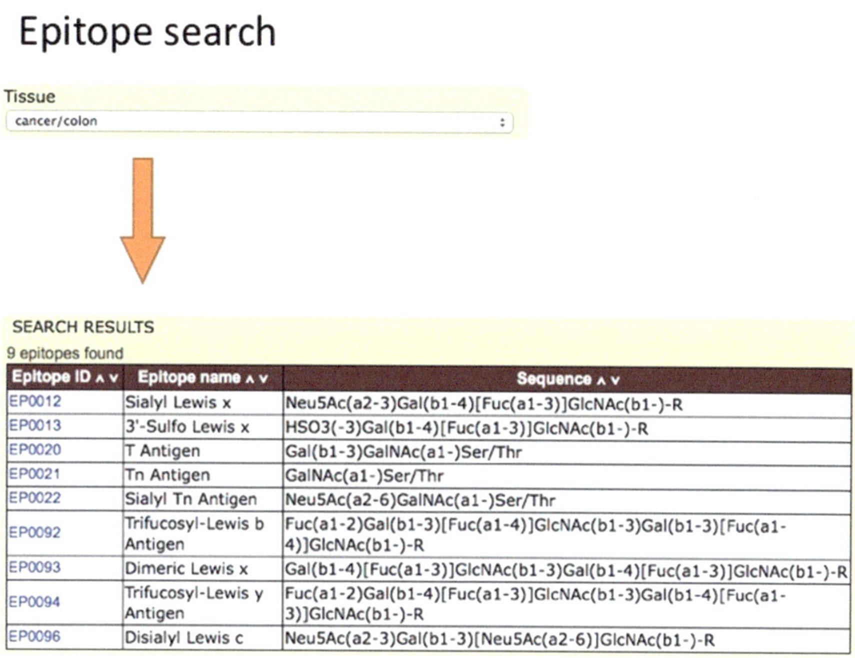

Epitope ID	Epitope name	Sequence
EP0012	Sialyl Lewis x	Neu5Ac(a2-3)Gal(b1-4)[Fuc(a1-3)]GlcNAc(b1-)-R
EP0013	3'-Sulfo Lewis x	HSO3(-3)Gal(b1-4)[Fuc(a1-3)]GlcNAc(b1-)-R
EP0020	T Antigen	Gal(b1-3)GalNAc(a1-)Ser/Thr
EP0021	Tn Antigen	GalNAc(a1-)Ser/Thr
EP0022	Sialyl Tn Antigen	Neu5Ac(a2-6)GalNAc(a1-)Ser/Thr
EP0092	Trifucosyl-Lewis b Antigen	Fuc(a1-2)Gal(b1-3)[Fuc(a1-4)]GlcNAc(b1-3)Gal(b1-3)[Fuc(a1-4)]GlcNAc(b1-)-R
EP0093	Dimeric Lewis x	Gal(b1-4)[Fuc(a1-3)]GlcNAc(b1-3)Gal(b1-4)[Fuc(a1-3)]GlcNAc(b1-)-R
EP0094	Trifucosyl-Lewis y Antigen	Fuc(a1-2)Gal(b1-4)[Fuc(a1-3)]GlcNAc(b1-3)Gal(b1-4)[Fuc(a1-3)]GlcNAc(b1-)-R
EP0096	Disialyl Lewis c	Neu5Ac(a2-3)Gal(b1-3)[Neu5Ac(a2-6)]GlcNAc(b1-)-R

Fig. 12.14 An example of how to find epitopes that are specifically expressed in a cancer tissue

You can search the keyword "cancer" from "All fields" In this case, the result page includes all entries related to the keyword "cancer"

To Find Epitopes Whose Structure Is Similar to a Certain Epitope

Go to "Search epitopes" enter the epitope structure, and conduct the search. The search result will be shown. Clicking an Epitope ID, you can find the epitope's General page and all related information listed under each tab on the top of the General page.

Example
Carbohydrate chains similar to "Lewis a" structure (Fig. 12.15):

1. Input the keyword "(Gal)1(GlcNAc)1(Fuc)1" in the "Composition" column in the epitope search page.
2. Start searching by clicking the "Search" button.
3. The search results are shown.
4. You can obtain the list of epitopes possessing the similar structure to "Lewis a" within their epitope structures.

Epitope search

Composition

(Gal)1(GlcNAc)1(Fuc)1

SEARCH RESULTS

18 epitopes found

Epitope ID ∧ ∨	Epitope name ∧ ∨	Sequence ∧ ∨
EP0005	O-Fucose Glycan /EGF Repeat	Neu5Ac(a2-3/6)Gal(b1-4)GlcNAc(b1-3)Fuc(a1-)Ser/Thr
EP0007	Lewis a	Gal(b1-3)[Fuc(a1-4)]GlcNAc(b1-)-R
EP0008	Sialyl Lewis a	Neu5Ac(a2-3)Gal(b1-3)[Fuc(a1-4)]GlcNAc(b1-)-R
EP0009	3'-Sulfo Lewis a	HSO3(-3)Gal(b1-3)[Fuc(a1-4)]GlcNAc(b1-)-R
EP0011	Lewis x	Gal(b1-4)[Fuc(a1-3)]GlcNAc(b1-)-R
EP0013	3'-Sulfo Lewis x	HSO3(-3)Gal(b1-4)[Fuc(a1-3)]GlcNAc(b1-)-R
EP0014	6'-Sulfo Sialyl Lewis x	Neu5Ac(a2-3)[HSO3(-6)]Gal(b1-4)[Fuc(a1-3)]GlcNAc(b1-)-R
EP0015	Sialyl 6-Sulfo Lewis x	Neu5Ac(a2-3)Gal(b1-4)[Fuc(a1-3)][HSO3(-6)]GlcNAc(b1-)-R
EP0016	6,6'- Disulfo Sialyl Lewis x	Neu5Ac(a2-3)[HSO3(-6)]Gal(b1-4)[Fuc(a1-3)][HSO3(-6)]GlcNAc(b1-)-R
EP0017	Cyclic Sialyl 6-Sulfo Lewis x	cyclicNeu5Ac(a2-3)Gal(b1-4)[Fuc(a1-3)][HSO3(-6)]GlcNAc-R
EP0142	6-Sulfo Lewis x	Gal(b1-4)[Fuc(a1-3)][HSO3(-6)]GlcNAc(b1-)-R
EP0211	Lewis a (b1-3)Gal	Gal(b1-3)[Fuc(a1-4)]GlcNAc(b1-3)Gal-R
EP0251	Blood Group H Type 1	Fuc(a1-2)Gal(b1-3)GlcNAc(b1-)-R
EP0252	Blood Group H Type 2	Fuc(a1-2)Gal(b1-4)GlcNAc(b1-)-R
EP0256	Blood Group A Type 1	GalNAc(a1-3)[Fuc(a1-2)]Gal(b1-3)GlcNAc(b1-)-R
EP0257	Blood Group A Type 2	GalNAc(a1-3)[Fuc(a1-2)]Gal(b1-4)GlcNAc(b1-)-R
EP0258	Blood Group A Type 3	GalNAc(a1-3)[Fuc(a1-2)]Gal(b1-3)GalNAc(a1-)-R
EP0259	Blood Group A Type 4	GalNAc(a1-3)[Fuc(a1-2)]Gal(b1-3)GalNAc(b1-)-R

Fig. 12.15 An example of how to find epitopes whose structure is similar to the "Lewis a" epitope

References

Benson DA, Clark K, Karsch-Mizrachi I, Lipman DJ, Ostell J, Sayers EW (2014) GenBank. Nucleic Acids Res 42:D32–D37

Feizi T (1985) Demonstration by monoclonal antibodies that carbohydrate structures of glycoproteins and glycolipids are onco-developmental antigens. Nature 314:53–57

Hakomori S (1984) Tumor-associated carbohydrate antigens. Annu Rev Immunol 2:103–126

Kanehisa M, Goto S, Sato Y, Kawashima M, Furumichi M, Tanabe M (2014) Data, information, knowledge and principle: back to metabolism in KEGG. Nucleic Acids Res 42:D199–D205

Kawasaki T, Nakao H, Takahashi E, Tominaga T (2006) GlycoEpitope: the integrated database of carbohydrate antigens and antibodies. Trends Glycosci Glycotechnol 18:267–272

Lombard V, Golaconda Ramulu H, Drula E, Coutinho PM, Henrissat B (2014) The carbohydrate-active enzymes database (CAZy) in 2013. Nucleic Acids Res 42:D490–D495

Rose PW, Bi C, Bluhm WF, Christie CH, Dimitropoulos D, Dutta S et al (2013) The RCSB Protein Data Bank: new resources for research and education. Nucleic Acids Res 41:D475–D482

Sell S (1990) Cancer-associated carbohydrates identified by monoclonal antibodies. Hum Pathol 21:1003–1019

Uniprot Consortium (2014) Activities at the Universal Protein Resource (UniProt). Nucleic Acids Res 42:D191–D198

Chapter 13
SugarBindDB

Julien Mariethoz, Khaled Khatib, Tiphaine Mannic, Davide Alocci, Matthew P. Campbell, Nicolle H. Packer, Elaine H. Mullen, and Frederique Lisacek

Abstract The SugarBind Database (SugarBindDB) covers knowledge of glycan binding of human pathogen lectin adhesins. It is a curated database; each glycan-binding pair is associated with at least one published reference. The usage of this database is illustrated through four case studies highlighting the various ways information can be found, explored and/or extracted.

Keywords Carbohydrate • Glycan binding • Lectin • Adhesin • Host-pathogen interactions • Glycoepitope • Data integration

13.1 Introduction

The SugarBind Database (SugarBindDB) covers knowledge of glycan binding of human pathogen lectin adhesins. It is a curated database; each glycan-binding pair is associated with at least one published reference. The interaction between pathogen lectins and host glycans has been published throughout the medical and biochemical literature beginning in the mid-1970s (Heyningen 1974).

J. Mariethoz
Proteome Informatics Group, SIB Swiss Institute of Bioinformatics, Geneva, Switzerland

K. Khatib
Glycoinformatics Inc., Great Falls, VA, USA

T. Mannic
Faculty of Sciences, University of Geneva, Geneva, Switzerland

D. Alocci • F. Lisacek (✉)
Proteome Informatics Group, SIB Swiss Institute of Bioinformatics, Geneva, Switzerland

Faculty of Sciences, University of Geneva, Geneva, Switzerland
e-mail: frederique.lisacek@isb-sib.ch

M.P. Campbell • N.H. Packer
Biomolecular Frontiers Research Centre, Macquarie University, North Ryde, NSW, Australia

E.H. Mullen
The MITRE Corporation, McLean, VA, USA

K.F. Aoki-Kinoshita (ed.), *A Practical Guide to Using Glycomics Databases*,
DOI 10.1007/978-4-431-56454-6_13

SugarBindDB was created in 2002 within the MITRE Corporation and publicly released in 2005. It was originally designed as a complement to a pathogen-capture technology based on the binding of viral, bacterial and biotoxin lectins to specific glycans displayed on glycoprotein films. This approach relied mainly on GlycoSuiteDB (Cooper et al. 2001) and UniProtKB/Swiss-Prot (UniProt Consortium 2012) to locate glycoproteins bearing a sugar sequence similar to a glycan ligand of a pathogen lectin. The database was transferred in 2010 to the SIB Swiss Institute of Bioinformatics where it was integrated in the ExPASy server, as described in (Shakhsheer et al. 2013). The transferred content was subsequently augmented, and the interface was completely modified to support user-friendly data browsing and searching as described in (Mariethoz et al. 2015). New features and content are regularly added and released quarterly on average.

The core data of SugarBindDB is a triple comprising a pathogenic agent, a lectin/adhesin and a glycan ligand. Each of these entities is named with as much precision as possible: taxonomic designation for agents, protein name for lectins and epitope name for ligands. Synonyms are listed whenever reported. When names are missing (which is frequent for lectins and pathogen strain names), then these entities are labelled N/S meaning "non-specified". The database includes additional information to supplement the core data, such as related diseases and affected tissues or organs.

The new version of SugarBindDB was developed within the same framework as UniCarbKB (see Chapter X), as the two projects are related. The layout and underlying philosophy promoting internal connectivity are common to the two resources. However, because the data and associated information are different, the result of searches is not output in the same way.

SugarBindDB content is displayed in views. For example, an agent view will show the taxonomic name linked to the National Center for Biotechnology Information (NCBI) Taxonomy database, agent properties (e.g. morphology, motility, etc.), the structures of stored ligands associated with this agent and the reference(s) supporting the evidence of the agent-ligand relationship linked to NCBI/PubMed. A view also lists a range of links connecting to further information, either internal or external to the database. As illustrated in Fig. 13.1 for *Burkholderia pseudomallei* H-99, links are displayed on the right side of the page. When a link is activated, it leads to a new page where all corresponding ligands will be shown. For instance, clicking on any of the agent properties (e.g. "flagellated") will list all stored ligands that are known to be bound by flagellated bacteria. This is a typical internal link. Other links such as "BURPE" below "HAMAP Proteome" connect to external resources, namely, in this case, the summary page of *Burkholderia pseudomallei* H-99 in High-quality Automated and Manual Annotation of microbial Proteomes (HAMAP), a bacteria-oriented subproject of the UniProtKB protein annotation scheme (Pedruzzi et al. 2015). Finally, lower to the right, more internal links appear in coloured blocks that indicate a type of association (colour code) and a number (in brackets).

The following illustrates in a series of four case studies the usage of search, navigation and visualisation tools that are implemented in SugarBindDB to consult or reveal information. Note that the default display of glycan structures is the

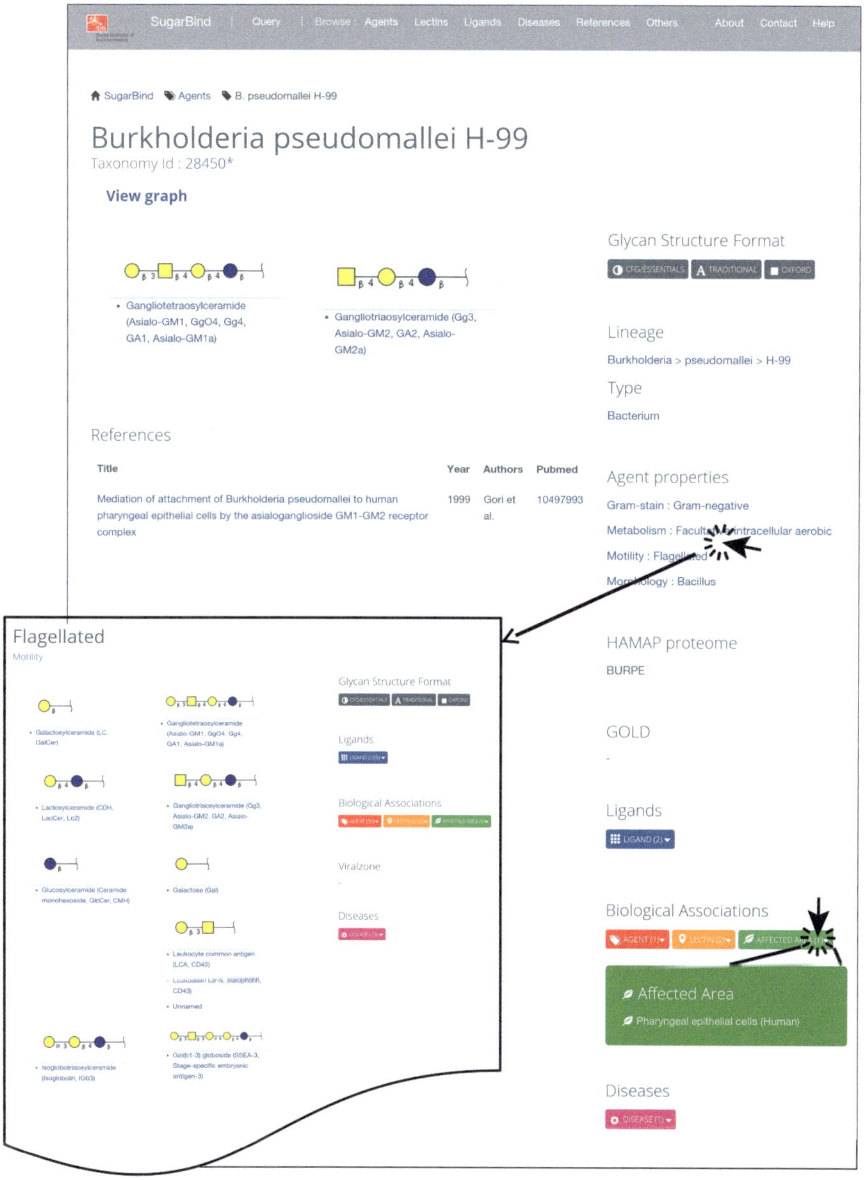

Fig. 13.1 Agent view for *Burkholderia pseudomallei* H-99, highlighting links to related pages in the database

most commonly used notation described in the textbook *Essentials in Glycobiology* (http://www.ncbi.nlm.nih.gov/books/NBK1908/) and promoted by the Consortium for Functional Glycomics (CFG). Two other options of popular notation are the two-dimensional "IUPAC condensed" and "Oxford Glycoscience" standards. Clicking on the corresponding dark grey buttons on the right-hand side under "Glycan Structure Format" changes the display.

13.2 Case Study 1

13.2.1 Browsing Versus Querying the Database for Pathogens

There are two options for accessing information regarding pathogens and their associated lectins: browsing or querying the database. These two approaches are available from the menu bar at the top of the homepage as shown in Fig. 13.2. Browsing is, of course, less targeted than querying, as exemplified in the next two sections—though in the end, the same information is reached.

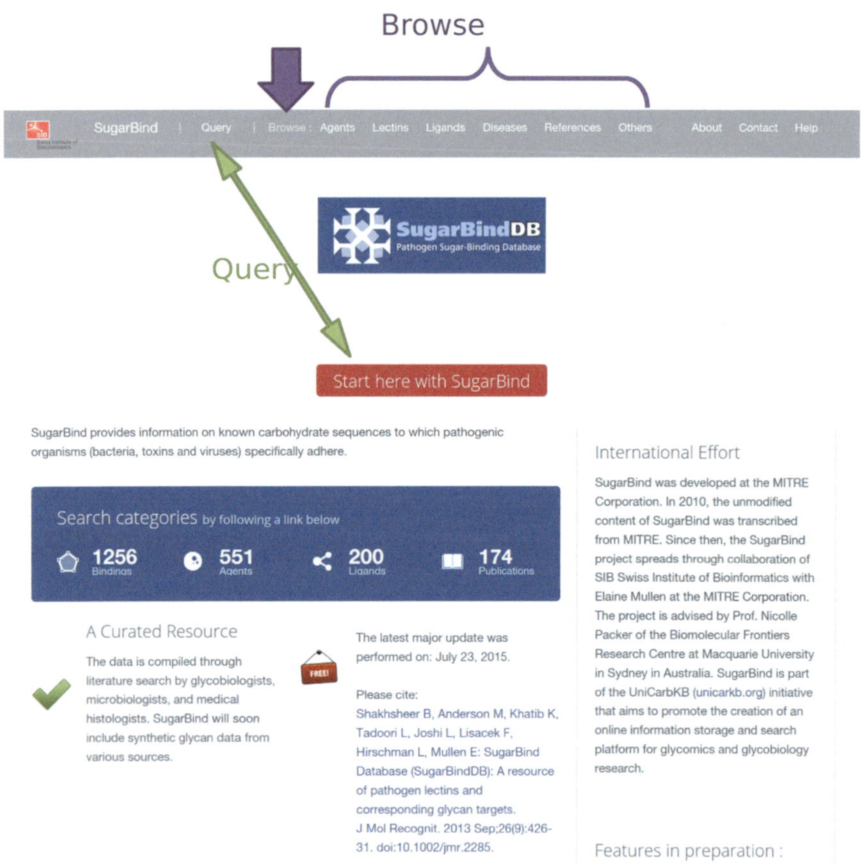

Fig. 13.2 SugarBindDB homepage showing the possible options for querying or browsing the database

13.2.1.1 Browsing

Step 1 Consider the upper light grey bar at the top of the homepage. Clicking on "Agents" prompts the complete set of pathogens included in SugarBindDB listed by name, type (bacterium, virus, toxin) and taxonomy ID (linked to NCBI taxonomy: http://www.ncbi.nlm.nih.gov/taxonomy).

Step 2 If a term is typed in the "Filter" window, the list will be reduced to the pathogen names including this term. In this example, typing "aeru" and clicking on filter will reduce the list to the *Pseudomonas aeruginosa* species as well as other species possibly described with comments referring to *Pseudomonas aeruginosa*. There are 15 *Pseudomonas aeruginosa* strains recorded in the database, though filtering also collects *Burkholderia cenocepacia* J2135 due to the extra comment referring to *Pseudomonas aeruginosa*. Each pathogen name is directly linked to the page that describes all carbohydrate ligands that it is known to bind to.

Step 3 For example, clicking on *Pseudomonas aeruginosa 19142* prompts the "Agents" page displaying the only reported glycan ligand and known agent properties. The default display is the most commonly used notation described in the textbook *Essentials in Glycobiology* (http://www.ncbi.nlm.nih.gov/books/NBK1908/) and promoted by the Consortium for Functional Glycomics (CFG). Two other options of popular notation are the 2D "IUPAC condensed" and "Oxford Glycoscience" standards. Clicking on the corresponding grey buttons on the right-hand side under "Glycan Structure Format" will change the display.

This sequence is shown in Fig. 13.3a. Despite the less-precise selection obtained from filtering the comprehensive list of pathogenic agent names, once this name can be spotted, then direct access to the pathogen page is possible.

13.2.1.2 Searching

Step 1 Clicking either on the red box "Start here with SugarBind" or on "Query" in the upper grey bar or on "Bindings" in the blue box of the homepage will prompt the query page. By default, the query is set on "Agents", but of course, any alternative radio button can be selected to search other categories (e.g. ligand). As a name is typed in the query window, a drop-down scrollable list of valid items is displayed. In this example, typing "aeru" prompts the term "Pseudomonas aeruginosa" that is selected.

Step 2 Clicking on the "Search" button opens a "Result" page that displays the complete set of lectin-ligand pairs recorded in the database for *Pseudomonas aeruginosa*. The output is structured in blocks, each corresponding to either a known or an unspecified (N/S) strain.

Step 3 Clicking on the "Ligand (1)" button to the right of *Pseudomonas aeruginosa 19142* prompts the "Agents" page displaying the only reported glycan structure and its known properties.

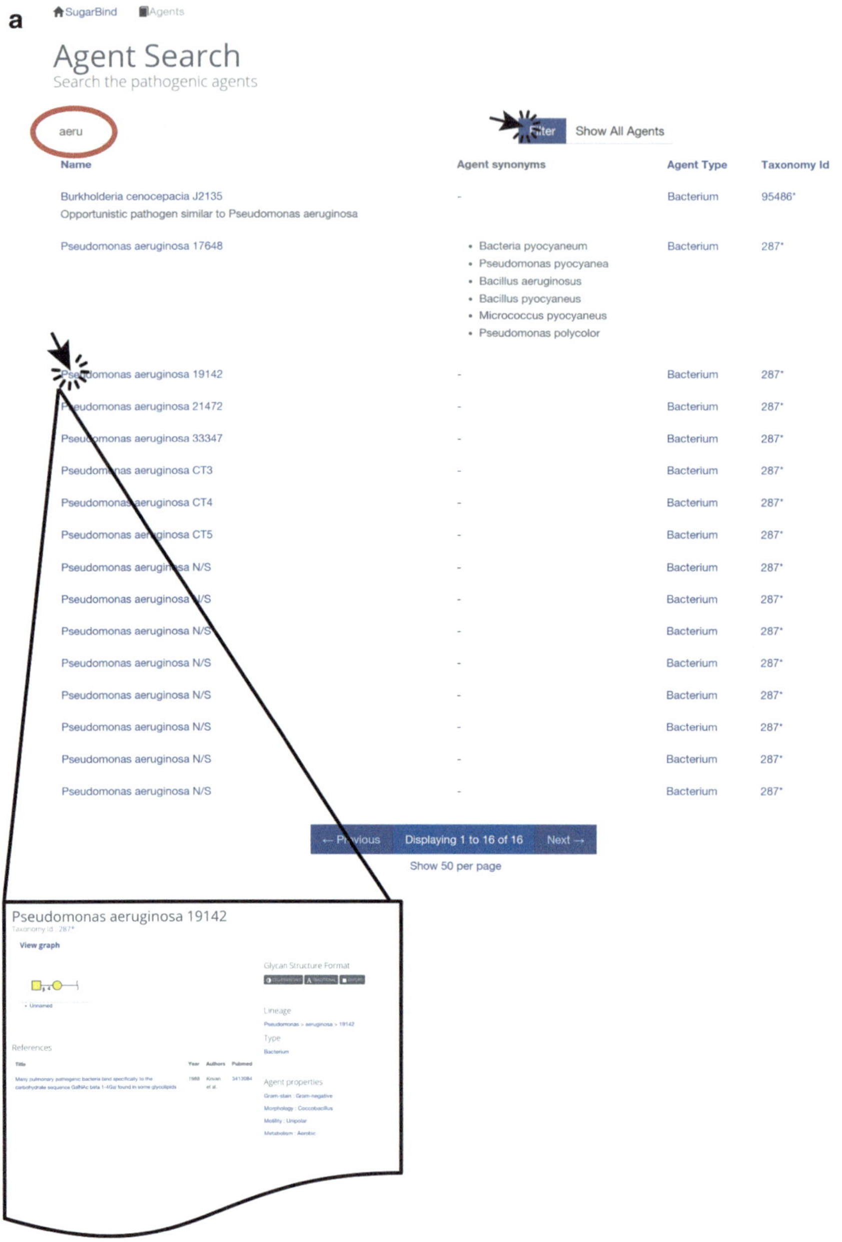

Fig. 13.3 (**a**) Example of agent *search* in the full agent list by entering partial words and filtering the list, (**b**) example of agent *query* by entering terms selected by auto-completion. Option (a) is an alternative to option (b), and the same agent page can be accessed either way (*Pseudomonas aeruginosa* 19142, in this example)

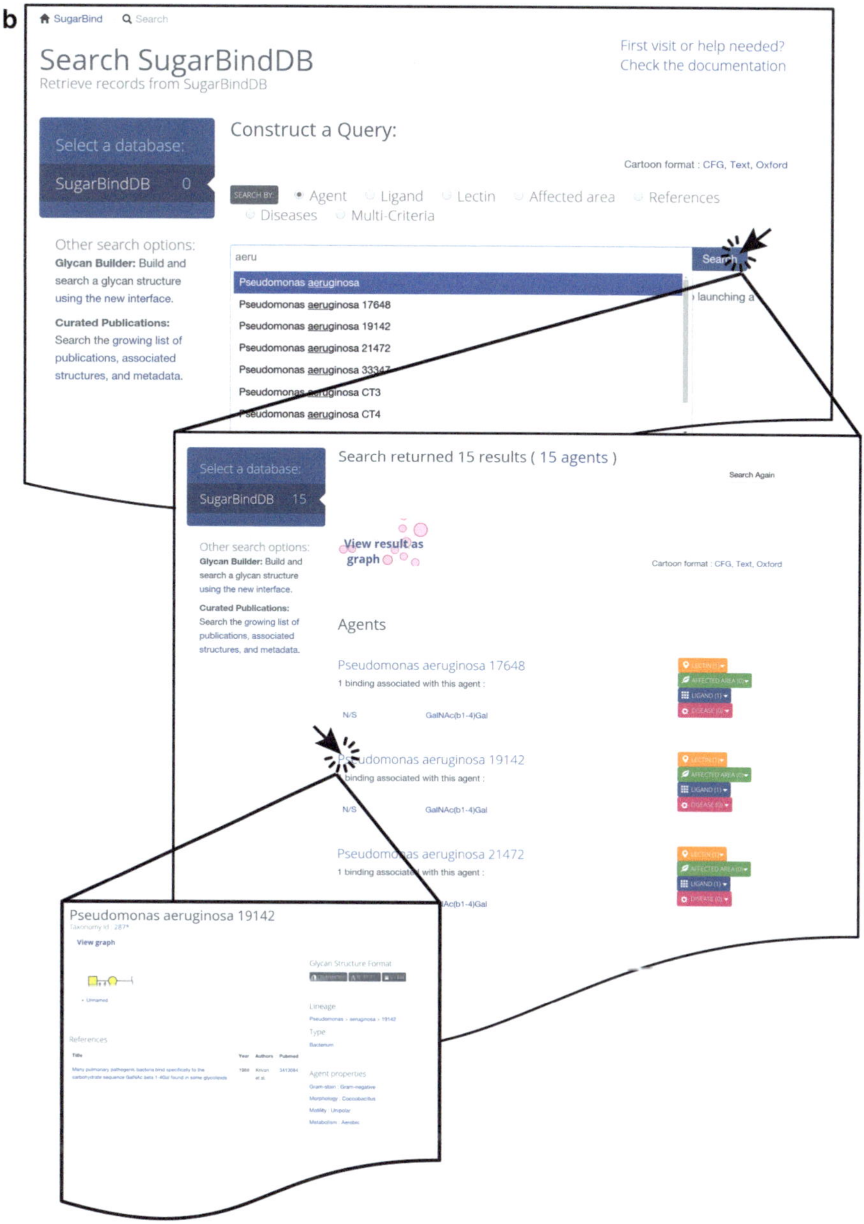

Fig. 13.3 (Continued)

This sequence is shown in Fig. 13.3b. The query option leads to precisely delineating the database coverage for a specific pathogenic agent. With this structured overview, the user is better equipped to select the relevant agent pages.

13.3 Case Study 2

13.3.1 *Search for a Pathogen, Visualise Related Structures and Explore Associations*

This case study highlights the connectivity of the database. For argument's sake, the starting point is a lectin.

Step 1 Clicking either on the red box "Start here with SugarBind" or on "Query" in the upper grey bar or on "Bindings" in the blue box of the homepage will prompt the query page. By default, the query is set on "Agents". In this example, the "lectin" radio button is selected to search the *Escherichia coli* lectin FedF. As a name is typed in the query window, a drop-down scrollable list of valid items is displayed. Typing "fe" prompts the term "FedF" that is selected.

Step 2 Clicking on the "Search" button opens a "Result" page that displays the complete set of lectin-ligand pairs related to FedF in the database. The list is structured in blocks, each corresponding to either a known or an unspecified (N/S) strain. The eight ligand structures associated with *Escherichia coli* strain 107/86 in the database are listed in linear form of the IUPAC encoding as a block of results. Mousing over each glycan sequence triggers the display of the corresponding selected graphic representation (CFG, traditional, Oxford). The ligand name list is prompted by clicking on the "Ligand (8)" button to the right of the lectin name.

Step 3 The "FedF | E. coli 107/86" page shows the corresponding lectin view and displays structures of the eight ligands with their possibly known common names, such as "A6 type 1" for the first one. Virtually all information contained in this page is linked to either further information internal or external to the database.

Step 4 The exploration of internally linked information is straightforwardly achieved through clicking on:

(i) Ligand names to prompt the corresponding ligand page (or ligand view) that provides the list of references where the ligand is mentioned, various encoding schemes of the structure such as GlycoCT (Herget et al. 2008) and more associations with other agents, lectins and diseases

(ii) Properties listed on the right-hand side of the page, such as "supporting structure", i.e. "Fimbria" in the case of FedF (see Fig. 13.4) that enable the user to examine all other ligands that are bound by a fimbriae-associated lectin/adhesin

(iii) Coloured association boxes that are ubiquitous in the database and instantiated in each page where they occur. The colour code is identical throughout: red for agents, blue for ligands, orange for lectins, green for tissue/affected area and pink for disease. The figure in brackets appearing in each box counts the number of existing links from the current page to each entity type. In the case of FedF, as illustrated in Fig. 13.4, both the red agent and

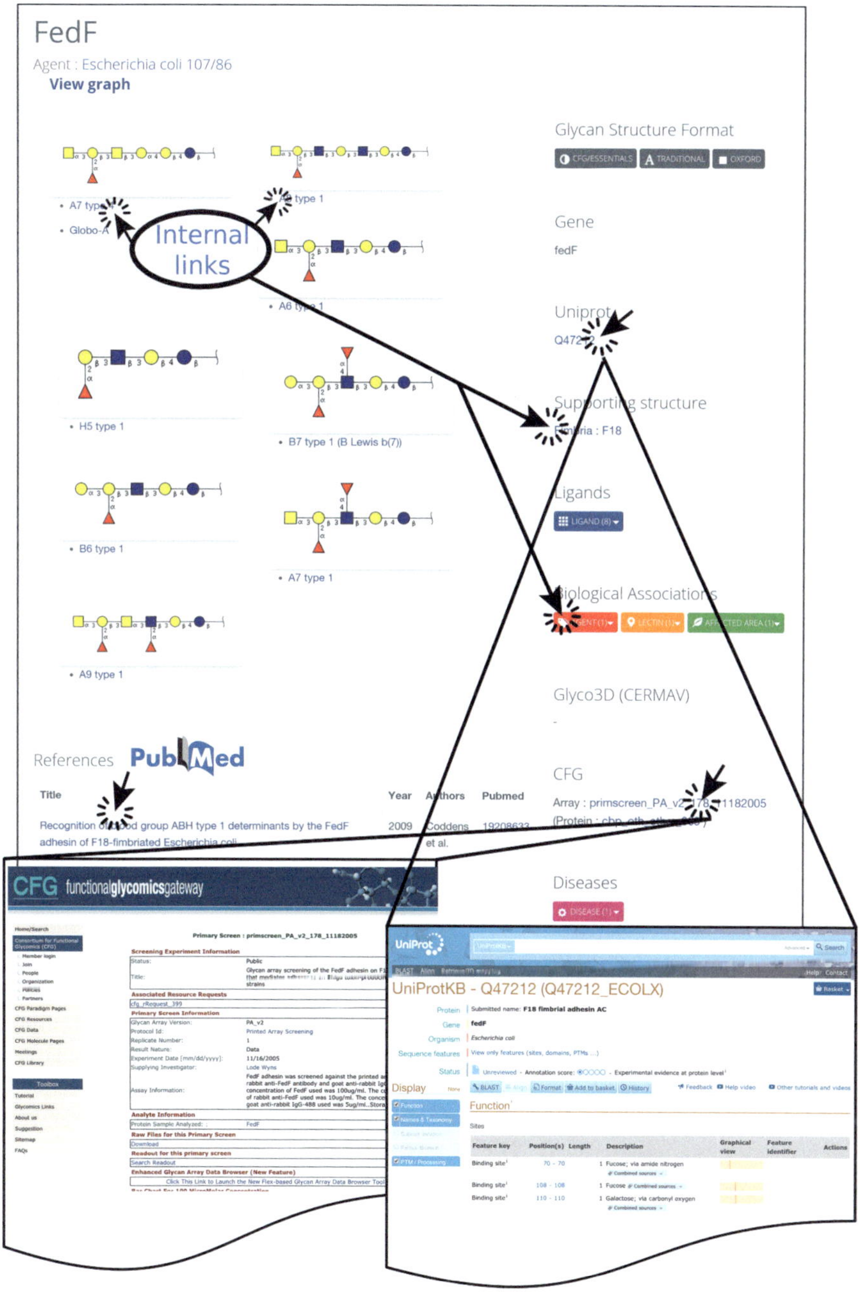

Fig. 13.4 Lectin view for FedF in *E. coli*, highlighting links to related content internal and external to the database

orange lectin boxes are set to 1, since the E. coli-FedF link is unique. The green or pink boxes are also set to 1 for FedF but can be multiple in cases where several studies across different tissues were reported or when various syndromes are known.

Step 5 The exploration of externally linked lectin information currently spans three related domains:

(i) Protein annotation as reported in the UniProtKB database (with a direct cross-reference to the database via the corresponding accession number, e.g. Q47212 for FedF as shown in Fig. 13.4)
(ii) Three-dimensional structure of the lectin when available in the Glyco3D database (http://glyco3d.cermav.cnrs.fr), which is not the case for FedF
(iii) Glycan array experiments in the CFG, when available, which is the case for FedF, as shown in Fig. 13.4.

Needless to say, all cited references are linked to PubMed. In summary, Fig. 13.4 highlights the different types of links that can be activated to access further information from a lectin page.

13.4 Case Study 3

13.4.1 Query with Multiple Criteria and Explore Associations

This case study emphasises the flexibility of the query tool. Indeed, more than one term can be typed in the query window and multiple terms need not be in a fixed category set by the radio buttons above the query window. The following explains how to perform this search:

Step 1 Clicking either on the red box "Start here with SugarBind" or on "Query" in the upper grey bar or on "Bindings" in the blue box of the homepage will prompt the query page. By default, the query is set on "Agents". In this example, the radio button is moved from "Agents" to "Multi-Criteria". Then, "BabA", "SabA" and "Sialyl-Lewis X" are successively entered into the query window, each via a drop-down scrollable list of valid items. The implicit logical connector between these input terms is "OR".

Step 2 Clicking on the "Search" button opens a "Result" page that displays the complete set of lectin-ligand pairs related to the two lectins and the one ligand listed in the database. The list is structured in blocks, each corresponding to either a known or an unspecified (N/S) strain. These blocks are grouped under "Lectins" and "Ligands" headers matching the categories of entities of which names were input in the query window. Each of the seven blocks shows ligand structures associated with individual lectins of several strains of *Helicobacter pylori* in the database.

Step 3 As previously noted in Case study 2, the exploration of internally linked information is straightforwardly achieved through clicks on ligand names or other properties listed on the right-hand side of the page, while the exploration of externally linked information is achieved via explicit links to other databases.

13.5 Case Study 4

13.5.1 Query with Multiple Pathogen Names, Visualise and Explore Binding Specificity

This case study introduces a visualisation tool designed to support comparative studies based on shared properties of ligands or lectins/adhesins.

Step 1 Clicking either on the red box "Start here with SugarBind" or on "Query" in the upper grey bar or on "Bindings" in the blue box of the homepage will prompt the query page. By default, the query is set on "Agents". In this example, several names are successively typed in the query window, each via a drop-down scrollable list of valid items. Typing "pneumonia" prompts a list of agent names that contain this term. Assume "Klebsiella pneumoniae" is selected. Retyping "pneumonia" next to "Klebsiella pneumoniae" prompts the same list of agent names that contain this term. This time, assume "Chlamydia pneumoniae" is selected. Finally, after typing one last time "pneumonia" next to "Klebsiella pneumoniae" "Chlamydia pneumoniae", assume "Streptococcus pneumoniae" is selected from the same list of agent names that contain the term.

Step 2 Clicking on the "Search" button opens a "Result" page that displays the complete set of lectin-ligand pairs related to all strains of the three selected pathogens as stored in the database. The 19 lectin-ligand pairs associated with these 14 pathogens can be visualised all together by clicking on the "view result as graph" box located above the list. A new window/tab is automatically opened, showing, first, a so-called Sankey graph, which maps all information on a hierarchical graph defined in the following order: agent-lectin-ligand. Each agent name (pathogen) is an initial node linked to its associated strains. Each strain is a node linked to its associated lectin(s), and each lectin is another node linked to the glycan structure that it binds. Each ligand is a final node. The links are visualised as grey connections. Clicking on any node highlights the path that goes through it via a colour change to orange. Note that clicking on any name leads directly to the corresponding entity page. Figure 13.5 shows the effect of clicking on the gangliotetraosylceramide ligand node and the subsequent orange colouring of the possible paths between ligands and pathogens. This graphic view allows tracing back which lectin of which strain actually binds gangliotetraosylceramide (alias Gal(b1-3)GalNAc(b1-4)Gal(b1-4)Glc(b1-1)). In this way, the user can investigate the comparable behaviour of distinct pathogens that generate similar clinical symptoms.

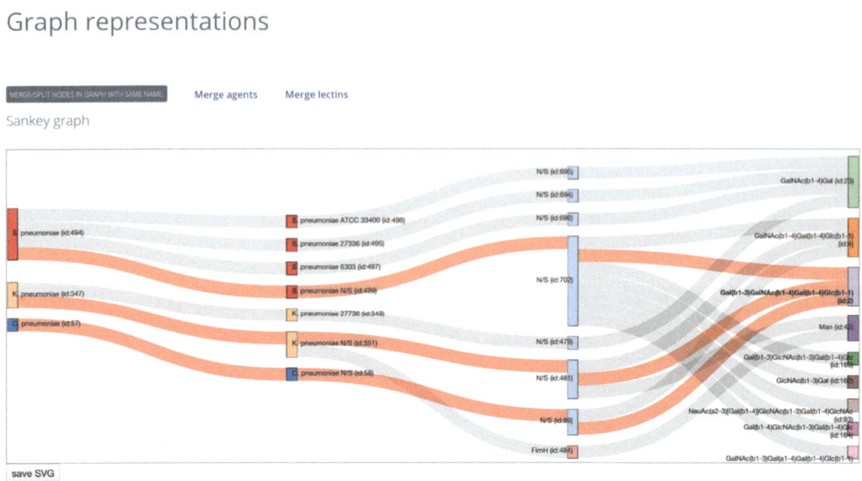

Fig. 13.5 Graph view of the comparison between *Klebsiella pneumoniae*, *Chlamydia pneumoniae* and *Streptococcus pneumoniae* in terms of glycan-binding specificity. The orange path emphasises Gal(b1-3)GalNAc(b1-4)Gal(b1-4)Glc(b1-1) as the ligand bound by all three species via their respective lectins

Step 3 Clicking again on a node of the selected path reverts to grey for all paths. Any colouring of the graph can be saved in SVG format simply by hitting the "save SVG" button. The SVG format preserves the image resolution when it is reduced or enlarged. SVG images can be opened in any web browser. Note that coloured boxes can be moved around in the page to modify the layout.

The four cases above illustrate information search and browsing options in SugarBindDB. At this stage, the database is stable but will improve as its content and cross-references are augmented, and more search and predictive tools are developed.

For newcomers, documentation is still limited, but, for a general overview of SugarBindDB, a new user can click on the "First visit or help needed?" link on the top right in the homepage and search result pages.

For basic knowledge of glycobiology, users are invited to visit the glycopedia.eu website.

13.6 Troubleshooting Tips

Sankey graph exploration may be confusing, since names and nodes are often very close together, and clicking on one or the other may be difficult. However, the underlying links have different purposes, as illustrated below:

- Clicking on a name opens the SugarBindDB page corresponding to the named entity in a new tab of the web browser.

- Clicking on a "node" represented as a coloured box (see Fig. 13.5) colours in orange all paths going through this node. Clicking again reverses colouring to initial grey.

We recommend initial mousing over the graph to visualise the distinction:

- Mousing over a name prompts a pointer—either a hand or unidirectional arrow ().
- Mousing over a node prompts a four-directional arrow () indicating possible interaction with the object in the graph.

This preliminary check will guarantee proper selection.

13.7 Conclusion

SugarBindDB is part of a series of data resources that were developed within the same framework and that complement one another. The main database UniCarbKB (Campbell et al. 2014) covers information on glycoproteins and associated glycans. SugarBindDB focuses on the recognition of these glycans by pathogen lectin adhesins. The last addition to this database collection is UniCarb-DB (Hayes et al. 2011), which stores experimental evidence for structures assigned to glycans and was recently restructured. This collection will co-evolve to give rise to broader capabilities.

References

Campbell MP, Peterson R, Mariethoz J, Gasteiger E, Akune Y, Aoki-Kinoshita KF, Lisacek F, Packer NH (2014) UniCarbKB: building a knowledge platform for glycoproteomics. Nucleic Acid Res 42(1):D215–D221

Cooper CA, Harrison MJ, Wilkins MR, Packer NH (2001) GlycoSuiteDB: a new curated relational database of glycoprotein glycan structures and their biological sources. Nucleic Acids Res 29:332–335

Hayes CM, Karlsson NG, Struwe W, Rudd PM, Lisacek F, Packer NH, Campbell MP (2011) UniCarb-DB: a database resource for glycomic discovery. Bioinformatics 27(9):1343–1344

Herget S, Ranzinger R, Maass K, von der Lieth CW (2008) GlycoCT-a unifying sequence format for carbohydrates. Carbohydr Res 343(12):2162–2171

Mariethoz J, Khatib K, Campbell MP, Packer NH, Mullen E, Lisacek F (2015) SugarBindDB, a resource of pathogen lectin-glycan interactions in glycoscience: biology and medicine Taniguchi N, Endo T, Hart G, Seeberger P, Wong CH (eds). Springer

Pedruzzi I, Rivoire C, Auchincloss AH, Coudert E, Keller G, de Castro E, Baratin D, Cuche BA, Bougueleret L, Poux S, Redaschi N, Xenarios I, Bridge A (2015) HAMAP in 2015: updates to the protein family classification and annotation system. Nucleic Acids Res 43(Database issue):D1064–D1070

Shakhsheer B, Anderson M, Khatib K, Tadoori L, Joshi L, Lisacek F, Hirschman L, Mullen E (2013) SugarBind database (SugarBindDB): a resource of pathogen lectins and corresponding glycan targets. J Mol Recognit 26(9):426–431

UniProt Consortium (2012) Reorganizing the protein space at the Universal Protein Resource (UniProt). Nucleic Acids Res 40(Database issue):D71–D75

Van Heyningen S (1974) Cholera toxin: interaction of subunits with ganglioside GM1. Science 183:656–657

Chapter 14
PAConto: RDF Representation of PACDB Data and Ontology of Infectious Diseases Known to Be Related to Glycan Binding

Elena Solovieva, Noriaki Fujita, Toshihide Shikanai, Kiyoko F. Aoki-Kinoshita, and Hisashi Narimatsu

Abstract The Pathogen Adherence to Carbohydrate Database (PACDB) has been developed by the Research Center for Medical Glycoscience (RCMG) and released in March 2010. Being the members of the "Life-Science Database Integration Project" of National Bioscience Database Center (NBDC) of Japan Science and Technology Agency (JST), we have decided to organize PACDB information and enrich its content by integration with other biomedical resources. We have designed and developed the ontology, called PAConto. This ontology is based on the Semantic Web technologies and is represented in Resource Description Framework (RDF) format. Using the Semantic Web approach with Web Ontology Language (OWL), Simple Knowledge Organization System (SKOS), and RDF standards, we have described the semantics of PACDB data, enriched this data with additional various classifications, and linked PACDB data with related biomedical resources. In addition to the ontology, we have developed a system with a user interface, using which the users can retrieve and search information from PAConto. The RDF files for PAConto with RDF/SPARQL-based user interface, documentation, and user's guide are available at http://acgg.asia/db/diseases/. In this chapter, we will introduce the topics of PACDB and PAConto and explain how the users can navigate through this system and search and display RDF data. Also, we will provide the contact information for feedback purposes. We hope that this PAConto ontology with PACDB database will help the users to better understand the etiology, pathogenesis, and manifestations of the diseases known to be related to glycan binding.

E. Solovieva • N. Fujita • T. Shikanai • H. Narimatsu (✉)
Research Center for Medical Glycoscience (RCMG), National Institute of Advanced Industrial Science and Technology (AIST), Tsukuba, Japan
e-mail: h.narimatsu@aist.go.jp

K.F. Aoki-Kinoshita
Research Center for Medical Glycoscience (RCMG), National Institute of Advanced Industrial Science and Technology (AIST), Tsukuba, Japan

Faculty of Science and Engineering, Soka University, Hachioji, Japan

K.F. Aoki-Kinoshita (ed.), *A Practical Guide to Using Glycomics Databases*,
DOI 10.1007/978-4-431-56454-6_14

Keywords Pathogen Adherence to Carbohydrate Database (PACDB) • Pathogen Adherence to Carbohydrate Ontology (PAConto) • Microbial glycan-binding proteins • Host glycan ligands • Lectin-glycan interactions • Pathogenic microorganisms • Infectious diseases • Semantic Web approach • RDF representation • RDF/SPARQL-based user interface • Data integration

14.1 Introduction

In this section, we will highlight the main topics (subjects) of the PACDB database that includes the following: the pathogenic microorganisms (pathogens), the microbial glycan-binding proteins, the target sources and glycan ligands of the host, the interactions between microbial glycan-binding proteins and host glycan ligands, and the diseases, in which lectin-glycan interactions play important roles in the diseases' pathogenesis. Also, we will mention about using the Semantic Web approach for description and integration of the data of a particular scientific field and about ontologies as a basis for domain knowledge representation. Finally, we will outline the reasons why we decided to develop PAConto.

14.1.1 About PACDB

PACDB has been created by Research Center for Medical Glycoscience (RCMG) and released in March 2010. At the present time, PACDB provides information on about 370 strains of 127 microorganisms and about 1700 lectin-glycan interactions of two types: binding and not binding. Also, PACDB provides information on 97 infectious diseases in which the interaction between adherence molecules of infectious agents and glycan ligands of the host cells plays an important role in the disease pathogenesis. All of the information for the creation of this database was obtained from 214 scientific articles. The Web-based user interface has been created, released, and is now available at http://jcggdb.jp/search/PACDB.cgi.

14.1.2 Overview of the Topics of PACDB

The PACDB data may be divided into several topics, as follows: pathogenic microorganisms (pathogens), microbial glycan-binding proteins, target sources and host glycan ligands, interactions between microbial glycan-binding proteins and host glycan ligands, and infectious diseases known to be related to glycan binding. In this subsection, we will briefly describe the basic characteristics of each of these topics.

14.1.2.1 Pathogenic Microorganisms (Pathogens)

Various types of microorganisms, such as bacteria, viruses, fungi, and parasitic protozoa, are infectious agents (pathogens) that cause diseases recorded in PACDB. In this subsection, we will discuss about these infectious agents, their pathogenic properties, and virulence factors, using information obtained from article written by Janeway et al. (2001).

In general, many microorganisms are pathogenic to mammals. Each pathogen is characterized by the following features: the ways of its transmission, the mechanism of its replication, and the mechanisms by which it causes disease or pathology. As is noted by Janeway et al. (2001), the degree of infectivity of the pathogenic microorganisms is determined by the following indicators: the number, route, mode of transmission, and stability of an infectious agent outside the host.

Some pathogens are resistant to drying and heating and can survive for long periods of time outside of their host. Other pathogens, such as the *human immunodeficiency virus* (*HIV*), which cannot exist outside of their hosts, are transmitted only by direct contact with bodily fluids or tissues of an infected person. In order to cause infectious disease, pathogenic microorganisms must first bind to the epithelial surface of the skin, gastrointestinal, respiratory or genitourinary tracts, and then must establish a focus of infection by overcoming the innate immune responses of the host.

The pathogenic microorganisms can be classified as extracellular and intracellular pathogens. Moreover, the latter can also be subdivided into two categories: (1) pathogens that grow and replicate in the cytoplasm of eukaryotic host cells, such as viruses and certain bacteria like *Chlamydia* and *Listeria*, and (2) pathogens that use membrane vesicles, such as the *Mycobacteria* (Janeway et al. 2001).

After the focus of infection was established, some pathogens can spread from the original site to other parts of the body. Also, some pathogens have the ability to produce specific exotoxins (protein toxins), and these toxic substances can spread to the whole of the body. The extracellular pathogens spread through the bloodstream or lymphatic system, while the intracellular pathogens must spread from cell to cell.

Most pathogenic microorganisms cause disease only in one or a few related host species. This high degree of specificity, which many pathogens have with their hosts, can be explained by the requirements of the pathogens for attachment to a particular cell-surface molecule, as well as the fact that the microorganism-host interactions are required to support replication of the pathogen (Janeway et al. 2001). In many cases, these particular cell-surface molecules, which are required for pathogen attachment, are glycoproteins, glycolipids, or polysaccharides.

In PAConto, the pathogenic microorganisms are the main entities in the ontology, and they are the main entries in the PAConto user interface.

14.1.2.2 Microbial Glycan-Binding Proteins

All of the types of infectious agents registered in PACDB interact with host glycans by adherence molecules. In most cases, these pathogen adherence molecules are glycan-binding proteins. Glycan-binding proteins can be divided into two main groups: lectins and glycosaminoglycan-binding proteins (GAG-binding proteins) (Varki et al. 2009). Microbial glycan-binding proteins can be a lectin or a glycosaminoglycan-binding protein. Below we will discuss about microbial glycan-binding proteins mentioning both their types.

Bacterial Glycan-Binding Proteins

There are various forms of bacterial glycan-binding proteins, such as proteins that are located on the tip or along the shaft of a fimbriae or pili, located directly on the bacterial surface or secreted as toxins (Holgersson et al. 2009). Most bacterial lectins are located on the fimbriae or pili and interact with glycoprotein and glycolipid ligands on host cells (Esko and Sharon 2009). *Helicobacter pylori* and *Pseudomonas aeruginosa* species are examples of bacteria that have bacterial surface lectins, which bind specific sugar molecules on the surface of host cells and mediate adhesion to them (Holgersson et al. 2009). An example of an exotoxin that has carbohydrate-binding activity is *cholera toxin* (Esko and Sharon 2009), which has high affinity to a GM1 molecule (Holgersson et al. 2009). Along with lectins, some bacteria have GAG-binding proteins, which have binding activity to heparan sulfate (Varki et al. 2009).

Viral Glycan-Binding Proteins

In viral infection, the adhesion of a virus to the host cell, which is mediated by protein-carbohydrate interactions, plays an important role in disease pathogenesis. Some viruses have lectins that bind to glycoproteins or glycolipids which contain sialic acid (Holgersson et al. 2009). For example, the hemagglutinin of *Influenza virus* binds to sialic acid-containing glycans. Other viruses have glycosaminoglycan-binding proteins that use heparan sulfate proteoglycans as receptors for adhesion (Esko and Sharon 2009). For example, the gD protein of *Herpes simplex virus*, which is GAG-binding protein, binds to heparan sulfate proteoglycans with high affinity and specificity to certain sulfate groups (Varki et al. 2009).

Protozoa Glycan-Binding Proteins

Some parasitic protozoa also use glycans as ligands for adhesion molecules. For example, the 260-kD heterodimeric lectin of *Entamoeba histolytica* binds to terminal galactose/N-acetylgalactosamine residues on glycoproteins and glycolipids. Another example is *Plasmodium falciparum* (*malaria*). The specific sialic acid-binding adhesin on merozoites of *P. falciparum*, such as EBA-175 (erythrocyte-binding antigen-175), interacts with glycoproteins of the erythrocyte membrane of the host and mediates the binding to erythrocytes, which leads to invasion of the host. The circumsporozoite form of *P. falciparum* also expresses GAG-binding proteins and binds to heparan sulfate on the surface of hepatocytes (Esko and Sharon 2009).

Fungal Glycan-Binding Proteins

Several fungal lectins from yeasts and molds, such as lectins of *Candida albicans*, *Histoplasma capsulatum*, *Paracoccidioides brasiliensis*, *Saccharomyces cerevisiae*, and others, have interactions with host glycoconjugates and can be involved in the pathogenesis of fungal infections (Varrot et al. 2013; Singh et al. 2011).

These pathogen adherence molecules will be discussed in Sect. 14.2.2.4, in connection with the Classification of Microbial Glycan-Binding Proteins.

14.1.2.3 Target Sources and Host Glycan Ligands

Target Sources

Pathogenic microorganisms can grow in various organs and tissues of the host organism. All pathogens can be divided into those that replicate inside the cells and those that replicate and grow in the extracellular spaces inside the body or on the surface of the epithelium. Intracellular pathogens must first invade the cells of the host and then replicate inside these cells. Many of the pathogenic microorganisms are extracellular pathogens. The site of the body in which pathogenic microorganism grows and replicates can affect the type and severity of the pathological process that is caused by this microorganism (Janeway et al. 2001).

As is noted by Esko and Sharon (2009), tropism to various tissues of host organisms can be explained by the binding specificity of glycan-binding proteins. Most bacteria, and perhaps other microorganisms, have several glycan-binding proteins with specificities to different carbohydrates, and these specificities can help define the range of tissues that are susceptible to these microorganisms (Esko and Sharon 2009).

There are some examples when the glycan-binding specificity can explain the tissue tropism of the pathogenic microorganisms.

Varki et al. (2009) describe the glycan-binding specificity for bacteria:

> Urinary tract infection by specific serotypes of Escherichia coli depends on binding mannose or P blood group structures.

Moreover, Esko and Sharon (2009) describe this specificity for viruses and parasitic protozoa:

> The specificity of the hemagglutinin correlates with the structures of sialylated glycans expressed on target epithelial cells in animal hosts. The circumsporozoite form of P. falciparum binds to heparan sulfate in a tissue-specific manner, with preferred binding to the basal surface of hepatocytes and the basement membrane of kidney tubules.

We will discuss tissue tropism of pathogenic microorganisms in Sect. 14.2.2.4, in connection with the Classification of Target Sources.

Host Glycan Ligands

In this subsection, by using as an example the mucin glycans, we will briefly discuss about the host glycan ligands based on information obtained from an article written by Juge (2012).

Mucins are high-molecular-mass glycoproteins. The human mucin (MUC) family consists of membrane-bound (including cell-surface) and secreted mucins. Mucins are produced in the intestine and are the main components of mucus and the apical surface components of all mucosal epithelia. As is noted by Juge (2012), the mucin glycans are the major receptors for gastrointestinal microbes, and the glycan-rich domains of mucins provide preferential binding sites for pathogens and commensal bacteria. Moreover, mucins are extensively O-glycosylated, and their O-glycans make up to 80% of the total mucin mass. In describing the characteristics of mucin glycans, Juge (2012) noted that:

> Human intestinal mucins display decreasing gradients of fucose and ABH blood group and an increasing acidic gradient from ileum to rectum. More than 100 complex oligosaccharides were identified, mostly mono-, di- or trisialylated.

Because carbohydrates play a crucial role in pathogen invasion of host cells, the identification of the structures of glycans is a very important discovery in this aspect of glycoscience.

14.1.2.4 Interactions Between Microbial Glycan-Binding Proteins and Host Glycan Ligands

In this subsection, we will briefly explain the interactions between microbial glycan-binding proteins and host glycan ligands using the example of viral pathogens, including influenza A viruses.

Various viruses use the interactions between their own viral lectins and glycans of host glycoconjugates for entry into host cells (Breedam et al. 2014).

As is noted by Breedam et al. (2014), the microbial glycan-binding proteins may be divided into glycosaminoglycan-binding proteins and lectins, and these two types are distinguished by the structural basis of their glycan recognition.

Also, these authors noted that, this difference is as follows:

> glycosaminoglycan-binding proteins interact with negatively charged glycosaminoglycans via clusters of positively charged aa residues.

In contrast, they also note that the carbohydrate recognition domains (CRDs) of the "most strict sense lectins" selectively recognize specific portions of N-glycans, O-glycans, or glycolipids (sometimes also glycosaminoglycans).

Viswanathan et al. (2010) discuss about the role that the influenza viral coat protein hemagglutinin (HA) plays in viral entry and infection. They note that HA functions on the viral surface as a glycan binding protein and enables viral attachment to host epithelia by binding to sialylated glycan receptors on the host cell surface.

Similarly to viral pathogens, other pathogens, like bacteria, fungi, and parasitic protozoa, also interact with host glycans, and these interactions act as primary keys to pathogen colonization of the host cells.

14.1.2.5 Infectious Diseases Known to Be Related to Glycan Binding

PACDB provides information about diseases in the pathogenesis of which the interaction of microbial glycan-binding proteins and host glycan ligands plays an important role.

Most of the organisms described in PACDB, such as *Bordetella pertussis*, *Influenza virus*, *Entamoeba histolytica*, *Cryptococcus neoformans*, and *Equine rhinitis A virus*, are disease-causing agents that cause infectious diseases in human or animal hosts. Also, some microorganisms, such as *Hepatitis B virus* (*HBV*), *Hepatitis C virus* (*HCV*), and *Helicobacter pylori*, are infectious agents that can be involved in cancer development. Some of the diseases that are recorded in PACDB, such as stomach ulcer, are caused by a combination of various factors, among which the primary cause is microorganisms such as *Helicobacter pylori*.

As mentioned above, pathogenic microorganisms must first adhere to epithelial or mucosal cells and then penetrate or colonize them. If it is successful, the pathogen establishes a site of infection and an infectious disease occurs (Janeway et al. 2001).

For many types of infection, if an effective primary response occurs, the pathological manifestation will be little, or there will be no residual pathology. In some cases of infection, the pathogenic microorganisms or the host defense responses lead to significant tissue damage. The mechanisms of damage are quite different for various infectious agents. Many pathogens have cytopathic effects and cause damage to infected cells. This is very common for intracellular infectious agents, directly damaging the cells that house them. Extracellular secretion of protein toxins (exotoxins) is the major mechanism by which many extracellular pathogens interact with host cells and cause disease. These exotoxins interact with the host cells by binding to surface receptors. Both the innate and the adaptive immune responses control infection and clear the infectious agents. At the same time, the immune response also can lead to tissue damage and cause pathologic processes (Janeway et al. 2001).

Some types of pathogenic microorganisms, such as *Cytomegalovirus*, after a primary infection can persist in a latent form as a subclinical, lifelong infection. These infections can be reactivated years later when the immune system is suppressed, for example, as it is in *AIDS* (*acquired immune deficiency syndrome*), and cause pathology (Janeway et al. 2001).

As is noted by Janeway et al. (2001), for some infections, the type of pathology also depends on the host area where pathogens grow. For example:

> *Streptococcus pneumoniae* in the lung causes pneumonia, whereas in the blood it causes a rapidly fatal systemic illness.

As mentioned above, there is a wide range of diversity among the infectious diseases known to be related to glycan binding in terms of disease pathogenesis, primary locations of pathological processes, severity of the diseases, manifestations, and other characteristics. We will discuss about these diseases again in Sect. 14.2.2.4, in connection with the disease classifications.

14.1.3 Semantic Web Approach for PACDB Data Representation and Integration

14.1.3.1 Ontologies as Basis for Integration of Information Resources

In the life science domain, the Semantic Web approach is widely used for integration of data from different biological databases or other types of information resources (Cheung et al. 2007). For data sources with similar syntactic and semantic representation, the integration of them is not a difficult process. If data from different data sets are represented in the same RDF format, it is unnecessary to map one format to another, and multiple data sets can be used by applications in the same way. Also, the queries for the different data sources can be performed through the same interface, such as a SPARQL (Simple Protocol and RDF Query Language) query end point. On the other hand, semantically related data, which are represented in the RDF format, can be easily linked to each other, regardless of the terminology used to describe them.

14.1.3.2 Ontologies as Knowledge Bases for Domain Knowledge Representation

The Semantic Web standards such as RDF Schema (RDFS), Web Ontology Language (OWL) (McGuinness and Harmelen 2004), Simple Knowledge Organization System (SKOS) (Miles and Brickley 2005), and others can be used to describe the knowledge of a particular domain in the form of ontologies. The domain knowledge can be characterized by the concepts that represent entities of the domain, by the relations between concepts, and by the properties of concepts that describe the characteristics of them. As is discussed by Lambrix et al. (2007), the ontologies can be classified into the following types: controlled vocabulary, taxonomy, thesaurus, data models, and knowledge bases. They also noted that:

> most biological ontologies are controlled vocabularies, taxonomies or thesauri, but there are also ontologies that are knowledge bases and use OWL as their representation language.

In an introduction of biological ontologies, Lambrix et al. (2007) discussed about the benefits of using ontologies for domain knowledge representation and noted that:

> ontologies lead to a better understanding of a field and to more effective and efficient handling of information in that field.

14.1.3.3 Semantic Web Technologies in PAConto Development

As highlighted before, the Semantic Web approach is now widely used to describe the complexity of biomedical information. Therefore, we have decided to organize PACDB information by representing its data in RDF format and to integrate its content with other biomedical resources by using Semantic Web technologies. Along with this, we have decided to create an ontology of infectious diseases known to be related to glycan binding. For this purpose PAConto has been designed and developed. PAConto is not only an RDF representation of PACDB data, but this ontology can also be considered as a knowledge base for the related biomedical domain. This can be argued because PAConto contains all of the required components for it, such as (1) concepts and instances that represent the actual entities in the domain, (2) many types of semantic relations between concepts, (3) properties (attributes) of concepts, and (4) axioms, which represent facts that we think is true in the domain area (Lambrix et al. 2007). Furthermore, the representation languages for PAConto are RDF Schema, OWL, and SKOS.

14.2 Development of PAConto

In this section, we will show in detail the structure of PAConto with information about the classes, relations, and properties. Next, we will discuss about using additional information from external biomedical resources, such as the Medical Subject Headings (MeSH) vocabulary (MeSH 2015), and others, to annotate PACDB data. Then, we will describe the integration of PAConto data with other glyco-related databases, such as the MonosaccharideDB database (MonosaccharideDB 2015) and the GlycoEpitope database (Okuda et al. 2015; GlycoEpitope 2015), which provide their data in RDF format. Finally, we will discuss the linkage of PAConto data with internal resources such as the Japan Consortium for Glycobiology and Glycotechnology Database (JCGGDB) (JCGGDB 2015).

14.2.1 Overview of PAConto

PAConto is the RDF representation of PACDB data and ontology of infectious diseases known to be related to glycan binding. PAConto contains not only data from PACDB, but also additional information like various classifications, link information to internal and external information resources, and some information extracted from these resources.

There are three components of PAConto. The main component is the data from PACDB that are represented in RDF format and recorded in the paconto.rdf file. This file is available at http://jcggdb.jp/rdf/diseases/paconto. The next component is the PAConto vocabulary that is recorded in the paconto-schema.rdf file and includes the ontology description and definition of classes, individuals, and properties for PAConto ontology. This RDF file for the PAConto vocabulary is available at http://jcggdb.jp/rdf/diseases/paconto-schema. The third component is the various classifications of PACDB topics that were developed on the basis of the scientific literature and information resources from the UMLS (Unified Medical Language System) (UMLS 2015). These classifications are defined in paconto-schema.rdf and used in paconto.rdf for describing PACDB data. In the user interface, these classifications are used for annotation of PACDB data and for data searching and filtering. We will introduce this functionality of our user interface in Sect. 14.3.2.

The first and third components make up the content of PAConto which is provided by the user interface.

14.2.2 Structure of PAConto Ontology

In this subsection, we will explain at first the classes and properties from publicly available RDF vocabularies and then the self-defined classes and individuals that we defined in our vocabulary for the PAConto ontology. All of these classes, individuals, and properties are used to describe the semantics of PACDB data and semantic relations among different topics of PACDB. Next, we will explain about how we use the SKOS concept class and semantic relation properties for RDF representation of PACDB data. After, we will describe the various classifications of PACDB topics that we have created and added to PAConto. Finally, we will mention about the integration of PAConto with other information resources.

14.2.2.1 Namespaces and Classes from the Public Ontologies and Vocabularies Used in PAConto

In the Semantic Web approach, all entities and relations between them need to be characterized by classes and properties from publicly available RDF vocabularies and ontologies or by self-defined classes and properties. The new vocabulary for

PAConto ontology (http://jcggdb.jp/rdf/diseases/paconto-schema) was created to respond to the need for detailed representation of PACDB data. In case there are already existing publicly available RDF vocabularies or ontologies that can be used for definition of our concepts and relations among them, we used the classes and properties from these vocabularies and ontologies. For using them in our ontology, we included the base URL of their namespaces in the RDF files. The namespaces, classes, and some of the properties that were used are listed below.

Namespaces

- skos: <http://www.w3.org/2004/02/skos/core#>
- rdfs: <http://www.w3.org/2000/01/rdf-schema#>
- rdf: <http://www.w3.org/1999/02/22-rdf-syntax-ns#>
- owl: <http://www.w3.org/2002/07/owl#>
- dct: <http://purl.org/dc/terms/>
- bibo: <http://purl.org/ontology/bibo/>
- prism: <http://prismstandard.org/namespaces/basic/2.0/>
- uniprot: <http://purl.uniprot.org/core/>
- glycan: <http://purl.jp/bio/12/glyco/glycan#>
- ggdsch: <http://jcggdb.jp/rdf/diseases/ggdonto-schema#>
- gdgsch: <http://jcggdb.jp/rdf/diseases/gdgdb-schema#>
- gmsch: <http://jcggdb.jp/rdf/diseases/gmncbi-schema#>
- pacsch: <http://jcggdb.jp/rdf/diseases/paconto-schema#>
- paconto: <http://jcggdb.jp/rdf/diseases/paconto#>

Classes and Properties from the Public Vocabularies and Ontologies

- <owl:Class rdf:about=“http://purl.jp/bio/12/glyco/glycan#component/”>
- <owl:Class rdf:about=“http://purl.jp/bio/12/glyco/glycan#compound/”>
- <owl:Class rdf:about=“http://purl.jp/bio/12/glyco/glycan#glycan_epitope/”>
- <owl:Class rdf:about=“http://purl.jp/bio/12/glyco/glycan#glycoconjugate/”>
- <owl:Class rdf:about=“http://purl.jp/bio/12/glyco/glycan#glycolipid/”>
- <owl:Class rdf:about=“http://purl.jp/bio/12/glyco/glycan#glycoprotein/”>
- <owl:Class rdf:about=“http://purl.jp/bio/12/glyco/glycan#saccharide/”>
- <owl:Class rdf:about=“http://purl.jp/bio/12/glyco/glycan#N-glycan/”>
- <owl:Class rdf:about=“http://purl.jp/bio/12/glyco/glycan#O-glycan/”>
- <owl:Class rdf:about=“http://purl.jp/bio/12/glyco/glycan#monosaccharide/”>
- <owl:Class rdf:about=“http://purl.jp/bio/12/glyco/glycan#polysaccharide/”>
- <owl:Class rdf:about=“http://purl.jp/bio/12/glyco/glycan#glycoconjugate_sequence/”>
- <owl:Class rdf:about=“http://purl.org/ontology/bibo/Journal/”>
- <owl:Class rdf:about=“http://purl.org/ontology/bibo/Article/”>
- <owl:Class rdf:about=“http://purl.org/ontology/bibo/AcademicArticle/”>
- <owl:ObjectProperty rdf:about=“http://purl.uniprot.org/core/strain/”>
- <owl:ObjectProperty rdf:about=“http://purl.uniprot.org/core/rank/”>
- <owl:ObjectProperty rdf:about=“http://purl.uniprot.org/core/organism/”>
- <owl:ObjectProperty rdf:about=“http://purl.jp/bio/12/glyco/glycan#has_affinity_to/”>

- <owl:ObjectProperty rdf:about="http://purl.jp/bio/12/glyco/glycan#has_motif/">
- <owl:ObjectProperty rdf:about="http://purl.jp/bio/12/glyco/glycan#has_epitope/">
- <owl:ObjectProperty rdf:about="http://purl.org/dc/terms/references/">
- <owl:DatatypeProperty rdf:about="http://purl.uniprot.org/core/scientificName/"/>">
- <owl:DatatypeProperty rdf:about="http://purl.jp/bio/12/glyco/glycan#has_sequence/">
- <owl:DatatypeProperty rdf:about="http://purl.org/ontology/bibo/pmid/">
- <owl:DatatypeProperty rdf:about="http://prismstandard.org/namespaces/basic/2.0/publicationName/">
- <owl:AnnotationProperty rdf:about="http://purl.jp/bio/12/glyco/glycan#has_strain/">
- <rdf:Property rdf:about="http://purl.org/dc/terms/date/">
- <rdf:Property rdf:about="http://purl.org/dc/terms/description/">

The namespaces of the specifications RDF, RDFS, OWL, and SKOS indicate the standards for the Semantic Web to represent thesauri, taxonomies, controlled vocabularies, ontologies, and so on. Therefore, these namespaces are always used in ontology description. The namespaces of the vocabularies DCT (Dublin Core Metadata Terms), PRISM, and BIBO (Bibliographic Ontology) are used in our ontology for ontology definition, and for description of the scientific articles that have been used as references for the creation of PACDB.

As can be seen from the above lists, many classes from the GlycoRDF ontology (Ranzinger et al. 2015; GlycoRDF 2015) are used in our PAConto ontology. GlycoRDF is an ontology to standardize glycomics data in RDF format, and we used many classes and properties from this ontology for describing concepts related to host glycan ligands, target sources, and microbial glycan-binding proteins. Also, we used properties for taxonomic data from the UniProt Knowledgebase (UniProtKB) (UniProtKB 2015) for describing concepts of our ontology related to pathogenic microorganisms. Below we will mention the classes and individuals that we newly defined in the PAConto ontology.

14.2.2.2 Classes and Their Instances Defined in the PAConto Ontology

In this subsection, we will briefly explain the main classes and some of their instances that are defined in the PAConto vocabulary to use for description of the semantic of PACDB data and semantic relations between topics of data and for creation of the classifications.

As shown in Table 14.1, the many classes of PAConto vocabulary are defined as subclasses of more general classes (superclasses) defined in publicly available RDF vocabularies and ontologies. Due to this, we wanted to make a connection with other RDF vocabularies, but also to create ways to accurately describe our data. As an

Table 14.1 Names and definitions for the main classes of the PAConto ontology

Class name	Label (rdfs:label)	Superclasses from public ontologies (rdfs:subClassOf)	Description (rdfs:comment)
pacsch:Microorganisms	Microorganisms	uniprot:Taxon	For describing information about microorganisms
pacsch:OrganismForms	Organism forms		Forms of microorganisms
pacsch:MicrobialGlycanBindingProteins	Microbial glycan-binding proteins	glycan:glycan_binder	For describing information about microbial glycan-binding proteins: lectins and glycosaminoglycan-binding proteins
pacsch:OccurInOrganismInReference	Occurrence in organism in reference		Occurrence of microbial lectins in microorganism reported in PubMed references
pacsch:TargetTissuesAndCells	Target tissues and cells	glycan:source_natural	For describing information about target tissues and cells in hosts
pacsch:GlycansPACDB	PACDB glycans	glycan:glycoconjugate glycan:saccharide	For describing information about glycan ligands recorded in PACDB
pacsch:LigandsStructuralFeatures	Carbohydrate ligands structural features	glycan:component	For describing information about structural features (characteristics) and structural parts of carbohydrate ligands; the individuals of this class are features (characteristics) or substructures of glycans or glycoconjugates
pacsch:GlycoEpitopesPACDB	GlycoEpitopes in PACDB	glycan:glycan_epitope	For describing information about glycoepitopes recorded as carbohydrate ligands in PACDB
pacsch:TropismInReference	Tropism in reference		Tissue/cell tropism of microorganisms reported in PubMed references
pacsch:AffinityInReference	Affinity in reference		Affinity of host glycans to microbial lectins reported in PubMed references
pacsch:InteractionInReference	Interaction in reference		Interaction of microbial lectins with host glycan ligands reported in PubMed references

pacsch:DiseasesPACDB	PACDB diseases	ncit:Diseases_and_Disorders	For describing information about diseases from PACDB
pacsch:DiseasesClassificationsMeSH	Classifications of diseases using MeSH	ncit:Diseases_and_Disorders	For describing classifications of diseases using MeSH vocabulary (MeSH 2015)
		uniprot:Annotation	
pacsch:ReferencesPACDB	PACDB references	glycan:citation	For describing information about references that were used in creation of PACDB
		bibo:Journal	
		bibo:Article	
		bibo:AcademicArticle	
pacsch:SemanticRelationsPAConto	Semantic relations PAConto	owl:ObjectProperty	For describing information about semantic relations between concepts in PAConto

ncit:Diseases_and_Disorders http://ncicb.nci.nih.gov/xml/owl/EVS/Thesaurus.owl#Diseases_and_Disorders (NCIt 2015)
MeSH Medical Subject Headings (MeSH 2015)

example, the pacsch:MicrobialGlycanBindingProteins class is defined as a subclass of the glycan:glycan_binder class, and there are some properties also defined in our ontology which have the pacsch:MicrobialGlycanBindingProteins class as the domain in their definition.

Many classes listed in this table are defined for describing the meaning of the topics of PACDB data, which were discussed above in Sect. 14.1.2. This applies to such classes as pacsch:Microorganisms, pacsch:TargetTissuesAndCells, pacsch:GlycansPACDB, pacsch:DiseasesPACDB, etc. To represent the information about semantic relations between topics of PACDB data, the pacsch:SemanticRelationsPAConto class with its subclass and instances was defined and added to the PAConto vocabulary.

The classes and their instances listed in Tables 14.1 and 14.2 also are used for creating various classifications of PACDB topics. For example, the pacsch:DiseasesClassificationsMeSH class from Table 14.1 and the instances of this class such as pacsch:DiseasesClassificationSystems, pacsch:DiseasesClassification Pathogens, and pacsch:DiseasesClassificationAnimal from Table 14.2 are used for the creation of three related classifications of diseases recorded in PACDB. Also, the instances of the ggdsch:ConceptType class such as pacsch:Microbial GlycanBindingProteinForms, pacsch:TargetsClassificationNCIT, and pacsch: GlycansTypes are used for creating the classifications of microbial glycan-binding proteins, target sources, and host glycan ligands, respectively. We will discuss about these classifications again in Sect. 14.2.2.4.

As shown in Table 14.2, the four instances of the pacsch:OrganismForms class are defined in our ontology. These instances indicate various types of pathogenic microorganisms, such as bacteria, viruses, fungi, and parasitic protozoa, and as we mentioned in Sect. 14.1.2.1, these pathogens cause diseases recorded in PACDB.

In the next subsection, we will describe the instances of the SKOS concept class that are the main elements of our PAConto ontology.

14.2.2.3 Concepts and Concept Relations

In the development of our ontology, we used the Semantic Web approach and Semantic Web technologies as RDF, RDFS, OWL, and SKOS. The various elements from these RDF specifications and RDF vocabularies are used together in combination, but the main structure of our ontology is defined by using the concept class from the SKOS vocabulary. All topics of PACDB are represented as instances of the skos:Concept class, and different types of topics are indicated by the rdf:type property, which shows that this concept is also the instance of classes such as pacsch:Microorganisms, pacsch:MicrobialGlycanBindingProteins, pacsch:GlycansPACDB, pacsch:DiseasesPACDB, etc. (see Table 14.1). These main concepts of the PAConto ontology with semantic relations between them are shown in Fig. 14.1.

Above, we mentioned that the pacsch:SemanticRelationsPAConto class is used for describing the semantic relations between concepts. For this purpose, we also

Table 14.2 Names and definitions for the main instances, which are used in PAConto for ontology description

Instance name	Class of instance (rdf:type)	Label (rdfs:label)	Description (rdfs:comment)	Resources with additional information (rdfs:seeAlso)
pacsch:MicrobialGlycan BindingProteinForms	ggdsch:ConceptType	Microbial glycan-binding protein forms	For classification of microbial glycan-binding proteins	pacsch:MicrobialGlycan BindingProteins
pacsch:TargetsClassification NCIT	ggdsch:ConceptType	Classification of targets in host-pathogen interactions	Classifications of target sources in host-pathogen interactions based on the NCIt, with adaptations to PACDB data (NCIt 2015)	pacsch:TargetTissuesAndCells
pacsch:GlycansTypes	ggdsch:ConceptType	Glycans types	Types of carbohydrates and glycoconjugates recorded in PACDB	pacsch:GlycansPACDB glycan:glycoconjugate glycan:saccharide
pacsch:Diseases Classifications	ggdsch:ConceptType	Classifications of diseases	Classifications of infectious diseases recorded in PACDB by using MeSH vocabulary (MeSH 2015)	pacsch:DiseasesPACDB pacsch:DiseasesClassifications MeSH
pacsch:DiseasesClassification Systems	pacsch:DiseasesClassifications MeSH	Classification of diseases by systems	Classification of infectious diseases by organ systems using MeSH vocabulary (MeSH 2015)	
pacsch:DiseasesClassification Pathogens	pacsch:DiseasesClassifications MeSH	Classification of diseases by pathogens	Classification of infectious diseases by pathogens using MeSH vocabulary (MeSH 2015)	

pacsch:DiseasesClassificationAnimal	pacsch:DiseasesClassificationsMeSH	Classification of animal infectious diseases	Classification of animal infectious diseases using MeSH vocabulary (MeSH 2015)	
pacsch:Bacteria	pacsch:OrganismForm	Bacteria	Superkingdom bacteria	taxonomy:2
pacsch:Viruses	pacsch:OrganismForms	Viruses	Superkingdom viruses	taxonomy:10239
pacsch:Fungi	pacsch:OrganismForms	Fungi	Kingdom fungi	taxonomy:4751
pacsch:ParasiticProtozoa	pacsch:OrganismForms	Parasitic protozoa		

ggdsch:ConceptType class of GGDonto ontology (http://jcggdb.jp/rdf/diseases/ggdonto-schema) that is used for describing types of concepts
NCIt National Cancer Institute Thesaurus (NCIt 2015)
MeSH Medical Subject Headings (MeSH 2015)
taxonomy:2 http://purl.uniprot.org/taxonomy/2
taxonomy:10239 http://purl.uniprot.org/taxonomy/10239
taxonomy:4751 http://purl.uniprot.org/taxonomy/4751

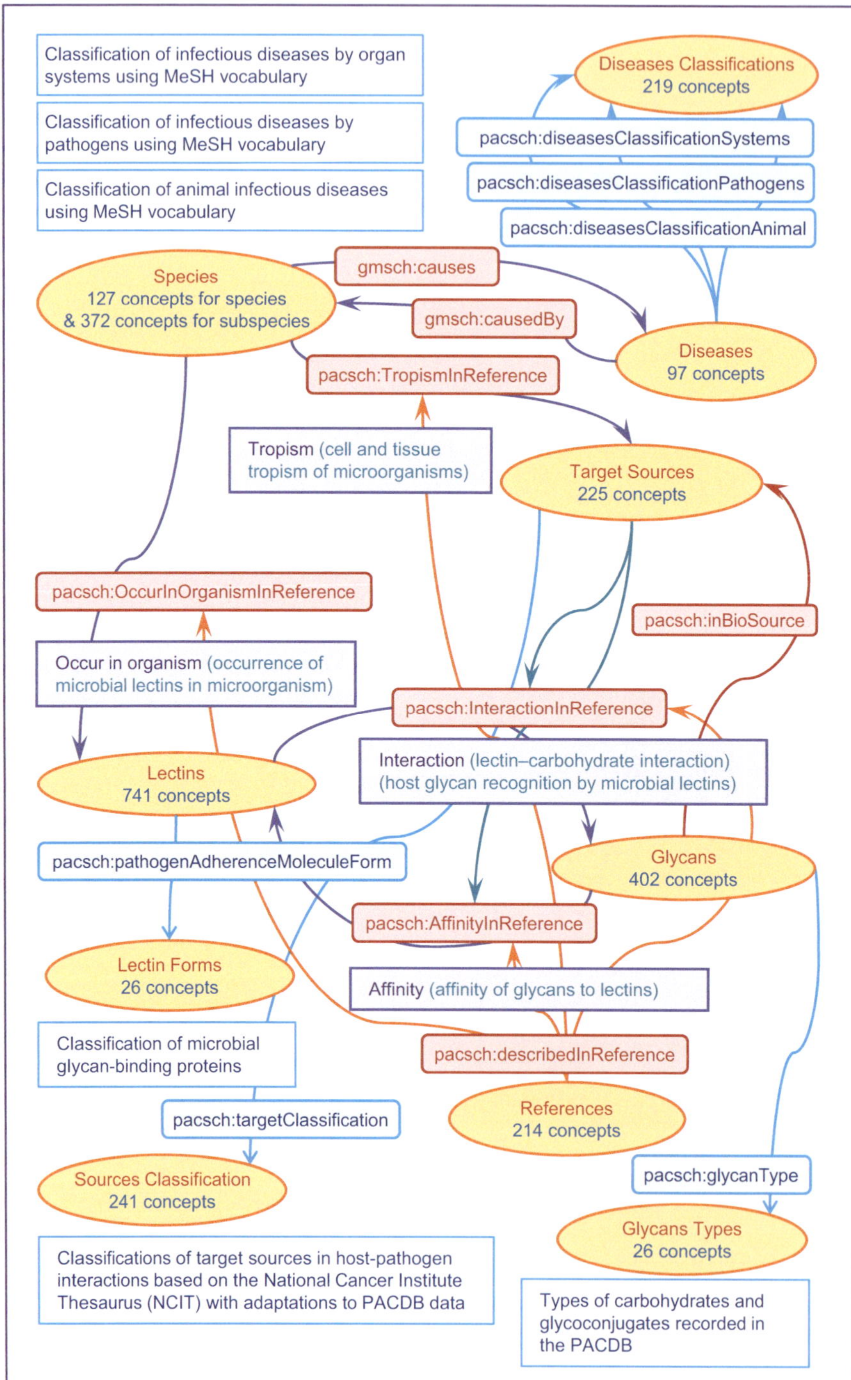

Fig. 14.1 Main concepts of ontology with semantic relations and classifications

defined other classes like pacsch:TropismInReference, pacsch:AffinityInReference, pacsch:InteractionInReference, and pacsch:OccurInOrganismInReference. These types of relations are listed in Table 14.1 and shown in Fig. 14.1. By using the term "InReference," we show that these semantic relations were reported in scientific articles, which we have used as references in the creation of PACDB.

Figure 14.1 also shows various classifications about which we will discuss in the next subsection.

14.2.2.4 Classifications of PACDB Topics

All of the classifications that we have defined in PAConto are shown in Fig. 14.1. The various sources of information that have been used for the creation of these classifications are listed below:

- Scientific literature (Esko and Sharon 2009; Holgersson et al. 2009; Singh et al. 2011; Varki et al. 2009; Varrot et al. 2013)
- Unified Medical Language System (UMLS) (UMLS 2015)
- Medical Subject Headings vocabulary (MeSH) (MeSH 2015)
- National Cancer Institute Thesaurus (NCIt) (NCIt 2015)

In the following list, we specify the classifications which we have created and added to our PAConto ontology:

- Classification of infectious diseases by organ systems using the MeSH vocabulary (MeSH 2015)
- Classification of infectious diseases by pathogens using the MeSH vocabulary (MeSH 2015)
- Classification of animal infectious diseases using the MeSH vocabulary (MeSH 2015)
- Classification of microbial glycan-binding proteins (Esko and Sharon 2009; Holgersson et al. 2009; Singh et al. 2011; Varki et al. 2009; Varrot et al. 2013)
- Classifications of target sources in host-pathogen interactions based on the National Cancer Institute Thesaurus (NCIt), with adaptations to PACDB data (NCIt 2015)

As an example, in Fig. 14.2 we give a classification of microbial glycan-binding proteins, which was created on the basis of the scientific literature (Esko and Sharon 2009; Holgersson et al. 2009; Singh et al. 2011; Varki et al. 2009; Varrot et al. 2013).

Not all of the classifications were defined in our ontology. For classification of pathogenic microorganisms, we used the RDF version of the UniProt taxonomy database (UniProt Taxonomy 2015) that is based on the NCBI taxonomy database (Sayers et al. 2009). In PAConto we included a property called pacsch:ncbiTaxonomyId with the values which specify the unique taxonomic identifiers (Taxonomy ID) from NCBI.

All of these classifications are used for annotation of various topics of PACDB. For this purpose, we linked concepts, which represent the main topics of PACDB

Classification of Microbial Glycan-Binding Proteins: Lectins and Glycosaminoclycan-binding proteins
Glycan-Binding Protein forms (groups)

- Bacteria
  - bacterial glycan-binding proteins
    - bacteria adhesins
      - fimbriae (e.g., fimbriae, type P fimbriae, type S fimbriae, type 1 fimbriae)
        - glycan-binding subunit of fimbriae (e.g., FimH - glycan-binding subunit FimH of type-1 fimbriae, PapG)
      - pili (e.g., Type IV Pili)
        - pilin (e.g., PilA)
      - bacterial surface lectins; membrane proteins (e.g., BabA, SabA, LecA, LecB)
      - bacterial GAG-binding proteins; sulfated glycosaminoglycan-binding proteins (for bacteria such as *Borrelia burgdorferi*, *Haemophilus influenzae*, *Mycobacterium tuberculosis*, *Neisseria gonorrhoeae*, *Staphylococcus aureus*)
    - toxins (exotoxins); secreted bacterial toxins (glycan-binding subunits & toxic subunits) (e.g., cholera toxin (CT), Toxin A, Bt toxins, Shiga toxin)
      - glycan-binding subunits of toxin (e.g., B subunits of CT)
- Viruses
  - viral glycan-binding proteins
    - viral lectins (e.g., VP1)
      - hemagglutinin protein (HA) (e.g., HA of *Influenza A virus*)
      - hemagglutinin-neuraminidase protein (HN) (e.g., HN of *Human parainfluenza virus 3*: hPIV 3)
      - hemagglutinin-esterase-fusion protein (HEF) (e.g., HEF of *Influenza C virus*)
    - viral GAG-binding proteins; sulfated glycosaminoglycan-binding proteins (e.g., glycoproteins gB, gC, gD, caspid proteins, protein E)
- Parasites (Protozoa); Parasitic Protozoa
  - protozoa glycan-binding proteins
    - parasite lectins; parasite adhesins (e.g., *Entamoeba histolytica* 260-kD heterodimeric lectin, EBA-175 (erythrocyte-binding antigen-175))
    - protozoa GAG-binding proteins; sulfated glycosaminoglycan-binding proteins (e.g. for *Plasmodium*)
- Fungi
  - fungal glycan-binding proteins; fungal lectins
    - fungal lectins from yeasts; yeast lectins
      - extracellular yeast lectins (e.g. for *Candida albicans*)
      - cell wall-associated yeast lectins (e.g. for *Histoplasma capsulatum*, *Paracoccidioides brasiliensis*, Saccharomyces cerevisiae)
    - fungal lectins from microfungi (molds) (e.g. for Sporothrix schenckii)

Fig. 14.2 Classification of microbial glycan-binding proteins

data, to concepts, which represent the information from created classifications. In the user interface, these classifications can be browsed, and they are used for searching and filtering information, for which users may search.

14.2.2.5 Integration of PAConto Ontology with Other Glyco-Related Databases

As is discussed by Aoki-Kinoshita et al. (2013), applying the Semantic Web approach to the representation of glycomics data is an important process in order to organize the semantics of these data and to integrate, not only glyco-related databases with each other but also to integrate glycomics data with genomics, proteomics, medical data, and other relevant information.

In the previous subsection, we explained how we use the information from external biological and medical resources. This annotation information is not related to carbohydrates, but is related to other topics of PACDB such as pathogenic microorganisms, diseases, lectins, etc., which we described above.

In this subsection, we will introduce the integration of the PAConto ontology with other glyco-related databases. There are two glyco-related databases, the MonosaccharideDB database (MonosaccharideDB 2015) and the GlycoEpitope database (Okuda et al. 2015; GlycoEpitope 2015), with which we integrate our ontology in the way that is provided by the Semantic Web approach. The MonosaccharideDB is a database that contains monosaccharide information. The GlycoEpitope is a database that provides information about carbohydrate antigens (glycoepitopes) and their recognizing antibodies. For our ontology we use these databases to specify the structures of host glycan ligands. To indicate resources from MonosaccharideDB and GlycoEpitope, we used properties defined in GlycoRDF and in PAConto. The properties from GlycoRDF are listed below: glycan:has_monosaccharide, glycan:has_alias, glycan:has_monosaccharide_notation_scheme, and glycan:has_epitope. The properties from PAConto are shown in Table 14.3.

The properties that indicate Glycan ID and Motif ID from JCGGDB (Japan Consortium for Glycobiology and Glycotechnology Database) (JCGGDB 2015) are also defined in PAConto and shown in Table 14.3. By using these properties with properties from GlycoRDF such as glycan:has_attached_glycan and glycan:has_motif, we created a link between our ontology and internal resources from JCGGDB in order to characterize structures of host glycan ligands and to view these structures as symbolic diagrams. When, through RDFizing JCGGDB, their contents will become available in RDF format, we will integrate JCGGDB resources with our ontology in the ways that are provided by Semantic Web technologies.

14.3 RDF/SPARQL-Based User Interface for PAConto

In this section, we will describe the system overview and system implementation of PAConto. Then, we will show in detail the RDF/SPARQL-based user interface with step-by-step instructions on how to use this system.

14.3.1 System Overview and System Implementation

Along with the PAConto ontology, which has been developed for PACDB, we have developed a system with a user interface, using which the users that are not familiar with the Semantic Web technologies can retrieve and search information of interest

Table 14.3 Names and definitions of properties, which are defined in PAConto and used for integration PAConto with other glyco-related databases

Property name	Data type (rdf:type)	Label (rdfs:label)	Description (rdfs:comment)
pacsch:isSameAsMonosaccharide	owl:ObjectProperty	Is same as monosaccharide	For specifying RDF resource describing this monosaccharide (usually MonosaccharideDB) (MonosaccharideDB 2015)
pacsch:epitopeID	owl:DatatypeProperty	Epitope ID	Epitope ID in GlycoEpitope database (GlycoEpitope 2015)
pacsch:epitopeName	owl:DatatypeProperty	Epitope name	Epitope name in GlycoEpitope database (GlycoEpitope 2015)
pacsch:jcggdbGlycanId	owl:DatatypeProperty	JCGGDB glycan ID	Glycan ID used in JCGGDB (JCGGDB 2015)
pacsch:jcggdbMotifId	owl:DatatypeProperty	JCGGDB motif ID	Motif ID used in JCGGDB (JCGGDB 2015)

JCGGDB Japan Consortium for Glycobiology and Glycotechnology Database (JCGGDB 2015)

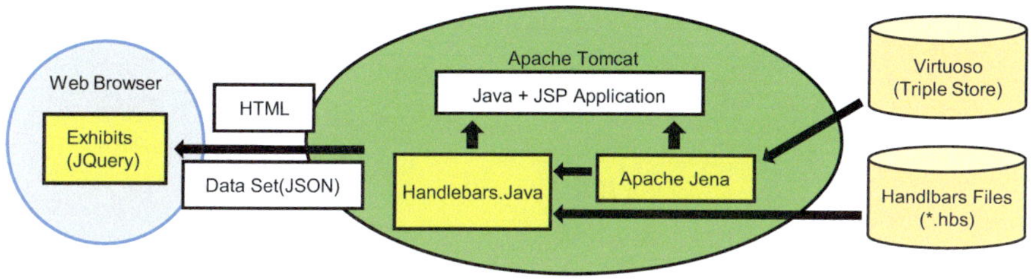

Fig. 14.3 System configuration, Handlebars.js (http://handlebarsjs.com/) is a JavaScript templating engine for the Semantic Web; Exhibit (http://simile-widgets.org/exhibit/) is a JavaScript framework that provides faceted browsing and text searching

from PAConto. This RDF/SPARQL-based user interface is available at http://acgg.asia/db/diseases/pacdb/.

In this user interface, by choosing the diseases, carbohydrate ligands, lectins, and so on from the respective classifications, the user can narrow down the subset of microorganisms of interest. On selecting one of the microorganisms of interest, the user can view detailed information about the diseases, interactions between pathogen adherence molecules and host glycan ligands, glycan structures, and other information related to the selected microorganism.

We have developed the basic structure of the Web system in Java + JSP. For better efficiency of programming, we used the Handlebars (http://handlebarsjs.com/) template as a definition of the Web pages. To process Handlebars files, we used the HandleBars.java which is a Java implementation of Handlebars. Also, we used Exhibit (http://simile-widgets.org/exhibit/) which is a JavaScript framework for large-scale data-rich interactive Web pages. Since Exhibit provides faceted browsing, sorting, and filtering functionality, structured data from multiple sets can be searched, filtered, and sorted by users. By using all of these implementations, the data from our ontology which contains rather complicated structure can be viewed, browsed, and searched, giving researchers the ability to find the information they are looking for very quickly. The system configuration with used applications and implementations is shown in Fig. 14.3.

14.3.2 Top Page of PAConto User Interface

The screenshot of the top page of the PAConto user interface is shown in Fig. 14.4. This top page is accessible at http://acgg.asia/db/diseases/pacdb/. All of the figures, which are presented in this section (Figs. 14.4, 14.5, 14.6, 14.7, 14.8, 14.9, 14.10, 14.11, 14.12, and 14.13), are the screenshots of the PAConto user interface.

As is mentioned above, the pathogenic microorganisms with their strains and subtypes are the main entries in the PAConto user interface, and detailed information that is shown in the page, which will be described in the following subsection, is

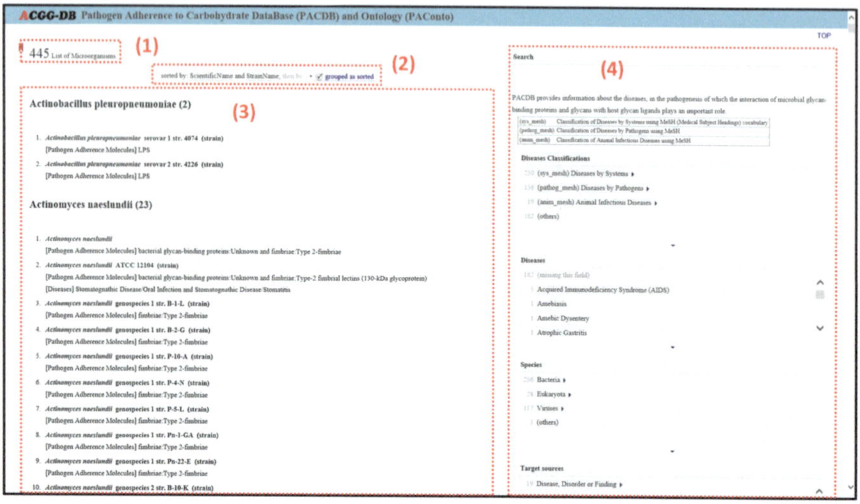

Fig. 14.4 Top page of PAConto user interface, list of pathogenic microorganisms. (*1*) Number of items in the list. (*2*) Menu for changing the layout of the list of pathogenic microorganisms. (*3*) List of pathogenic microorganisms. (*4*) Text searching and multiple facets for faceted navigation

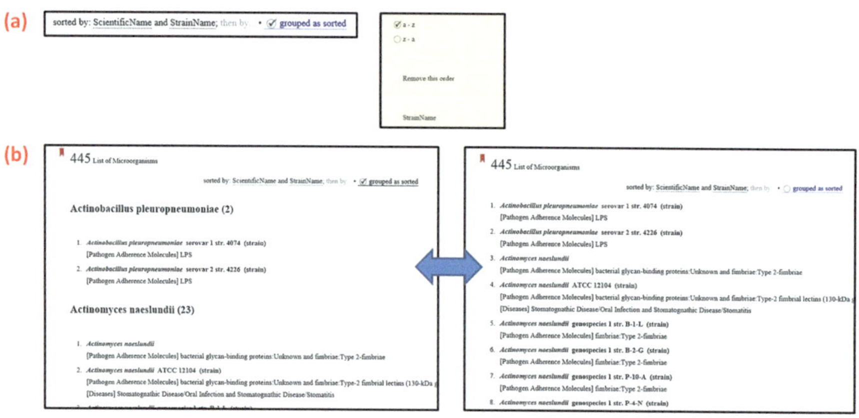

Fig. 14.5 Changing the layout of the list of microorganisms. (**a**) Menu for changing the layout of the list. (**b**) Different layouts of the list (grouped and ungrouped)

associated with one of the listed microorganisms. The list of pathogenic microorganisms and the number of items in this list are shown in Figs. 14.4(3) and 14.4(1), respectively. The layout of the list of the pathogenic microorganisms may be changed by selecting the items in the menu for changing the layout of the list. This menu is shown in Figs. 14.4(2) and 14.5(a). The different layouts of the list of microorganisms, which are sorted by [ScientificName and StrainName], are shown in Fig. 14.5(b) and differ by grouping condition (grouped or ungrouped).

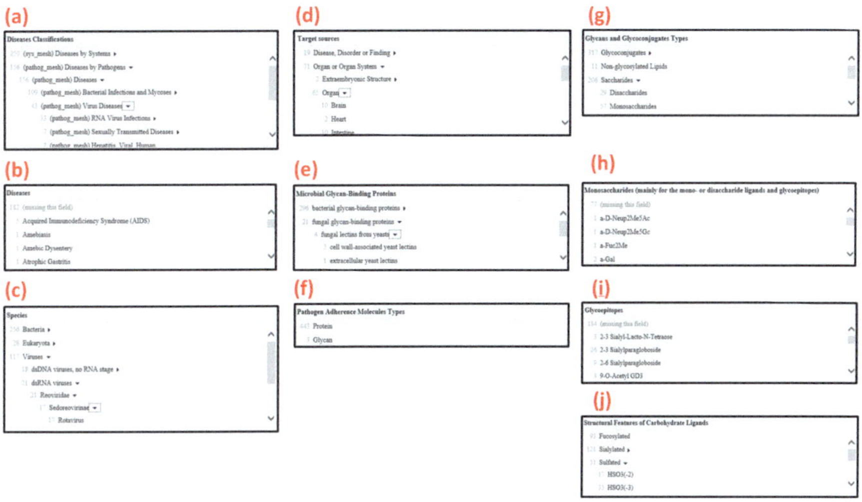

Fig. 14.6 Classifications and groups of topics on the top page; each classification or group corresponds to one facet set. (**a**) Diseases classifications. (**b**) Diseases. (**c**) Species. (**d**) Target sources. (**e**) Microbial glycan-binding proteins. (**f**) Pathogen adherence molecules types. (**g**) Glycans and glycoconjugates types. (**h**) Monosaccharides (mainly for the mono- or disaccharide ligands and glycoepitopes). (**i**) Glycoepitopes. (**j**) Structural features of carbohydrate ligands

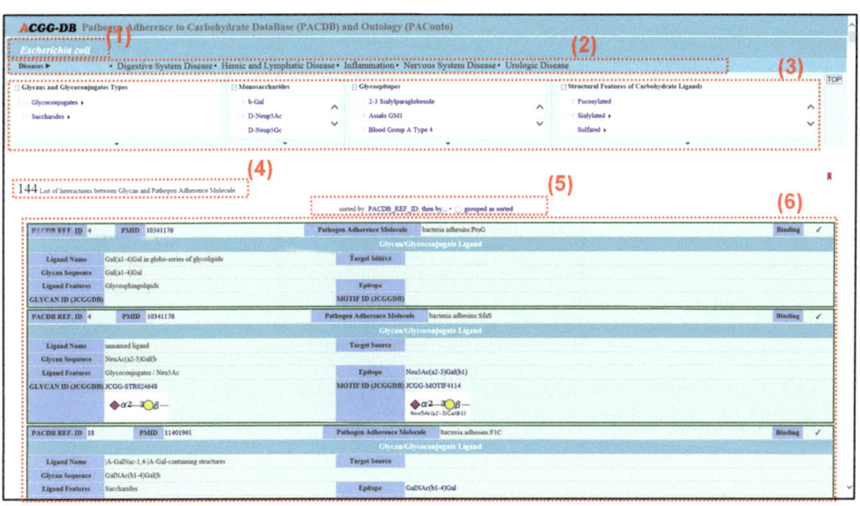

Fig. 14.7 Page of detailed information about the one pathogenic microorganism, *Escherichia coli*, is used as an example. (*1*) Microorganism. (*2*) Diseases. (*3*) Multiple facets for faceted navigation. (*4*) Number of items in the list of lectin-glycan interactions. (*5*) Menu for changing the layout of the list of lectin-glycan interactions. (*6*) List of lectin-glycan interactions

As is mentioned in the previous subsection, the text searching and faceted browsing (faceted navigation) are the functionalities that are provided in our system.

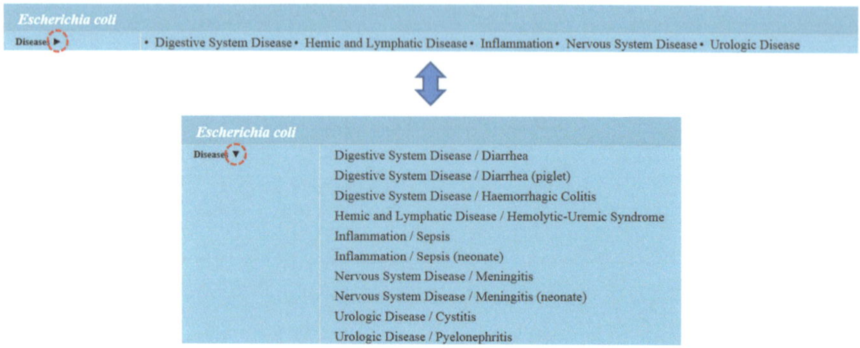

Fig. 14.8 Changing the layout of the view for diseases related to selected microorganism (list of categories of diseases only or drop-down list of categories of diseases with diseases names)

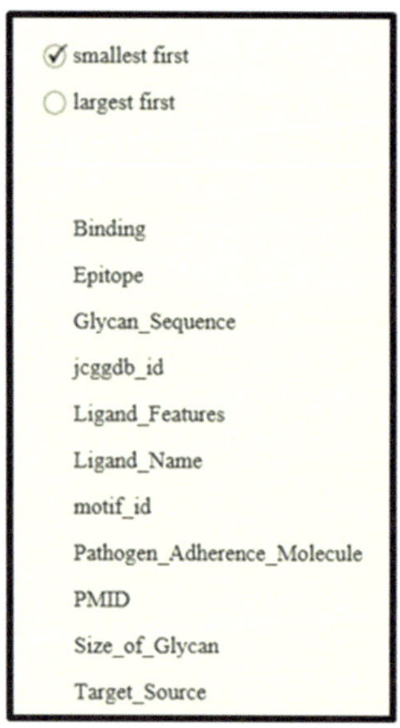

Fig. 14.9 Characteristics of the microbial glycan-binding proteins and host glycan ligands that are used as items for sorting the list of lectin-glycan interactions

Figure 14.4(4) shows the field for text searching and multiple facets for faceted navigation, which correspond to the data from various classifications and groups of topics.

The value of the input field [Search] in the upper-right corner of the top page is applied to the list of pathogenic microorganisms to generate the set of pathogens that are associated with this input string. The disease's names, symptoms, lectins names and types, host targets, host glycans and their characteristics, and so on may be used as keywords for text searching.

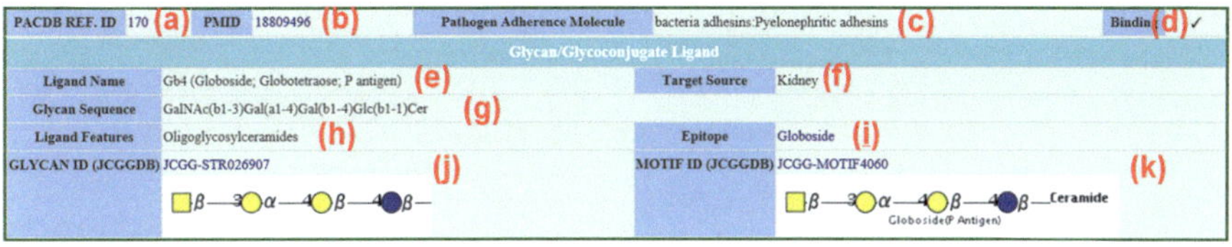

Fig. 14.10 Detailed information about microbial glycan-binding protein and host glycan ligand that correspond to one lectin-glycan interaction related to selected microorganism. (**a**) PACDB REF.ID (ID from PACDB). (**b**) PMID. (**c**) Pathogen adherence molecule: type and name. (**d**) Binding or not binding. (**e**) Ligand name. (**f**) Target source. (**g**) Glycan sequence. (**h**) Ligand features. (**i**) Epitope. (**j**) GLYCAN ID (JCGGDB). (**k**) MOTIF ID (JCGGDB)

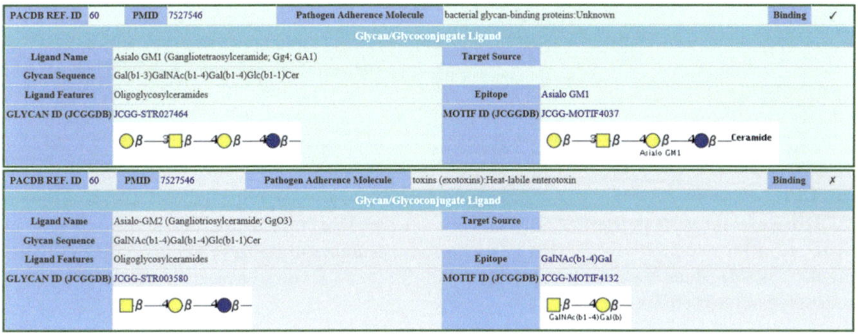

Fig. 14.11 Example of lectin-glycan interactions (binding or not binding)

By using various properties of data elements, the faceted browsing allows users to narrow down the data from large content and to explore the information, which is filtered by multiple filters selected by users themselves. In our user interface, for faceted browsing we use the various classifications that are publicly available or were defined in our ontology, as well as other groups of topics such as diseases, monosaccharides, glycoepitopes, and so on. Figure 14.6 and the list below show these classifications and groups of topics. Each classification or group corresponds to one facet list:

- (a) Diseases classifications
  - Classification of diseases using MeSH vocabulary
- (b) Diseases
  - List of names of diseases recorded in PACDB
- (c) Species
  - Species classification from UniProt taxonomy database

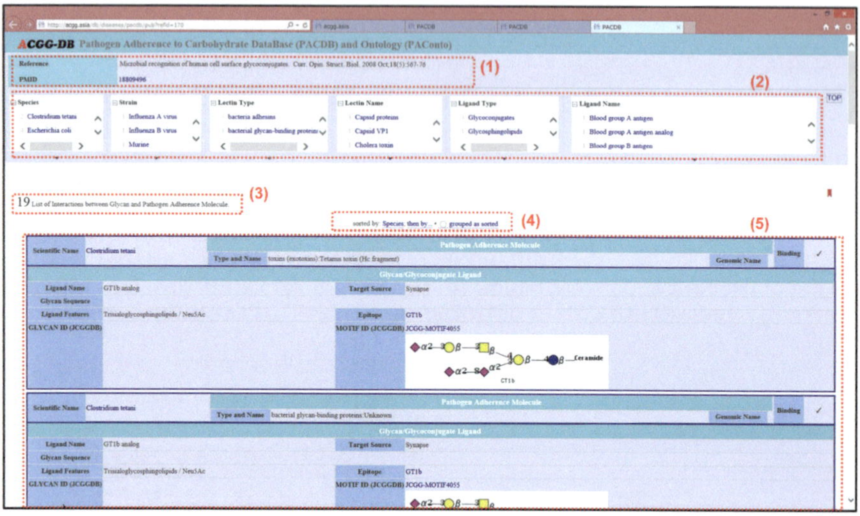

Fig. 14.12 Page of detailed information obtained from one scientific article; article with the PACDB REF. ID: 170 is used as an example. (*1*) Detailed information about this article with PMID. (*2*) Multiple facets for faceted navigation. (*3*) Number of items in the list of lectin-glycan interactions. (*4*) Menu for changing the layout of the list of lectin-glycan interactions. (*5*) List of lectin-glycan interactions

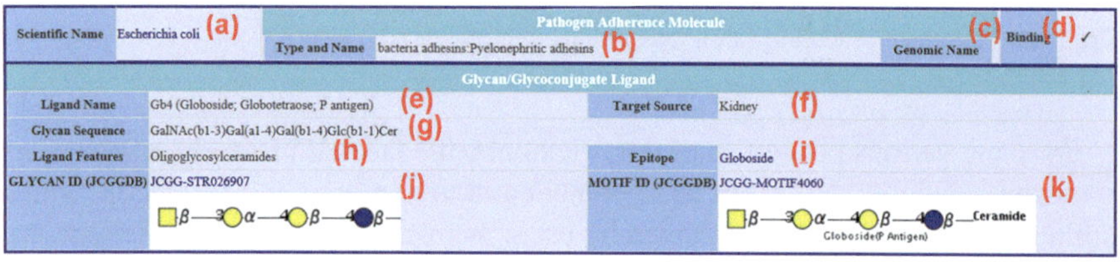

Fig. 14.13 Detailed information about microbial glycan-binding protein and host glycan ligand that correspond to one lectin-glycan interaction obtained from selected article. (**a**) Scientific name of pathogenic microorganism. (**b**) Pathogen adherence molecule: type and name. (**c**) Pathogen adherence molecule: genomic name. (**d**) Binding or not binding. (**e**) Ligand name. (**f**) Target source. (**g**) Glycan sequence. (**h**) Ligand features. (**i**) Epitope. (**j**) GLYCAN ID (JCGGDB). (**k**) MOTIF ID (JCGGDB)

- (d) Target sources
  - Classifications of targets based on the NCIt with adaptations to PACDB data
- (e) Microbial glycan-binding proteins
  - Classification of microbial glycan-binding proteins based on scientific literature
- (f) Pathogen adherence molecules types
  - Protein or glycan

- (g) Glycans and glycoconjugates types
  - List with hierarchical structure for types of glycans and glycoconjugates based on scientific literature
- (h) Monosaccharides (mainly for the mono- or disaccharide ligands and glycoepitopes)
  - List of names of monosaccharides from MonosaccharideDB database
- (i) Glycoepitopes
  - List of names of glycoepitopes from GlycoEpitope database
- (j) Structural features of carbohydrate ligands
  - List of structural features of carbohydrate ligands defined in PAConto in order to characterize them in detail

By choosing the facet filter items from multiple facet filter lists, the users can display only the pathogens that are associated with these selected items. This is an "OR" relationship between multiple facet filters. Because we use various classifications and groups of topics for faceted navigation, this allows users to better understand the structure of our content and our ontology, as well as the semantics of the data from PACDB, and allows them to quickly explore the information which they are looking for.

By selecting one of the pathogens shown in the pathogenic microorganisms list, users can access the page of detailed information, which we will describe in the following subsection.

14.3.3 Page of Detailed Information About a Pathogenic Microorganism

As described above, in the top page, the list of all pathogenic microorganisms, or that which is filtered by faceted filters, is displayed, and a particular microorganism can be selected to jump to a page of detailed information related to selected pathogen. To explain the items and information that are displayed in this page, we use as an example *Escherichia coli* (*E. coli*), and all of the figures, which are presented in this subsection (Figs. 14.7, 14.8, 14.9, 14.10, and14.11), are related to this pathogen.

The screenshot of the page of detailed information about a pathogenic microorganism is shown in Fig. 14.7. Figure 14.7(1) shows the selected microorganism, in this example *Escherichia coli*. Fig. 14.7(2) shows the list of categories of diseases for which *E. coli* is a disease-causing agent. If the user clicks the right-arrow [▶] at the right of the [Diseases] facet, a drop-down list of categories of diseases with disease names is displayed. These two layouts of the diseases list are demonstrated in Fig. 14.8. Figure 14.7(3) shows the multiple facets for faceted navigation, and all of these correspond to different characteristics of host glycan ligands. In this page, we use the following facet lists: [Glycans and Glycoconjugates Types], [Monosaccharides], [Glycoepitopes], and [Structural Features of Carbohydrate Ligands]. This is an "AND" relationship between multiple facets, and the explanation of them was given above.

The main information of this page is given in blocks corresponding to one lectin-glycan interaction. This information is shown in the next three parts of Fig. 14.7: Fig. 14.7(4) the number of items in the list of lectin-glycan interactions, Fig. 14.7(5) the menu for changing the layout of the list of lectin-glycan interactions, and Fig. 14.7 (6) the list of lectin-glycan interactions.

The drop-down menu can be used for changing the layout of the list of lectin-glycan interactions (Fig. 14.7(5)), the items for sorting the list of lectin-glycan interactions are displayed, as shown in Fig. 14.9. The characteristics of the microbial glycan-binding proteins and host glycan ligands are used for this sorting, and these characteristics are as follows: Binding, Epitope, Glycan_Sequence, jcggdb_id, Ligand_Features, Ligand_Name, motif_id, PACDB_REF_ID, Pathogen_Adherence_Molecule, PMID, Size_of_Glycan, and Target_Source. The meaning of these items will be mentioned in the explanation in Fig. 14.10.

As an example, in Fig. 14.10, we show the detailed information about microbial glycan-binding protein and host glycan ligand that correspond to one lectin-glycan interaction related to the selected microorganism. Below we will explain the parts of Fig. 14.10:

- (a) PACDB REF.ID
  - A unique ID that is assigned to every literature in PACDB; if the user clicks this ID, the corresponding [Page of Detailed Information obtained from one Scientific Article] will be displayed (See Sect. 3.4)
- (b) PMID
  - PubMed Unique Identifier; if the user clicks this PMID, the corresponding page at PubMed [http://www.ncbi.nlm.nih.gov/pubmed/] will be displayed
- (c) Pathogen adherence molecule
  - Name of the microbial glycan-binding protein from PACDB and its type from the classification of microbial glycan-binding proteins
- (d) Binding or not binding
  - Lectin-glycan interactions: binding or not binding
- (e) Ligand name
  - Name of host glycan ligand from PACDB
- (f) Target source
  - Name of host target source from PACDB

- (g) Glycan sequence
  - Sequence of host glycan ligand from PACDB
- (h) Ligand features
  - Types of glycans and glycoconjugates and structural features of carbohydrate ligands
- (i) Epitope
  - Name(s) of related glycoepitope(s) from the GlycoEpitope database; if the user clicks this name, the corresponding page at the [http://www.glycoepitope.jp] will be displayed
- (j) GLYCAN ID (JCGGDB)
  - For host glycan ligands: ID of corresponding glycans in JCGGDB and the structures of them using CFG (Consortium for Functional Glycomics) symbols; if the user clicks this ID, the corresponding page at JCGGDB [http://jcggdb.jp] will be displayed
- (k) MOTIF ID (JCGGDB)
  - For host glycan ligands: ID of corresponding motifs in JCGGDB and the structures of them using CFG symbols; if the user clicks this ID, the corresponding page at JCGGDB [http://jcggdb.jp] will be displayed

As shown in Fig. 14.11, the background color of the items in the list of lectin-glycan interactions depends on the type of these interactions: [✔ binding] or [✘ not binding] and makes their comparison easier.

By selecting one of the PACDB REF.ID shown in the list of lectin-glycan interactions, users will be able to access the page of detailed information obtained from one scientific article, which we will describe in the next subsection.

14.3.4 Page of Detailed Information Obtained from One Scientific Article

The screenshots (Figs. 14.12 and 14.13), which are presented in this subsection, are taken from the page of detailed information obtained from one scientific article. The information is provided on this page may be related to multiple pathogenic microorganisms in the case that they were all described in the selected scientific article. That is in contrast to the page which was explained in the previous subsection.

To explain the items and information that are displayed in this page, we use as example the scientific article with PACDB REF. ID 170, and both figures, which are presented in this subsection, show the information that was obtained from this scientific article. The detailed information about this article (including article title, publication date, journal title, volume and issue, starting and ending pages, PMID, and so on) is displayed in Fig. 14.12(1).

Figure 14.12(2) shows multiple facets for faceted navigation, and they correspond to the characteristic of the pathogenic microorganisms, pathogen adherence molecules, and the host glycan ligands. In this page, we use the next facet lists: [Species (Scientific Name of pathogen)], [Strain], [Lectin Type (Pathogen Adherence Molecule Type)], [Lectin Name (Pathogen Adherence Molecule Name)], [Ligand Type (from classification of Glycans and Glycoconjugates Types)], and [Ligand Name]. This is an "AND" relationship between multiple facets.

Similarly to the page which was described in detail in Sect. 14.3.3, the main information of this page is also given in blocks corresponding to one lectin-glycan interaction. This information is shown in the next three parts of Fig. 14.12: Fig. 14.12(3) the number of items in the list of lectin-glycan interactions, Fig. 14.12(4) the menu for changing the layout of the list of lectin-glycan interactions, and Fig. 14.7(5) the list of lectin-glycan interactions.

If the user clicks the drop-down menu for changing the layout of the list of lectin-glycan interactions, which is shown in Fig. 14.12(4), the items for sorting this list are displayed. The characteristics of the pathogen, microbial glycan-binding proteins, and host glycan ligands are used for this sorting, and these characteristics are as follows: Binding, Epitope, Glycan_Sequence, jcggdb_id, Ligand_Features, Ligand_Name, motif_id, Pathogen_Adherence_Molecule, Size_of_Glycan, Strain, and Target_Source. The meanings of these items for sorting (except for "strain," that is, the name of strain of pathogenic microorganism) were mentioned in the explanation of Fig. 14.10 in the previous subsection.

One example of lectin-glycan interactions from the list, which is shown in Fig. 14.12(5), is given in Fig. 14.13. In Fig. 14.13 we show the detailed information about microbial glycan-binding proteins and host glycan ligand that corresponds to one lectin-glycan interaction. This displayed information was recorded in PACDB on the basis of data from the selected scientific article (PACDB REF. ID: 170). Below we will explain the parts of Fig. 14.13.

- (a) Scientific name of pathogenic microorganism
  - Scientific name of pathogenic microorganism from PACDB
- (b) Pathogen adherence molecule: type and name
  - Type of microbial glycan-binding protein from classification of microbial glycan-binding proteins and its name from PACDB
- (c) Pathogen adherence molecule: genomic name
  - Genomic name of microbial glycan-binding protein from PACDB

The meanings of the items, which are marked on Fig. 14.13 as (d) ~ (k), correspond with the meaning of the items in Fig. 14.10(d) ~ (k), respectively, and were described in the previous subsection.

In this Sect. 14.3, we have explained how the RDF/SPARQL-based user interface can be used to easily access information from the PAConto ontology. We hope that these step-by-step instructions will allow the users to quickly access all of the features of our system and help them quickly start using this system.

14.4 Summary and Conclusions

In conclusion, we have developed an ontology with a SPARQL-based Web system named PAConto, which is an RDF representation of PACDB data and ontology of infectious diseases known to be related to glycan binding. The content of our system includes data from PACDB and newly created classifications of topics of PACDB data. These classifications were created on the basis of various sources of information, including scientific literature and information resources of biological and medical institutes and organizations. The content of our system is represented as an ontology which is described using standards for the Semantic Web, such as RDF, OWL, and SKOS. Our ontology can be used as a knowledge base for the biomedical domain of infectious diseases known to be related to glycan binding and for a better understanding of this field. Also, we have integrated our ontology with other glyco-related databases and relevant information from external biological and medical resources. By this, we have enriched the content of our system with additional information from related biomedical resources, and it can be helpful for the effective retrieval of information which is stored in different sources and databases. For this integration, we have used the Semantic Web approach for resources in RDF format and other techniques, such as the making of a simple link between two resources and the creation of our own RDF representation of information, for resources which have no RDF format. When, through RDFizing, the content of these resources will become available in RDF format, we will integrate them with our ontology in the ways that are provided by Semantic Web technologies.

For retrieving and searching PAConto ontology, we have developed the RDF/SPARQL-based user interface. SPARQL statements are used for querying the RDF data from our ontology. By using this user interface, the data from PAConto ontology with complicated structure can be quickly viewed, browsed, and searched by the users, without requiring knowledge of SPARQL.

To help users get the best out of our system, we have created a user's guide that introduces the topics of the content and explains the features of the user interface. As previously mentioned, this user's guide is available at http://acgg.asia/db/diseases/. Any users that have any problems or questions about PAConto can contact our development team through email at jcggdb-ml@aist.go.jp.

We would like to hope that our system will be helpful for all researchers interested in the aspects of glycoscience in better understanding infectious diseases known to be related to glycan binding and in effectively obtaining the information in this field.

Acknowledgments This work has been supported by National Bioscience Database Center (NBDC) of Japan Science and Technology Agency (JST).

References

Aoki-Kinoshita KF, Bolleman J, Campbell MP, Kawano S, Kim JD et al (2013) Introducing glycomics data into the Semantic Web. J Biomed Semant 4(1):39

Breedam WV, Pöhlmann S, Favoreel HW, de Groot RJ, Nauwynck HJ (2014) Bitter-sweet symphony: glycan-lectin interactions in virus biology. FEMS Microbiol Rev 38(4):598–632

Cheung KH, Smith AK, Yip KYL, Baker CJO, Gerstein MB (2007) Semantic Web approach to database integration in the life sciences. In: Baker CJO, Cheung KH (eds) Semantic Web: revolutionizing knowledge discovery in the life sciences. Springer, New York, pp 11–30

Comprehensive Monosaccharide Database (MonosaccharideDB) (2015) http://monosaccharidedb.org/. Accessed 25 Nov 2015

Esko JD, Sharon N (2009) Microbial lectins: Hemagglutinins, Adhesins, and Toxins. In: Varki A, Cummings RD, Esko JD, Freeze HH, Stanley P, Bertozzi CR, Hart GW, Etzler ME (eds) Essentials of glycobiology, 2nd edn. Cold Spring Harbor Laboratory, New York, Chapter 34

GlycoEpitope (2015) GlycoEpitope database. http://www.glycoepitope.jp/. Accessed 25 Nov 2015

GlycoRDF (2015) Current version. https://github.com/ReneRanzinger/GlycoRDF/blob/master/ontology/glycan.owl. Accessed 25 Nov 2015

Holgersson J, Custafsson A, Caunitz S (2009) Bacterial and viral lectins. In: Gabius HJ (ed) The sugar code: fundamentals of glycosciences. Wiley-VCH, Weinheim, pp 279–300

Janeway CA Jr, Travers P, Walport M, Shlomchik MJ (2001) Adaptive immunity to infection. In: Immunobiology: the immune system in health and disease, 5th edn. Garland Science, New York, Chapter 10

Japan Consortium for Glycobiology and Glycotechnology DataBase (JCGGDB) (2015) http://jcggdb.jp/index_en.html. Accessed 25 Nov 2015

Juge N (2012) Microbial adhesins to gastrointestinal mucus. Trends Microbiol 20(1):30–39

Lambrix P, Tan H, Jakoniene V, Strömbäck L (2007) Biological ontologies. In: Baker CJO, Cheung KH (eds) Semantic Web: revolutionizing knowledge discovery in the life sciences. Springer, New York, pp 85–99

McGuinness DL, Harmelen FV (eds) (2004) OWL web ontology language overview. https://www.w3.org/TR/2004/REC-owl-features-20040210/. Accessed 25 Nov 2015

Medical Subject Headings (MeSH) (2015) MeSH vocabulary. https://www.nlm.nih.gov/mesh/. Accessed 25 Nov 2015

Miles A, Brickley D (eds) (2005) SKOS core vocabulary specification. https://www.w3.org/TR/2005/WD-swbp-skos-core-spec-20051102/. Accessed 25 Nov 2015

National Cancer Institute Thesaurus (NCIt) (2015) NCIthesaurus. https://ncit.nci.nih.gov/ncitbrowser/. Accessed 25 Nov 2015

Okuda S, Nakao H, Kawasaki T (2015) GlycoEpitope: database for carbohydrate antigen and antibody. In: Taniguchi N, Endo T, Hart GW et al (eds) Glycoscience: biology and medicine. Springer, Tokyo, pp 267–273

Ranzinger R, Aoki-Kinoshita KF, Campbell MP, Kawano S, Lütteke T et al (2015) GlycoRDF: an ontology to standardize glycomics data in RDF. Bioinformatics 31(6):919–925

Sayers EW, Barrett T, Benson DA, Bryant SH, Canese K et al (2009) Database resources of the National Center for Biotechnology Information. Nucleic Acids Res 37(Database issue):D5–D15

Singh RS, Bhari R, Kaur HP (2011) Characteristics of yeast lectins and their role in cell–cell interactions. Biotechnol Adv 29(6):726–731

Unified Medical Language System (UMLS) (2015) UMLS knowledge sources. https://www.nlm.nih.gov/research/umls/. Accessed 25 Nov 2015

Universal Protein Resource (UniProt) Taxonomy (2015) http://www.uniprot.org/taxonomy/
Universal Protein Resource (UniProt) UniProt Knowledgebase (UniProtKB) (2015) http://www.uniprot.org/uniprot/. Accessed 25 Nov 2015
Varki A, Etzler ME, Cummings RD, Esko JD (2009) Discovery and classification of glycan-binding proteins. In: Varki A, Cummings RD, Esko JD, Freeze HH, Stanley P, Bertozzi CR, Hart GW, Etzler ME (eds) Essentials of glycobiology, 2nd edn. Cold Spring Harbor Laboratory, New York, Chapter 26
Varrot A, Basheer SM, Imberty A (2013) Fungal lectins: structure, function and potential applications. Curr Opin Struct Biol 23(5):678–685
Viswanathan K, Chandrasekaran A, Srinivasan A, Raman R, Sasisekharan V, Sasisekharan R (2010) Glycans as receptors for influenza pathogenesis. Glycoconj J 27(6):561–570

Part VI
Glycan Data Analysis

Chapter 15
RINGS: A Web Resource of Tools for Analyzing Glycomics Data

Kiyoko F. Aoki-Kinoshita

Abstract This chapter introduces the web resource called RINGS which can be accessed at http://www.rings.t.soka.ac.jp/ and provides freely available tools for analyzing and data mining glycomics data. These include DrawRINGS, an applet to draw glycan structures and obtain the KCF (KEGG Chemical Function)-formatted representation of the structure; ProfilePSTMM, a tool based on a probabilistic model which can extract patterns within groups of glycan structures; Glycan Miner, a tool for finding the common glycan substructures within a group of glycan structures; MCAW, a multiple alignment tool for glycans; Kernel Tool, a tool to find glycan substructures that are particular to one glycan structure data set compared to another; GPP, glycan pathway predictor for generating *N*-glycan biosynthesis pathways; and utilities for converting between various glycan structure representations. Each tool will be described in its usage and application, along with tips for using the tools most effectively. Moreover, RINGS provides a data management system as well as a feedback system to allow users to store their data on the RINGS server as well as to interact with developers to improve RINGS functionality.

Keywords Glycans • Glycobiology • Glycoinformatics • Data mining • Web resource

15.1 Introduction

Due to the development of semi-high-throughput technologies for generating glycomics data in the glycosciences, such as mass spectrometry (MS), nuclear magnetic resonance (NMR), and glycan arrays, numerous glycomics databases and algorithms have been developed to analyze this data in the past decade. The latter (algorithms), however, have been mainly published in the informatics literature, and few have been implemented as easy-to-use tools for the wet-lab glycobiologist. Therefore, RINGS was developed since 2006 in order to provide a

K.F. Aoki-Kinoshita (✉)
Faculty of Science and Engineering, Soka University, 1-236 Tangi-machi, 192-8577, Hachioji, Tokyo, Japan
e-mail: kkiyoko@soka.ac.jp

K.F. Aoki-Kinoshita (ed.), *A Practical Guide to Using Glycomics Databases*,
DOI 10.1007/978-4-431-56454-6_15

user-friendly interface to use these tools. In the current version of RINGS, the main text representation of glycan structures used is KCF (KEGG Chemical Function Format) (Akune et al. 2010). As new formats are being developed, utilities to translate between KCF and other formats have also been incorporated into RINGS. In the future, however, RINGS tools will be designed to allow any format to be used as input so that the user need not be concerned with the glycan representation to use.

15.2 DrawRINGS

DrawRINGS is an easy-to-use, web-based drawing tool for searching glycans using KCaM (Aoki et al. 2004). Users can also obtain KCF text for glycans by drawing structures on the canvas.

15.2.1 DrawRINGS Input

The DrawRINGS interface is a Java applet that runs in web browsers. Because of security risks, many browsers these days do not permit applets to run by default. However, it is still possible to run them with the appropriate browser settings, which are described in the *Tips* Sect. 15.2.3. The main drawing interface for DrawRINGS is illustrated in Fig. 15.1. There are nine buttons across the top, which

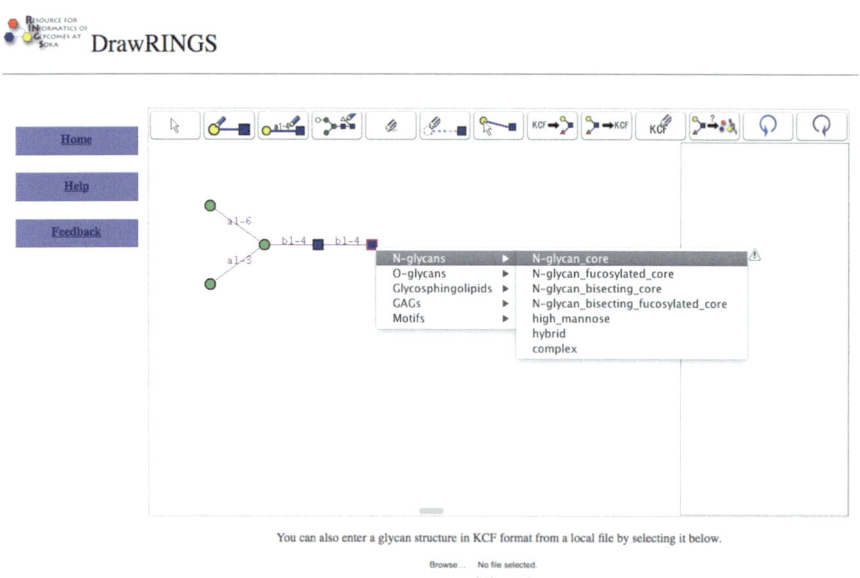

Fig. 15.1 The DrawRINGS input screen, illustrating the menu displayed when using the *core structure* button

can be used for drawing monosaccharides (called nodes), glycosidic bonds (called edges), erasing nodes, making queries, etc. Table 15.1 describes each button and its functionality.

Table 15.1 Each button in DrawRINGS allows users to perform the following tasks

Button	Function	Description
	Select	Select monosaccharides on the canvas. Multiple monosaccharides can be selected at once by dragging the mouse over the monosaccharides. Selected items can be moved around on the canvas
	Draw monosaccharide	A selection menu of monosaccharide names appears when the drawing area is clicked, and the selected monosaccharide will be drawn. Moreover, a pop-up window to input a monosaccharide name appears when "Other" is selected such that names of components not appearing in the list can be entered
	Draw glycosidic bond	A glycosidic bond can be drawn between two monosaccharides by clicking on the corresponding monosaccharides. A selection menu of glycosidic conformations appears when a bond is clicked. Moreover, a pop-up menu to input user-defined conformations appears by selecting "Other"
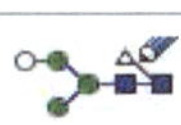	Core structure	Core structures include the common *N*-glycan core structures (including bisecting GlcNAc and core fucosylation), *O*-glycan structures, glycosaminoglycans (GAGs), and common motifs such as Lewis structures, glycolipids, lactosamine repeats, etc. Any of these basic structures can be drawn at once by selecting this button and clicking on the canvas, where a hierarchical menu of structures will be displayed
	Clear canvas	The entire canvas will be cleared of monosaccharides and bonds
	Delete monosaccharide	Delete a particular monosaccharide and its attached bond(s)
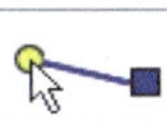	Move monosaccharide	A selected monosaccharide can be dragged across the canvas

(continued)

Table 15.1 (continued)

Button	Function	Description
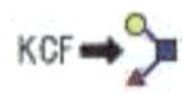	Draw KCF	The structure corresponding to the KCF in the text area will be drawn on the canvas
	Output KCF	The structure being drawn will be output into the text area in KCF format
	Clear KCF	The KCF in the text area will be deleted
	Run query	A pop-up menu of options to use when querying the RINGS databases for the drawn structure will appear. Options to select a score matrix, the database to search (RINGS, GlycomeDB, or both), and similarity score options are available (similarity to sort results by percentage matched, or matched to sort by number of matching components). If a score matrix is selected, the algorithm will use the values from the score matrix to weight the linkages appropriately. Please refer to the text for details
	Undo	The last action performed on the canvas will be undone
	Redo	The last action undone will be re-performed

15.2.1.1 Drawing/Inputting Structures onto the Canvas

To draw on the canvas, first use the buttons for either drawing monosaccharides or drawing structures. Then additional monosaccharides and glycosidic bonds can be added as necessary. Adjustments to the topology of the glycan can be made by using the select button and dragging the necessary components around on the canvas. Selected components can also be copied using Ctrl-C (or Command-C on Mac OS) and pasted using Ctrl-V (or Command-V on Mac OS).

A file containing a glycan in KCF format can also be loaded into DrawRINGS by selecting it using the *Browse...* button located below the canvas. Once selected, the structure can be drawn on the canvas by clicking the *Upload* button.

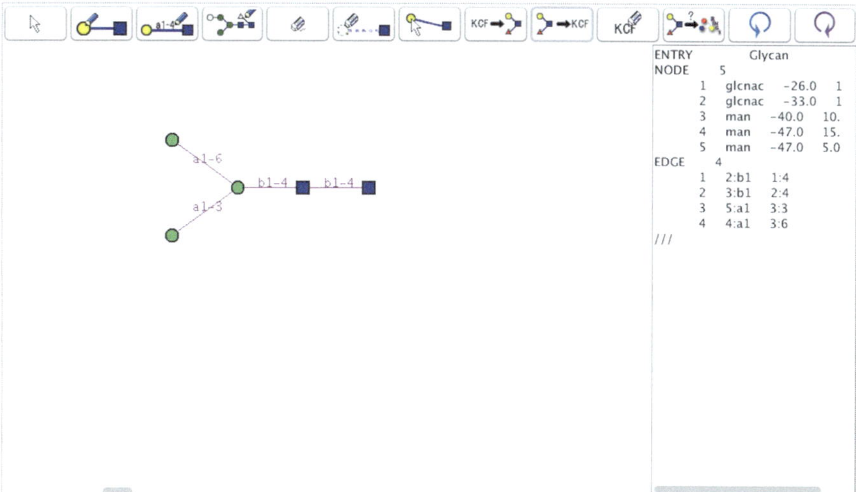

Fig. 15.2 The DrawRINGS screen after KCF has been outputted. The text area containing the KCF data can be edited manually as needed. The updated glycan image can then be obtained by clicking the *Draw KCF* button, which will update the canvas with the glycan represented by the KCF text

15.2.1.2 Obtaining KCF

Once a structure is drawn on the canvas, the KCF-formatted text representing it can be obtained by clicking the *Output KCF* button. Figure 15.2 illustrates how DrawRINGS looks when the KCF text is shown in the text area on the right-hand side of the canvas. This text area can be edited manually to modify the glycan displayed as well. The updated glycan figure can then be obtained by clicking the *Draw KCF* button, which will update the canvas with the glycan represented by the KCF text.

15.2.1.3 Running Queries

The RINGS database stores all glycan structures derived from KEGG GLYCAN (Aoki-Kinoshita and Kanehisa 2015) and GlycomeDB (Ranzinger et al. 2011), in addition to several structures that were manually curated from the literature, up to 2009. The structures in these data can be queried using the *Query* button in DrawRINGS.

The algorithm behind this query tool is KCaM (Aoki et al. 2004), which implements an efficient tree structure matching algorithm. Three options are provided by this query tool: (1) score matrix, (2) database, and (3) score type. A glycan score matrix can be analogized to BLOSUM or PAM for protein sequence alignment, where particular glycosidic linkages and monosaccharides that should be considered *similar* to one another, in particular within a certain glycan class, can be scored more highly to obtain higher-scoring glycans within the same class. Predefined score matrices are available, including *N*-glycans, *O*-glycans, and sphingolipids. A fourth matrix, called Link_similarity, is a hand-generated score matrix based on expert knowledge of glycosidic linkages and monosaccharides that may be substituted more frequently with other glycosidic linkages and monosaccharides, respectively. For example, β1-3 may be considered more similar to β1-4 than α1-2.

For the target database, users may choose between querying either the data RINGS, which includes KEGG GLYCAN and manually curated structures, or GlycomeDB, or both of these databases. Because the latter contains almost 40,000 structures, it takes a few minutes longer to query it.

Finally, the score type allows users to specify the scoring system to use, and the results are always displayed in descending order of score. The *Similarity* option gives a maximum score of 100, whereas the *Matched* option sums the score of matching monosaccharides and linkages, where each matching glycosidic linkage (and its monosaccharide) receives a score of 100. In fact, the *Similarity* option actually divides the *Matched* score by the larger of the two structures. Because each glycosidic linkage can receive a maximum score of 100, the maximum possible score would be 100 times the maximum number of linkages. For example, if a glycan having nine glycosidic linkages is compared against one of four, the maximum possible score is 100 times the larger of the structures (in this case, $100 \times 9 = 900$). Thus, if the number of matching linkages (and their monosaccharides) is, say, five, then the *Similarity* score between the two structures would be 56 %($= 500 \div 900$).

15.2.2 DrawRINGS Query Output

As a result of running a query, a new browser window listing the matching results in order of score will be displayed, as in Fig. 15.3. This figure is a screenshot of a query using the structure drawn in Fig. 15.2 and using the Link_Similarity score matrix on the GlycomeDB data. The results list the structures in order of similarity, where it is apparent that the most similar structures are displayed at the top. Each result is linked to the Glycan Entry page, shown in Fig. 15.4, which displays the various text representations of the structure. This page also provides a link to the KEGG GLYCAN database and reaction information about the glycan if available.

15.2.3 DrawRINGS Tips

The following are some tips regarding using DrawRINGS:

- Java needs to be enabled in the browser with appropriate security settings (usually at low levels).

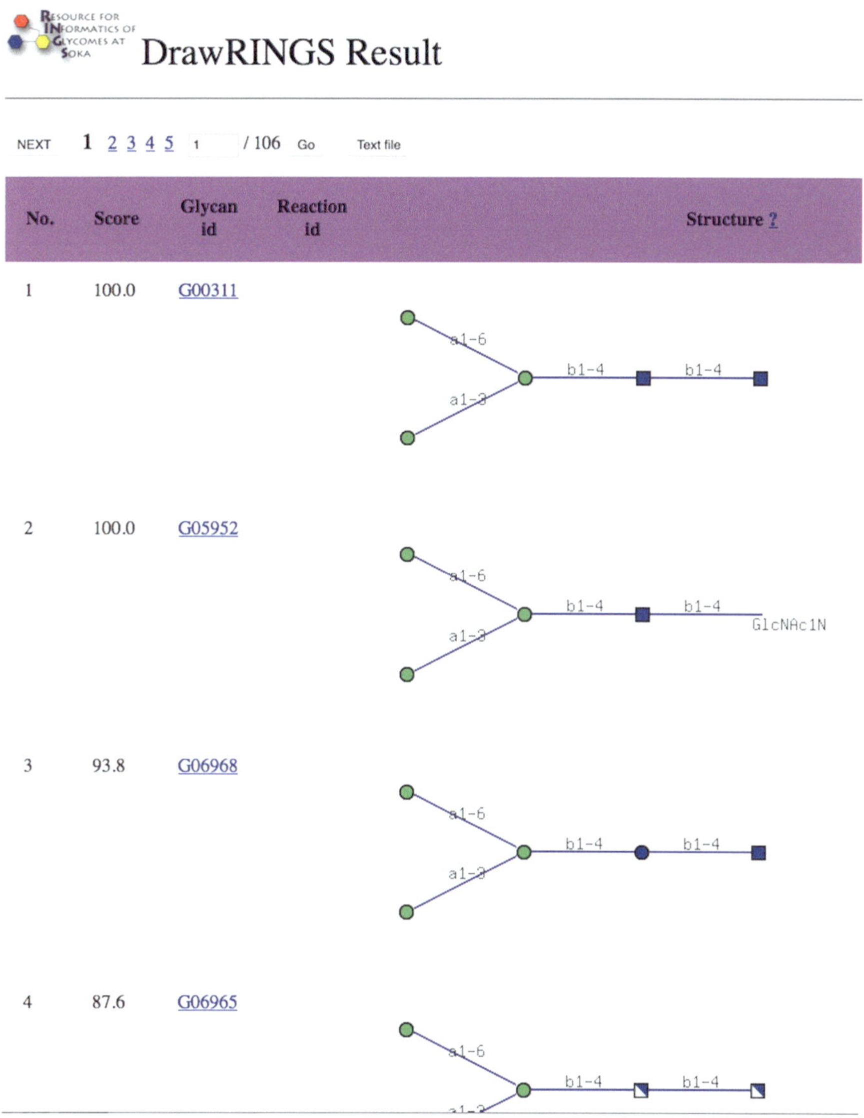

Fig. 15.3 The DrawRINGS result screen using the structure drawn in Fig. 15.2 and using the Link_Similarity score matrix on the GlycomeDB data. The results list the structures in order of similarity, where it is apparent that the most similar structures are displayed at the top. Each result is linked to the Glycan Entry page of RINGS (Fig. 15.4), which displays detailed information about the structure

RESOURCE FOR INFORMATICS OF GLYCOMES AT SOKA

Glycan Entry

Entry

G06928

Interaction ID

Class

N-Glycan

KCF

```
ENTRY       G06928                        Glycan
COMPOSITION (GlcNAc)2 (Man)3
MASS        910.8
CLASS       Glycoprotein; N-Glycan
DBLINKS     CCSD: 32909
            GlycomeDB: 21902
            JCGGDB: JCGG-STR012821
NODE        5
            1   GlcNAc      0      0
            2   GlcNAc    -10      0
            3   Man       -20      0
            4   Man       -30      5
            5   Man       -30     -5
EDGE        4
            1     2:b1    1:4
            2     3:a1    2:4
            3     4:b1    3:6
            4     5:b1    3:3
///
```

LINUCS Code

[][D-GlcNAc]{[(4+1)][b-D-GlcNAc]{[(4+1)][a-D-Man]{[(3+1)][b-D-Man]{}[(6+1)][b-D-Man]{}}}}

GlycoCT

```
RES
1b:x-dglc-HEX-1:5
2s:n-acetyl
3b:b-dglc-HEX-1:5
4s:n-acetyl
5b:a-dman-HEX-1:5
6b:b-dman-HEX-1:5
7b:b-dman-HEX-1:5
LIN
1:1d(2+1)2n
2:1o(4+1)3d
3:3d(2+1)4n
4:3o(4+1)5d
5:5o(3+1)6d
6:5o(6+1)7d
```

Structure

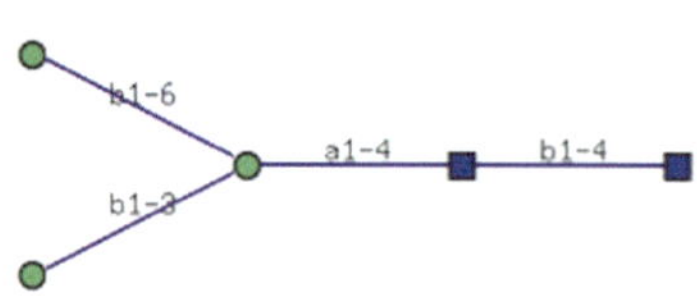

Fig. 15.4 A Glycan Entry page for the fourth matching structure in the DrawRINGS query results shown in Fig. 15.3. The various text representations of this structure are listed, along with a link to the original KEGG GLYCAN database

- Google Chrome may not work, but DrawRINGS should work on Safari, Firefox, and Internet Explorer.
- Set the URLs *http://www.rings.t.soka.ac.jp* and *http://rings.t.soka.ac.jp* in the list of allowed sites in Java settings.
- At the time of this writing, selecting the text area and using Ctrl-C (or Command-C on Macs) to copy the KCF are disabled on many browsers. Thus, in order to download the KCF text, it is recommended that DrawRINGS be used as a registered user having logged in and then a query be made on the RINGS database, in order to store the KCF in the user's data library. The User Management system is described later in Sect. 15.9 and instructions on running queries are described in the next subsection.

15.3 ProfilePSTMM

The ProfilePSTMM Tool was built based on the probabilistic model proposed called PSTMM (probabilistic sibling-dependent tree Markov model) (Ueda et al. 2004), which was then modified such that profiles could be directly output from the model (Aoki-Kinoshita et al. 2006). In simple terms, this model *learns* patterns from the input data in a process called *training*. Once a model is *trained* on a particular data set, it can essentially be used to *test* other glycans to determine how well it matches the pattern that has been learned by the model. In other words, during the *training* process, ProfilePSTMM attempts to find patterns inherent within a given glycan data set containing many disparate glycan structures. These patterns are then visualized in the results as distributions of monosaccharides at particular positions of a branched topology. Once a pattern has been learned, it can potentially be used to assess, or test, whether a given glycan structure that was not in the original data set should belong to the set or not. That is, it assesses the glycan to see how well it matches the pattern learned.

As for the ProfilePSTMM Tool, it performs the *training* process and simply outputs the *trained* pattern that was *learned* from the data. The model is currently not stored for later *testing* on other data. However, such advanced analyses are being planned at the time of this writing.

The ProfilePSTMM model was originally developed to aid in the understanding of glycan patterns being recognized by glycan-binding agents such as proteins, viruses, and antibodies, whose data can be obtained from glycan array databases. It uses a probabilistic model to overcome the noise that can be found in many of the glycan-binding data available, as well as weak-binding patterns which may skew analytical results. This is in comparison to tree alignment algorithms, which do not allow for such flexibility.

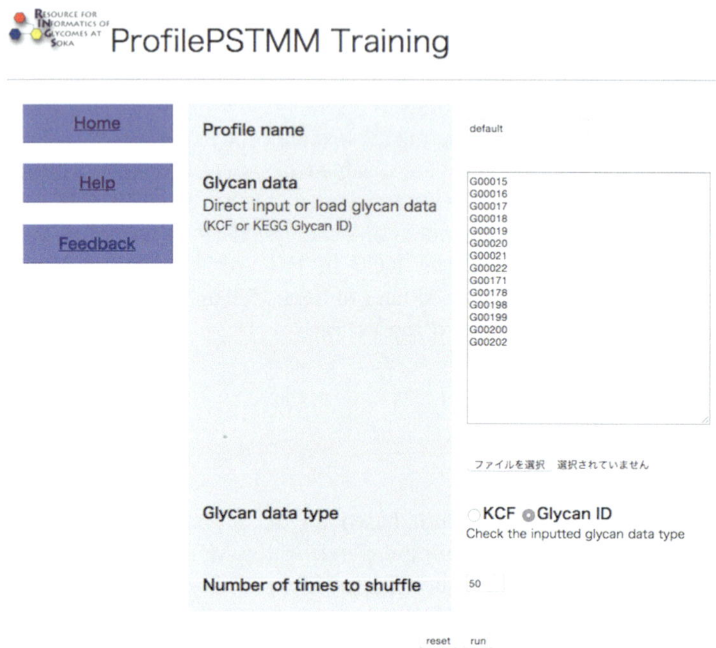

Fig. 15.5 A snapshot of the default ProfilePSTMM Tool input page. A list of KEGG GLYCAN IDs or KCF data can be specified

15.3.1 ProfilePSTMM Input

The ProfilePSTMM Tool takes as input a data set of glycans. The intention of this tool was to allow for the analysis of glycans that exhibit binding to a particular agent. Thus, data such as that from lectin and glycan arrays can be analyzed with this tool. A snapshot of the input screen is given in Fig. 15.5. A list of KEGG GLYCAN IDs or a concatenated list of KCF data can be specified. The *shuffle number* refers to the number of times the computation should be repeated to avoid local optima. Because this tool is a probabilistic model, it is suggested that it is run at least five times in order to find the optimal solution. A file can also be specified containing the input data, either as KEGG GLYCAN IDs or KCF data, instead of cutting and pasting into the text area.

15.3.2 ProfilePSTMM Output

As a result of training on the default glycan structures using shuffle number 5, the resulting glycan profiles will be displayed, as in Fig. 15.6. Looking closely at these results, one can see that there were basically two distinct profiles that

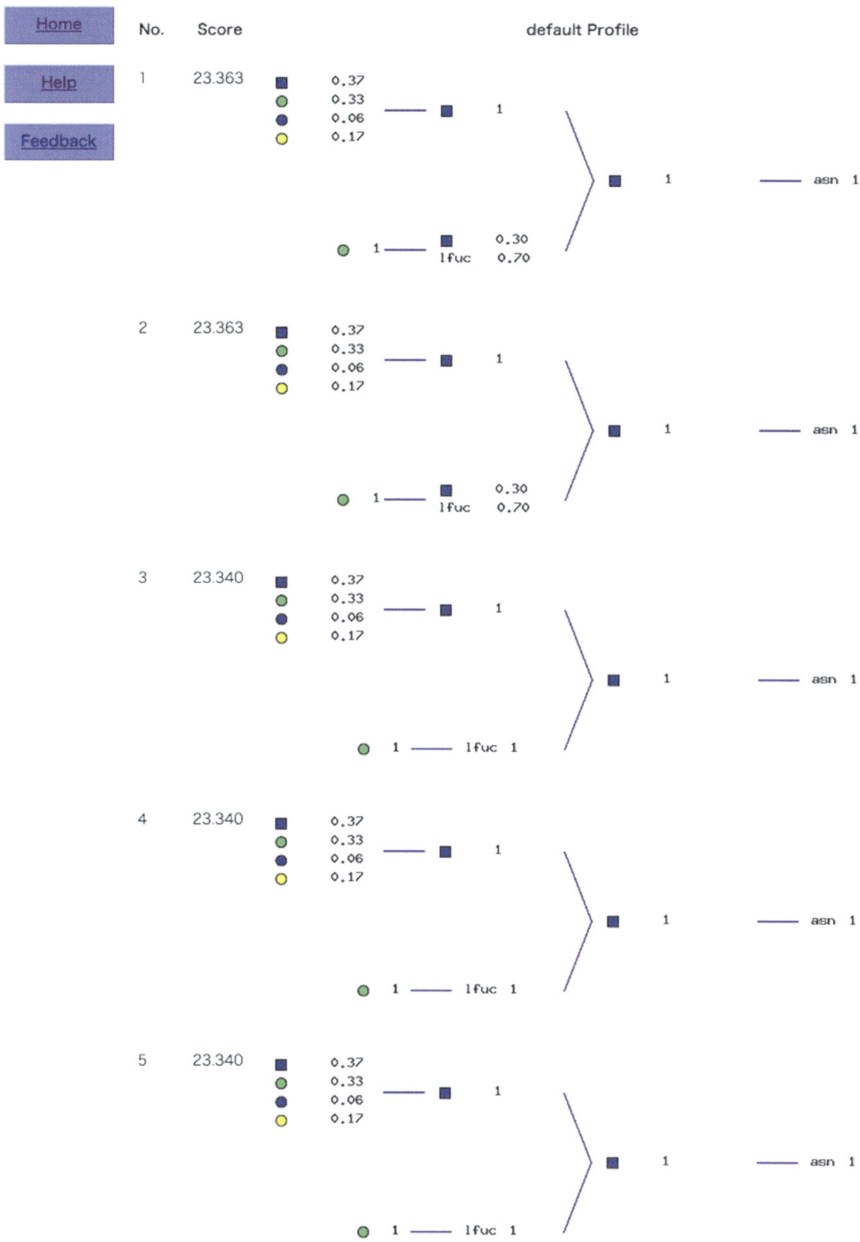

Fig. 15.6 Result screen after running ProfilePSTMM

were trained based on the data set, one with score 23.363 and one with 23.340. In both profiles, the biantennary pattern learned contained the same distribution of monosaccharides at the terminating ends. Mannose appeared 100 % on one end, and

GlcNAc, mannose, glucose, and galactose appeared with a ratio of 37:33:6:17. The single difference between these profiles appears at the neighboring position, where fucose either appears alone (100 %) or may be replaced with a 30 % probability with GlcNAc. The computed score is actually computed based on the *likelihood* score computed when learning the patterns in the glycan data set. Future work will entail assessing these scores to obtain significant values that can also be displayed to the user.

Note that the shape, or topology, of the profile is currently based on the concept of *maximum common subtree* of the input data, meaning that the shape of the profile that is output will be based on the smallest glycan structure that exists in the input data set. This causes the learning process to combine monosaccharide distributions into a smaller space, causing the results to be misleading. This is a currently known issue that is being updated and will be made more flexible in future versions, when it will be easier to ascertain the results.

15.3.3 ProfilePSTMM Tips

- At the time of this writing, the ProfilePSTMM Tool can be used for glycan data sets that do not contain very small glycan structures, but rather contain glycans that are of comparable size.
- In order to specify stronger binding glycans compared to others in the data set, weights can be implied by specifying multiple copies of heavily binding glycans in the input.
- The computation time may increase with larger shuffling number and larger number of glycans in the input.

15.4 Glycan Miner

The Glycan Miner Tool is based on a mining algorithm called α-closed frequent subtree mining (Hashimoto et al. 2008). This tool takes as input a large data set of glycans and attempts to find significant subtrees among the data. In a sense, it can find unique overly expressed glycan substructures within a set of glycan structures. It is suited for the analysis of glycan profiling data by mass spectrometry or glycan-binding data such as that from glycan arrays.

15.4.1 Glycan Miner Input

The input screen for the Glycan Miner Tool is shown in Fig. 15.7. The default values allow users to test the tool to see what kind of results can be obtained. The data

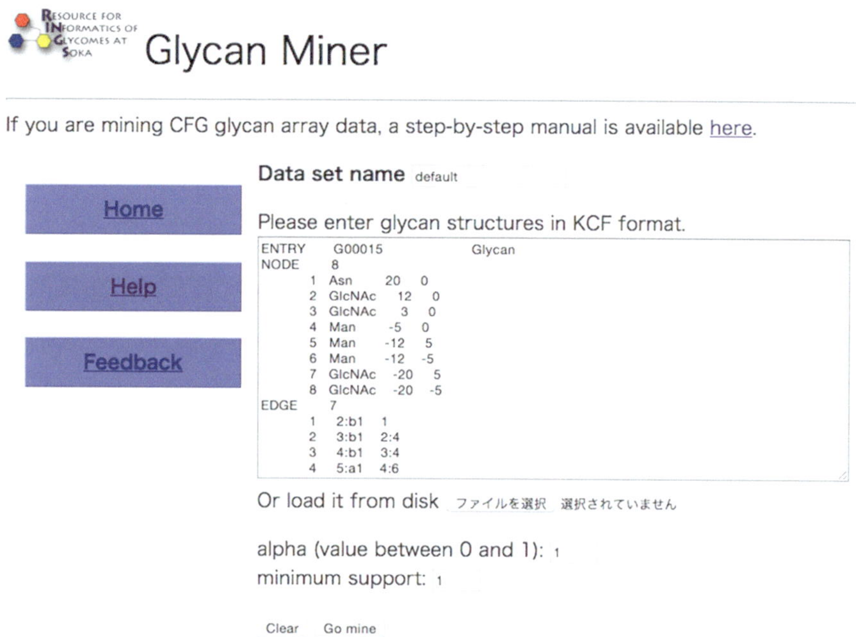

Fig. 15.7 Input screen for Glycan Miner with default values

should be specified in KCF format and can be inputted into the text area or specified with a filename containing the data. There are two options available: alpha and *minimum support*. These parameters refer to how a candidate subtree (extracted from the data set) should be handled. The *support* for any candidate subtree refers to the number of glycans in the input data set in which it appears. For example, if a data set of 80 structures, 50 *N*-glycans and 30 *O*-glycans, is given, the subtree consisting of the $Man_3GlcNAc_2$ core structure of *N*-glycans would have a support of 50. Candidate subtrees are computed from the input data set; they consist of all possible subtrees among the data.

The second parameter, alpha, essentially specifies the uniqueness of the subtrees to be output. An example would best illustrate this parameter, so two output examples are provided in Sect. 15.4.2. In simple terms, a larger value of alpha, closer to 1, will output a larger range of glycan subtrees, whereas a smaller value, closer to 0, will filter out redundant glycan subtrees.

15.4.2 Glycan Miner Output

Figures 15.8 and 15.9 are snapshots of the Glycan Miner results using the same support = 1 but different values for α. As illustrated, the results of $\alpha = .9$ contain

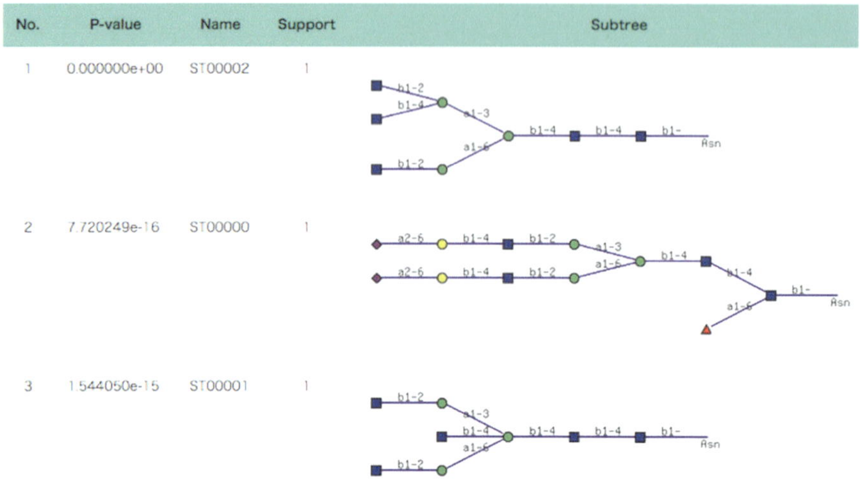

Fig. 15.8 Results of running Glycan Miner on the default data set, using a value of 1 for support and .3 for α

seven glycans, whereas those of $\alpha = .3$ contain just three. The difference lies in the fact that several of the seven glycans are rather similar to one another. More specifically, their support values are rather close, whereas the three selected using $\alpha = .3$ are less similar to one another.

The α-closed frequent mining algorithm compares the *support* between every glycan subtree pair in which one contains the other. Let us assume that there are two subtrees, the *N*-glycan core $Man_3GlcNAc_2$ which we call *largerT* and a subtree of this structure $Man_2GlcNAc_2$ which we call *smallerT*. The α-closed frequent mining algorithm will compare each of their respective supports (denoted as $supp(largerT)$ and $supp(smallerT)$) and will discard the larger if its support does not satisfy the equation $supp(largerT) < max(\alpha \times supp(smallerT), minsup)$, where *minsup* is the *minimum support* specified by the user. Since it can be assumed that subtrees with larger values of *support* will be less overlapping, α can be used to make this differentiation.

The Glycan Miner Tool also sorts the resulting subtrees based on p-value, which is generated based on the data set inputted. That is, a random data set of subtrees are generated, with which the support of the subtrees in the results are compared. The smaller the p-value, the fewer the chance that the same results are obtained randomly, and so the subtrees with smaller p-value are listed first.

15.4.3 Glycan Miner Tips

- p-values are generated based on the input, so the p-value on smaller data sets will not be very meaningful. Moreover, because it is generated for each run,

No.	P-value	Name	Support	Subtree
1	0.000000e+00	ST00004	1	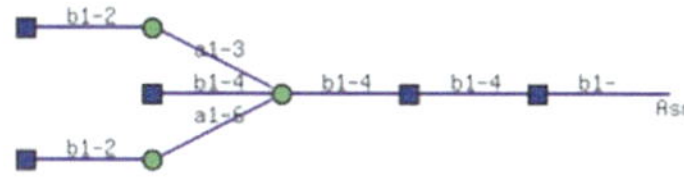
2	7.720249e-16	ST00001	6	
3	1.544050e-15	ST00003	1	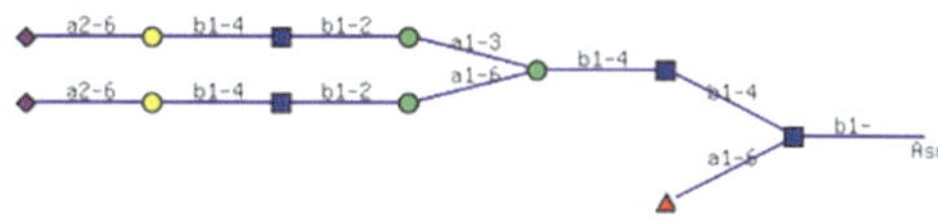
4	2.316075e-15	ST00000	3	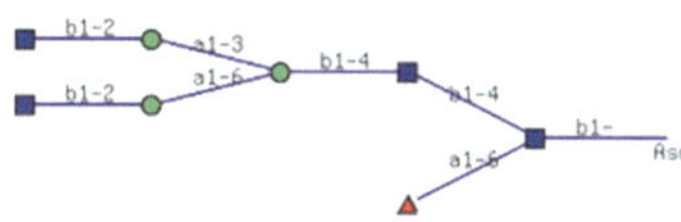
5	3.088100e-15	ST00005	1	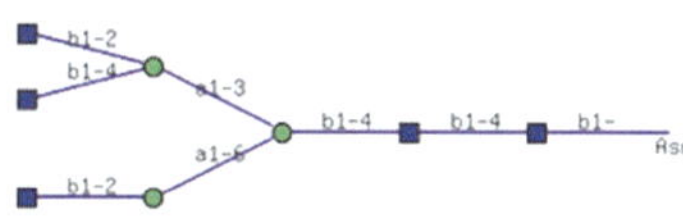
6	3.860124e-15	ST00006	2	
7	4.632149e-15	ST00002	2	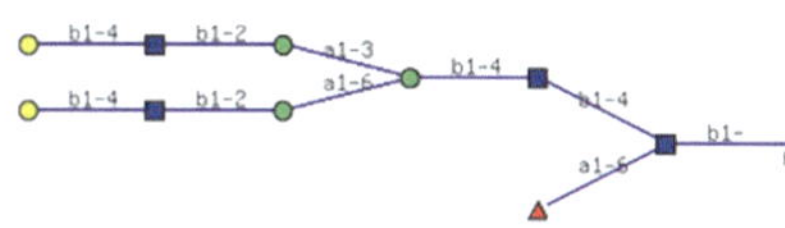

Fig. 15.9 Results of running Glycan Miner on the default data set, using a value of 1 for support and .9 for α

the order of the results may change, especially for small data sets. Such results would imply that there is no significant difference between the results and that they should all be treated equally.

- Glycosidic bond information is very specific: unspecified linkage information will not be matched with structures whose linkages are specified.
- IUPAC-formatted data can be used for Glycan Miner by following the link at the top of the input page for using CFG data.

15.5 MCAW

MCAW, or multiple carbohydrate alignment with weights, was originally developed as a temporary tool to determine the shape of profiles for ProfilePSTMM. However, it was decided that MCAW could be used on its own to align multiple glycans to find consensus sequence patterns across a set of glycans, so it was made available as a web tool. MCAW generates easy-to-understand glycan profile images as results. Moreover, detailed parameters are available for advanced users who wish to fine-tune the results.

Similar to the ProfilePSTMM Tool, MCAW can be used to find patterns across a set of glycans, such as from glycan-array or lectin-array data. Note, however, that it is not a probabilistic model, meaning that all the input data will be used at face value, meaning that all the data will be used in the results. In contrast, a probabilistic model will account for noise in the data, and so discard data that may be outliers.

15.5.1 MCAW Input

The MCAW input screen is shown in Fig. 15.10. Multiple KCF data can be pasted into the text area, or a file containing such data can be selected. The advanced parameters are described in Table 15.2. The default values can be used normally, but these options allow users to fine-tune the results if specific characteristics of the data inputted are known.

15.5.2 MCAW Output

The output of MCAW is a figure, as in Fig. 15.11, of the glycans aligned into a profile, but the results in PKCF (profile KCF format) can also be downloaded by the link provided. A legend for the figure is also provided as a link.

In the figure, similar to ProfilePSTMM, distributions of glycans and their linkages are given at various positions of glycans. Highly consistent portions of the alignment will have higher distributions of a particular monosaccharide at those positions. Figure 15.11 illustrates an example. One can see that at the positions numbered 7 through 10, highly *conserved* patterns of monosaccharides are found. This is a sialyl-Lewis X structure, which is a common motif among glycans in mammals. Moreover, position 18, which is attached to position 7, also contains a 6-sulfate approximately 16 % of the time, indicating that 6-sulfated sialyl-Lewis X structures can also be found in the data.

Fig. 15.10 The MCAW input screen, where multiple KCF data can be specified, and a variety of parameters are available for fine-tuning the results

Table 15.2 The advanced parameters in MCAW

Parameter	Description
Gap penalty	The penalty value to use when inserting a gap where a monosaccharide should align. Used to shift matching subportions of glycans when necessary to align the most similar parts
Monosaccharide	The score for matching monosaccharides
Anomer	The score for matching anomers
Nonreducing side carbon number	The score for matching the carbon number of monosaccharide on the nonreducing side of glycosidic bonds
Reducing side carbon number	The score for matching the carbon number of monosaccharides on the reducing side of glycosidic bonds (usually 1 or 2)

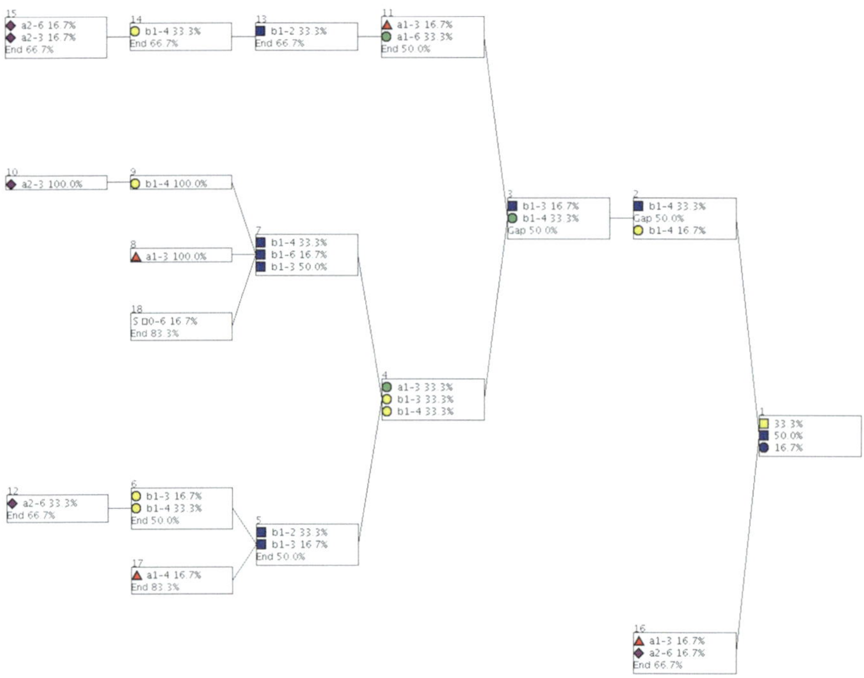

Fig. 15.11 A snapshot of the MCAW output screen, where distributions of monosaccharides and their glycosidic linkages are given at various positions

15.5.2.1 PKCF Format

The output of MCAW can also be downloaded as a PKCF (Profile KCF) file from the result page. The PKCF for the result in Fig. 15.11 is as follows.

```
ENTRY G04804-G04183-G04845-G05121-G05108-G04329 GlycanProfile
NODE 18
1 1=GlcNAc 2=GlcNAc 3=Glc 4=GalNAc 5=GalNAc 6=GlcNAc 0 0
2 1=GlcNAc 2=GlcNAc 3=- 4=- 5=- 6=Gal -8 6
3 1=Man 2=Man 3=- 4=- 5=- 6=GlcNAc -16 6
4 1=Man 2=Man 3=Gal 4=Gal 5=Gal 6=Gal -24 1
5 1=GlcNAc 2=GlcNAc 3=GlcNAc 4=GlcNAc 5=GlcNAc 6=GlcNAc -32 5
6 1=LFuc 2=LFuc 3=LFuc 4=LFuc 5=LFuc 6=LFuc -40 5
7 1=Gal 2=Gal 3=Gal 4=Gal 5=Gal 6=Gal -40 7
8 1=Neu5Ac 2=Neu5Ac 3=Neu5Ac 4=Neu5Ac 5=Neu5Ac 6=Neu5Ac -48 7
9 1=GlcNAc 2=GlcNAc 3=GlcNAc 4=0 5=0 6=0 -32 -3
10 1=Gal 2=Gal 3=Gal 4=0 5=0 6=0 -40 -2
11 1=Man 2=Man 3=0 4=0 5=0 6=LFuc -24 11
12 1=Neu5Ac 2=Neu5Ac 3=0 4=0 5=0 6=0 -48 -2
13 1=GlcNAc 2=GlcNAc 3=0 4=0 5=0 6=0 -32 11
14 1=Gal 2=Gal 3=0 4=0 5=0 6=0 -40 11
15 1=Neu5Ac 2=Neu5Ac 3=0 4=0 5=0 6=0 -48 11
16 1=0 2=0 3=0 4=Neu5Ac 5=0 6=LFuc -8 -6
17 1=0 2=0 3=LFuc 4=0 5=0 6=0 -40 -4
18 1=0 2=0 3=0 4=0 5=S 6=0 -40 3
EDGE 17
1 2-1:b1 1-1:4
1 2-2:b1 1-2:4
1 2-3 1-3
1 2-4 1-4
1 2-5 1-5
1 2-6:b1 1-6:4
2 3-1:b1 2-1:4
2 3-2:b1 2-2:4
2 3-3 2-3
2 3-4 2-4
2 3-5 2-5
2 3-6:b1 2-6:3
.
.
17 18-3 5-3
17 18-4 5-4
17 18-5 5-5:6
17 18-6 5-6
///
```

The general format of PKCF follows that of KCF, in that there are three major sections in order: `ENTRY`, `NODE`, and `EDGE`. The first section lists the glycan names as inputted into the MCAW tool, separated by hyphens, read in from the KCF files. The order of the names indicates the order of the items listed in the `NODE` and `EDGE` sections. The `NODE` and `EDGE` labels are followed by numbers, indicating the number of positions and connected positions, respectively. Usually, the number of edges is one less than the number of positions. The lines following the `NODE` heading are the details regarding each position. They are formatted as follows:

```
<pos#> <glycan#1>=<residue> <glycan#2>=<residue> ...<glycan#n>=
<residue> <x-coord> <y-coord>
```

where `pos#` indicates the position number within the tree topology and `x-coord` and `y-coord` represent their x- and y-coordinates, respectively, as drawn. Thus at each position, the residue names derived from each glycan are listed in order of the glycans listed in the `ENTRY` line. This is followed by x- and y-coordinates, similar to the KCF format.

In contrast, the `EDGE` section consists of lines corresponding to each glycosidic bond in each glycan. Thus the number of lines in this section equals the number of connected positions times the number of glycans. In the example above, there are 17 edges in the tree topology and 5 glycans, so there is a total of 85 lines, which have been cut to save space. Each line in this section is formatted as follows:

```
<link#> <pos#>-<glycan#>:<anomer><carbon#> <pos#>-<glycan#>:
<carbon#>
```

where `link#` is a unique number given to each edge. However, if the anomer and/or carbon number information is unavailable, these are left out. If both are unavailable, then the colon ":" is omitted. This is, again, a similar format to the KCF representation, except with additional position information included.

15.5.3 MCAW Tips

- The input data should be in KCF format, but with the extra restriction that a name must be included in the ENTRY line and that the name should not have a hyphen in it.
- MCAW is still in alpha-stage and may not give results for complex data sets, such as those having a large variety of structures. It may work better on data sets with similar glycan structures.
- Glycans with no glycosidic bonds (i.e., single monosaccharides) cannot be included in the input data.
- As of May 2016, new functionality has been made available such that weights can be specified with IUPAC-formatted data: follow the link for using CFG array data, available at the top of the input screen of MCAW.

15.6 Kernel Tool

The Glycan Kernel Tool in RINGS is an implementation of the biochemically weighted Kernel (Jiang et al. 2011). This tool implements a kernel that utilizes a score matrix for weighting similar glycosidic linkages and monosaccharides more highly in order to extract biochemically meaningful features from the data. Combined with a support vector machine (SVM), this tool can be used to detect biologically significant motifs across two data sets of glycans. It has been shown to extract the same motifs as those that were manually curated, which were consistent with the literature. This detection procedure consists of using the SVM to classify the two input data sets based on the kernel, and then feature extraction is performed based on the classification to detect the most significant features, or glycan substructures, that most distinguish the two data sets. In layman's terms, given two data sets of glycan structures to compare, this tool will find those glycan substructures that distinguish one data set from the other. That is, it will find substructures that are unique to the target data set, compared to the control.

15.6.1 Kernel Tool Input

Two data sets of glycans are required as input for comparison. The *target* data set of glycans is compared against the *control* data set to find unique substructures of glycans, or motifs. Figure 15.12 is a snapshot of the input screen for the Glycan

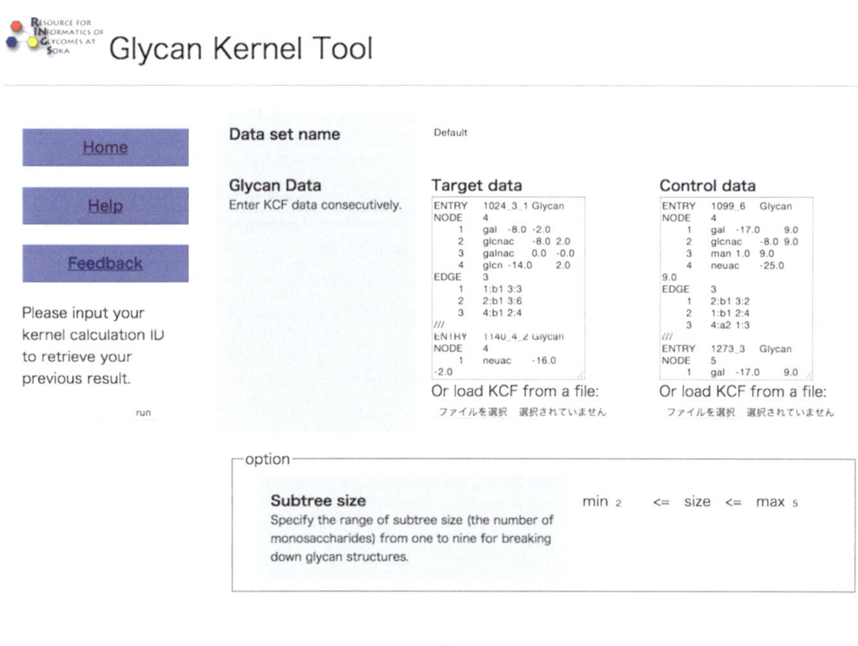

Fig. 15.12 A snapshot of the input screen for the Glycan Kernel Tool. Two text areas are provided for the user to provide KCF-formatted data of two data sets of glycans to compare. Because of the potentially long computation time, previously executed runs of this tool are assigned computation IDs, which can be retrieved from the text field on the left-hand side

Kernel Tool. Two text areas are provided for the user to provide KCF-formatted data of two data sets of glycans to compare. However, large numbers of glycans in the data sets will take longer to compute. Therefore, computation IDs are assigned to each run of this tool. Thus, users can close this browser window and retrieve the results later once they are complete by returning to this page and inputting the computation ID in the text field on the left-hand side of the page.

15.6.2 Kernel Tool Output

The output of this tool is a list of the glycan features and their *layers*, along with their scores, as shown in Fig. 15.13. Details are described in the original

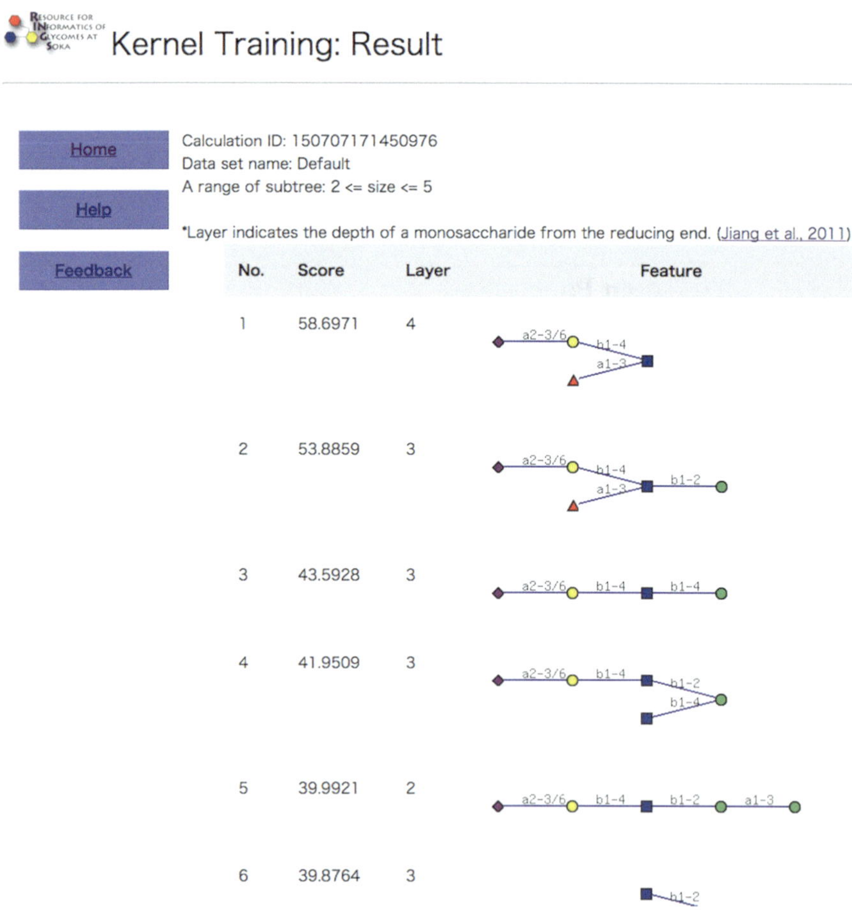

Fig. 15.13 A snapshot of the output screen for the Glycan Kernel Tool. The list of glycans, their scores, and their layers are listed in order of significance

manuscript (Jiang et al. 2011), but in general terms, the *layer* of the glycan feature is the distance of the given feature from the root node, or reducing end, of the original glycan within which it was found. That is, the layer indicates the number of glycosidic linkages between the feature and the reducing end. Thus, features with larger layers are those that are found closer to the nonreducing end of the input glycan(s). In Fig. 15.13, the top feature is actually a subtree of the second highest feature, which has an additional mannose (green circle) at its reducing end. Accordingly, their layers differ by one. Other features are also subsets of these top two features, indicating that this top feature is probably the single motif that most distinguishes the two data sets.

15.6.3 Kernel Tool Tips

- Data sets of over 40 structures may take over an hour to compute. Please be sure to save the computation ID for larger data sets.
- Since substructures unique to the target and not in the control will be extracted by this tool, swapping target and control data sets may result in interesting results, where structures unique to the control and not the target can be found.

15.7 GPP: Glycan Pathway Predictor

The glycan pathway predictor (GPP) is a tool that computes the *N*-glycans that could be theoretically biosynthesized given a set of glyco-enzymes involved with *N*-glycan biosynthesis. This generation procedure is based on the mathematical model proposed by Krambeck et al. in (2009). In short, the substrate specificities of each enzyme are specified using a mathematical model, and it is assumed that there are no limitations in the amount of each enzyme, sugar donors, or substrates. Then, GPP computes all possible glycan that can be synthesized using the input glycan structure and selected enzymes. The output results in a network of the synthesized glycans.

15.7.1 GPP Input

The GPP input screen is shown in Fig. 15.14. A single *N*-glycan formatted in KCF should be specified in the text area on the left. Then one or more enzymes should be selected from the list on the right. Finally, because this mathematical model could potentially continue indefinitely, a limit for the largest glycan structure to generate, indicated by molecular mass, should be specified. Table 15.3 gives a description of each enzyme that is available in GPP.

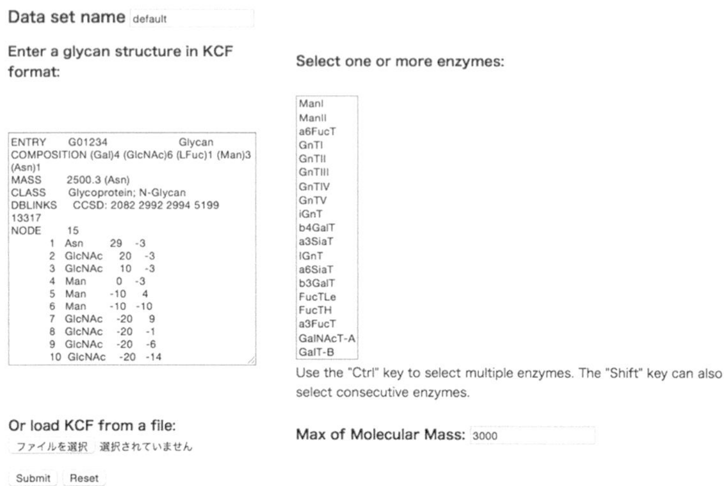

Fig. 15.14 A snapshot of the input screen for the GPP Tool. A single *N*-glycan formatted in KCF should be specified. Then one or more enzymes should be selected from the list on the right. Finally, the mass limit for the generated glycans should be entered in order to prevent the computation from continuing indefinitely

15.7.2 GPP Output

The output screen for GPP is a Java applet. Therefore, similar to the DrawRINGS applet, Java security permissions of the web browser used will need to be adjusted accordingly in order to obtain the results. An open-source applet called ZGRViewer (http://zvtm.sourceforge.net/zgrviewer.html) is used to display the resulting pathway. A snapshot of the output screen is shown in Fig. 15.15. The functionality available in ZGRViewer includes zoom, swipe, and double-clicking of nodes. Figure 15.16 is a snapshot of the zoomed in screen. Nodes represent each glycan structure synthesized, and they are connected by edges representing the enzyme involved in the biosynthesis reaction. Nodes can be double-clicked to open a browser window displaying the detailed information of the glycan, as in Fig. 15.17.

15.7.3 GPP Tips

- The glycan structure used as the first structure must be an *N*-linked glycan and a substrate for at least one of the enzymes selected.
- The resulting pathway map may take some time to load. Clicking on the canvas once may refresh the map and display it.

Table 15.3 The glycogenes available in GPP

Abbreviation	Name	EC number	Corresponding genes
ManI	α2-mannosidase I	3.2.1.113	MAN1A1, MAN1A2, MAN1B1, MAN1C1
ManII	α3/6-mannosidase II	3.2.1.114	MAN2A1, MAN2A2
GnTI	β2-*N*-acetylglucosaminyltransferase I	2.4.1.101	MGAT1
GnTII	β2-*N*-acetylglucosaminyltransferase II	2.4.1.143	MGAT2
GnTIII	β4-*N*-acetylglucosaminyltransferase III	2.4.1.144	MGAT3
GnTIV	β4-*N*-acetylglucosaminyltransferase IV	2.4.1.145	MGAT4A, MGAT4B
GnTV	β6-*N*-acetylglucosaminyltransferase V	2.4.1.155	MGAT5
b3GalT	β3-galactosyltransferase		B3GALT1, B3GALT2, B3GALT5
b4GalT	β4-galactosyltransferase	2.4.1.38	B3GALT1, B4GALT2, B4GALT3
a3SiaT	α3-sialyltransferase	2.4.99.6	ST3GAL3
a6SiaT	α6-sialyltransferase	2.4.99.1	ST6GAL1
a6FucT	α6-fucosyltransferase	2.4.1.68	FUT8
FucTLe	α3/4-fucosyltransferase III	2.4.1.65	FUT3, FUT5, FUT6
FucTH	α2-fucosyltransferase Se, H	2.4.1.69	FUT1, FUT2
a3FucT	α3-fucosyltransferase	2.4.1.152	FUT4, FUT7, FUT9
iGnT	Blood group i β3-*N*-acetylglucosaminyltransferase	2.4.1.149	B3GNT1, B3GNT6
IGnT	Blood group I β6-*N*-acetylglucosaminyltransferase	2.4.1.150	GCNT2
GalNAcT-A	Blood group A α3-*N*-acetylglucosaminyltransferase	2.4.1.40	ABO
GalT-B	Blood group B α3-galactosyltransferase	2.4.1.37	ABO

- The detail screen may not display in newer browsers due to Java applet security issues. GPP will be updated to include more enzymes in the near future, and the web display will also be updated accordingly.

15.8 Converter Utility

The Convert Tool in RINGS is a tool that aims to combine and automate the various glycan text format conversion utilities currently available in RINGS. It takes as input any of the major glycan text formats and provides options to convert the structures in the supported output format.

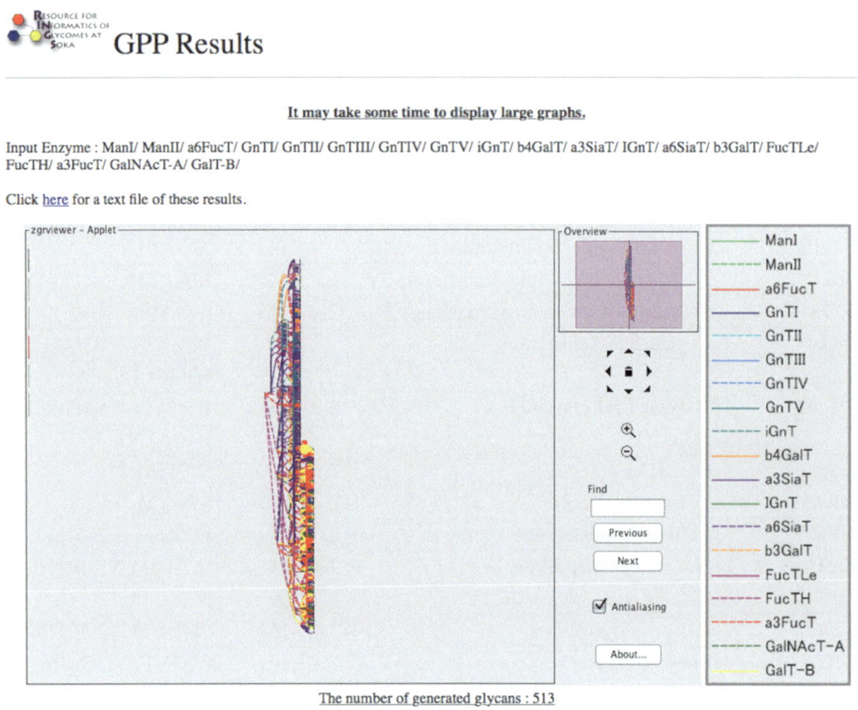

Fig. 15.15 A snapshot of the output screen of the GPP Tool

15.8.1 Convert Utility Input

The Convert Tool takes as input glycan structures in GLYDE-II, $\mathrm{GlycoCT}_{\mathrm{condensed}}$, KCF, IUPAC, LinearCode, and LINUCS formats. If multiple glycans are inputted, multiline formats such as GlycoCT and KCF need to be inputted with newlines in between. After clicking the "Submit" button, the representation format of the input is automatically recognized, and a list of possible output formats is provided as a pull-down menu. Figure 15.18 is a snapshot of the input screen for the Convert Tool, using GLYDE-II as an example. Note that only one structure in GLYDE-II can be converted at a time due to formatting issues in XML. Figure 15.19 shows how the input format of the text in Fig. 15.18 was recognized and that the possible representation formats to which it can be converted are KCF, $\mathrm{GlycoCT}_{\mathrm{condensed}}$, IUPAC, LinearCode, and MDL Mol. Moreover, the output file format can be selected from HTML, Text, JSON, and DL. HTML will display the results graphically, with images of the glycan structures that were converted. Text will display the results in plain text format, which is easier to process by computer and can be copied as text. JSON is a special format for computers to process the results, and DL refers to download, allowing users to download the results as a text file instead of displaying all the results in the browser, saving network bandwidth.

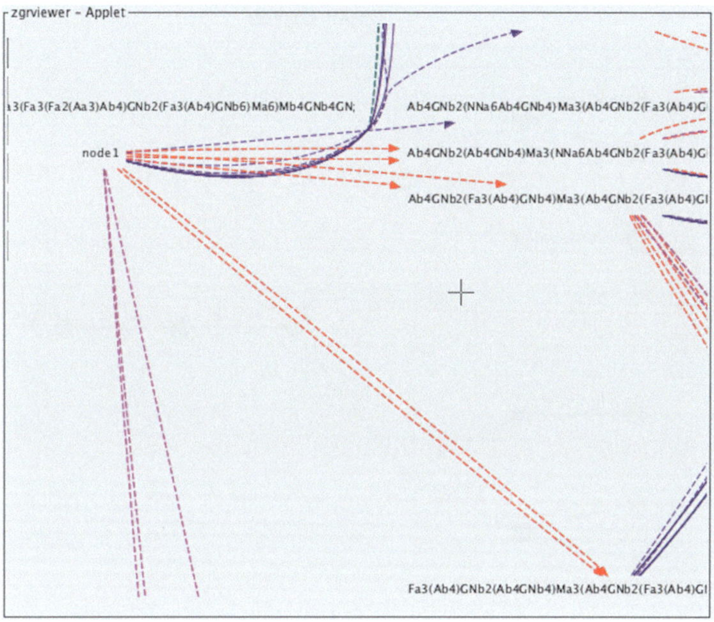

Fig. 15.16 A snapshot of the output screen zoomed in

15.8.2 Convert Utility Output

The output screen depends on the output formatted selected from the input page. Figure 15.20 is a snapshot of when the HTML format is selected. If either KCF or GlycoCT is selected as the output representation format, then a figure of each converted structure is also provided. This allows the user to visually confirm that the converted structure is correct.

15.8.3 Convert Utility Tips

- Depending on the conversion path via which input formats are converted to the selected output formats, some information may be lost in the process of conversion. This may further cause errors during the conversion. In that case, it is suggested that a format that contains maximal structural information, such as GLYDE-II or GlycoCT, be used as the input format. To do so, the existing input format can be translated to GLYDE-II or GlycoCT, and any missing information should be added manually before being translated to the target format.
- GLYDE-II-formatted structures can only be converted one structure at a time, due to formatting issues of XML.

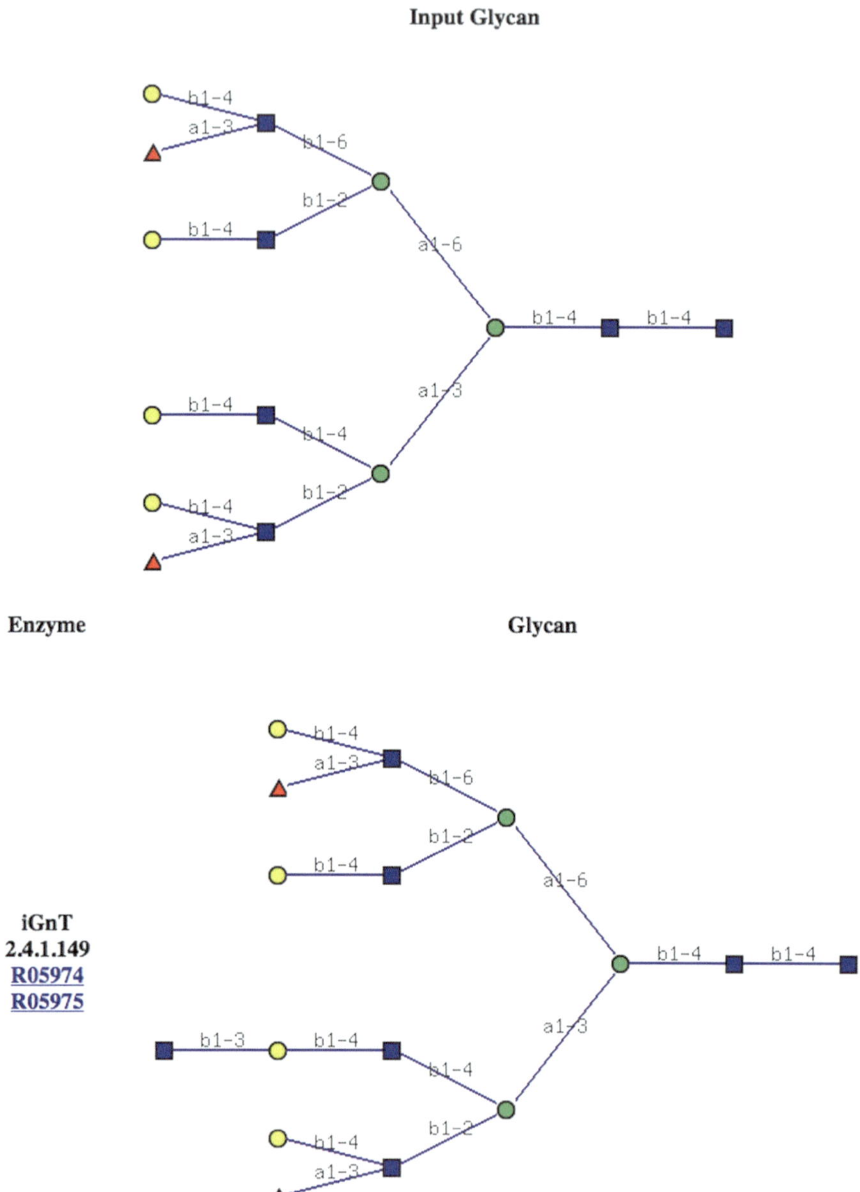

Fig. 15.17 A snapshot of the detail screen for a glycan selected from the output of the GPP Tool

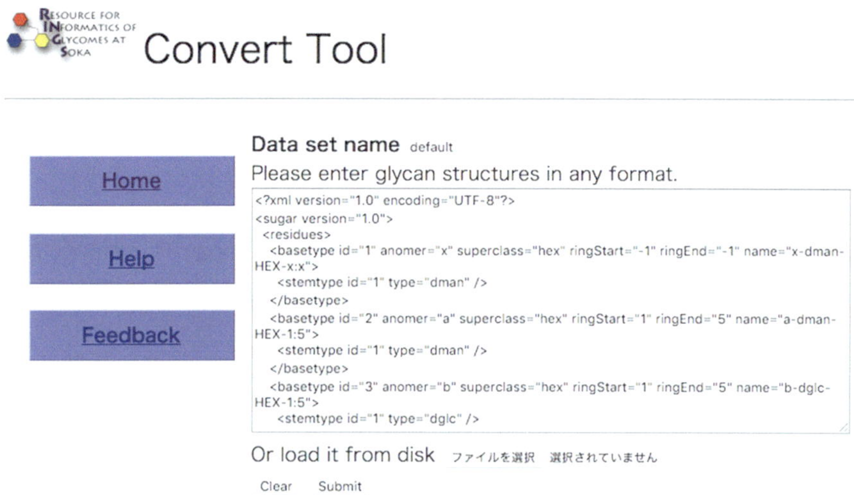

Fig. 15.18 A snapshot of the input screen for the Convert Tool. In this example, GLYDE-II is used as input. Note that only one structure at a time can be converted from GLYDE-II format

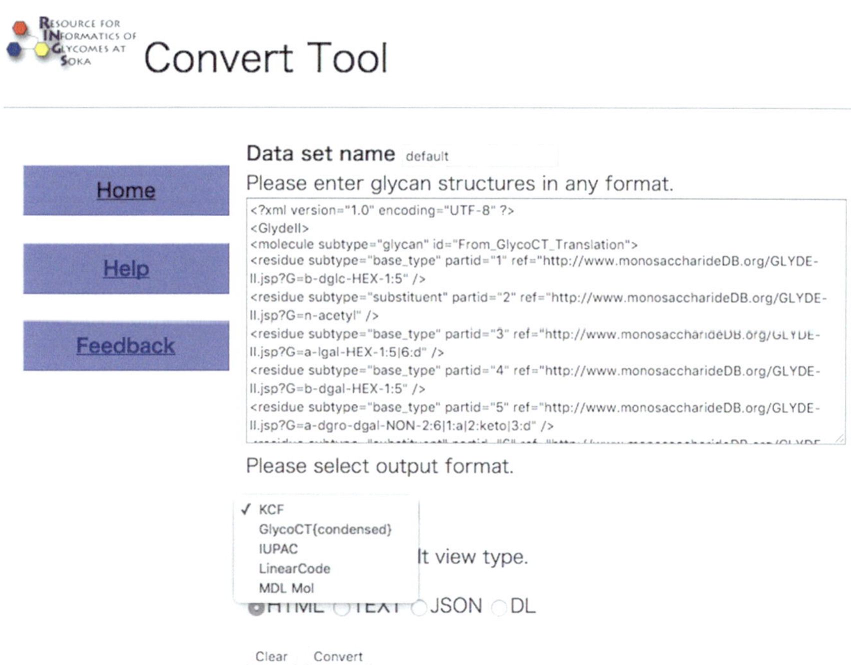

Fig. 15.19 The output format selection screen of the Convert Tool after a structure(s) has been inputted. The input format is automatically recognized, and a pull-down menu of the available output formats is displayed

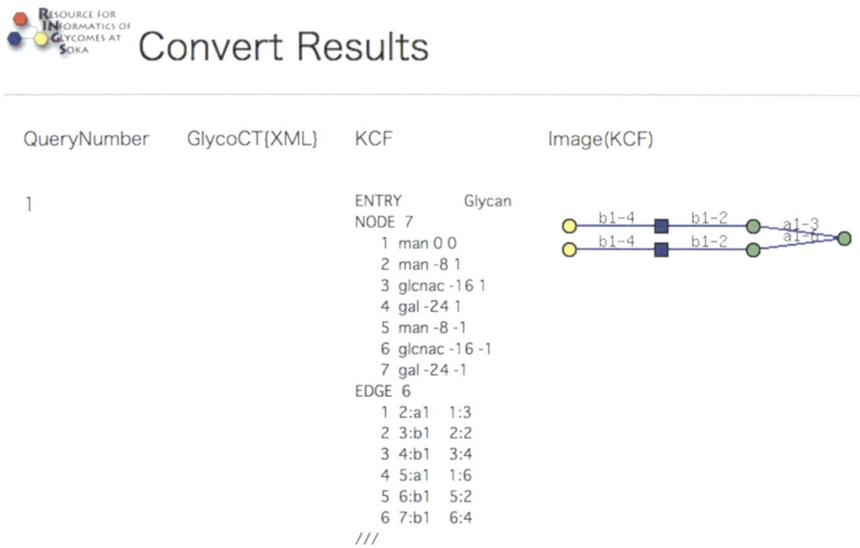

Fig. 15.20 A snapshot of the Convert Tool results page, displayed in HTML format. Each converted structure is displayed along with its figure so that users can visually confirm the structure

15.9 Data Management System

A data management system is available for users who have registered an account on RINGS. Registration is free of charge, and all data is kept private on the RINGS server. The purpose of this system is to allow users to store their past analysis results online without having to download the results of each tool individually. Moreover, data and results stored on RINGS from one tool can be used as input to other tools when the data formats are compatible. Thus, this system enables users to use the RINGS tools more efficiently.

15.9.1 Data Management System Usage

Users can create an account with their email address by clicking the "User registration form" link located at the top of the main RINGS page (Fig. 15.21). After successfully creating an account, users can then log-in with their registered email address and password via the Login form at the top right of this page.

Once users have logged in with their registered email address and password, the main page displays a data tree at the left, as in Fig. 15.22. As shown in this figure, by clicking on the name of the data set itself, the name of the data set and comments regarding the data set can be inputted.

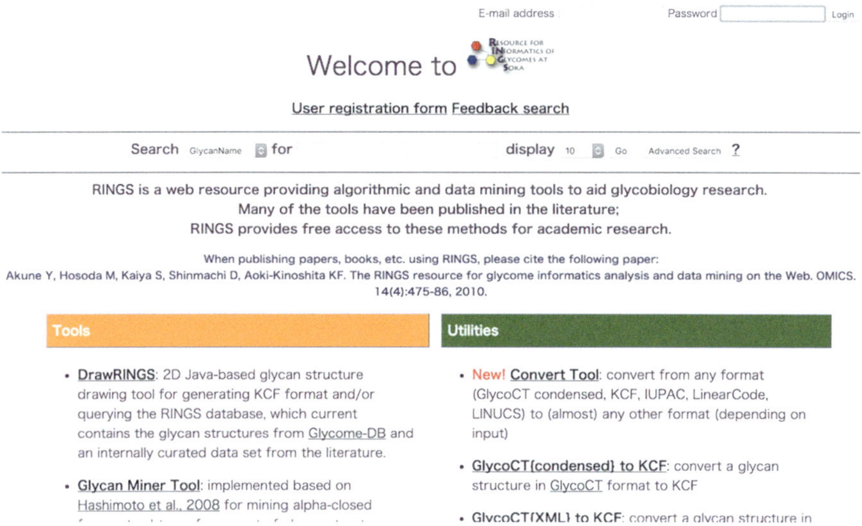

Fig. 15.21 Top of RINGS home page, where links to create a user account are available via the "User registration form" link. Registered users can then log-in with their registered email address and password, located at the top right of this page

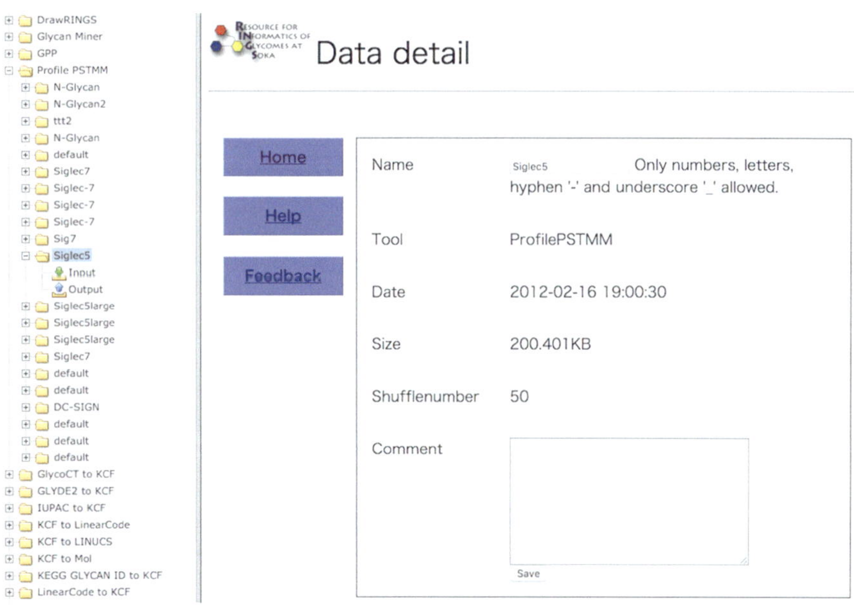

Fig. 15.22 Details regarding a data set can be inputted by clicking the name of the data set. The name can be modified and comments can be stored with each data set

Figure 15.23 is a view of the input data of the data set in Fig. 15.22. Having clicked the “Input” folder for this data set, the glycan structures used for input to ProfilePSTMM on this data set is shown on the right. A pull-down menu is available at the top left for other tools that can also take this data as input. The data set itself can also be downloaded via the “Download” link at the top right.

By clicking on the “Output” folder of a data set, the results of the analysis using the input data are shown. In Fig. 15.24, the output view for ProfilePSTMM is shown using the Profile Viewer. An option to view the results in text format is also provided on this page.

Finally, it is highly recommended that users log out using the “Logout” button at the bottom of the data tree when analysis is complete. This will ensure that the data is kept private and cannot be viewed by others. Note that the size of all the data is also indicated at the bottom. Although there is currently no quota, users can check their data usage and delete any unnecessary data as needed. Data can be deleted by clicking on a tool name in the data tree and then checking the data they want to delete. The “Delete” button is shown at the bottom of the page.

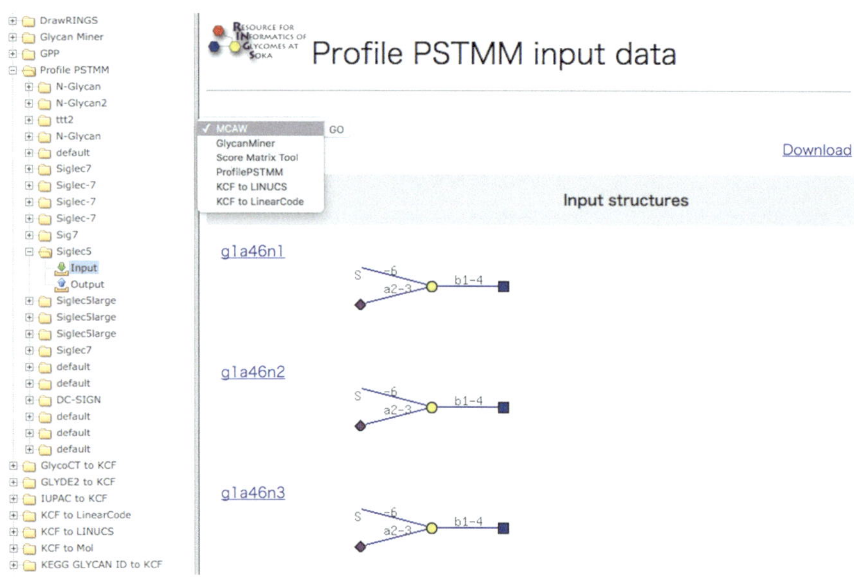

Fig. 15.23 Snapshot of RINGS after user has logged in and selected a data set. Here, the input data used for the analysis of Siglec 5 data is selected and can be viewed graphically on the right. Options to rerun this data using a tool in the pull-down menu are shown. The data itself can also be downloaded using the “Download” link at the right

Fig. 15.24 The output view of the analytical results using ProfilePSTMM on the data in Fig. 15.22. The results can also be viewed as text by clicking the button at the top of this page

15.9.2 Data Management System Tips

- Sometimes the data tree does not display after logging in, due to cached cookies and having not logged out a previous time. This can usually be solved by logging in again, and the tree should be displayed.
- It is highly recommended that a data set name be specified whenever running a tool. The default name can be used of course, but managing the data can become difficult as more data is accumulated.
- If for some reason the data tree is not updated with the latest results, the "Reload" button located at the bottom of the data tree can be used to refresh the data tree.

15.10 Feedback System

A feedback and bug report system is also available via the "Feedback search" link at the top of the main RINGS page, as shown in Fig. 15.21. Although anyone can view others' feedback without logging in, users are required to register and log in order to report any new bugs or feedback.

15.10.1 Usage

Figure 15.25 shows the top part of the Feedback Form. The "Report Form" link will open a new form for users to enter feedback after logging in. This form provides three fields: the tool regarding which feedback is being entered, a title to summarize the problem or suggestion, and the details of how the problem occurred or regarding the suggestion. All feedback can be viewed by selecting "all" and clicking the "Go" button using the main Feedback Form (Fig. 15.25). Feedback regarding a specific tool can also be selected in this form. The results of this form show the list of feedback, including Title, Report Data, Status (open, closed, or duplicate), the tool name, the number of votes for the feedback, any comments from developers, and a priority indicator. The bottom of the Feedback Form also shows the latest feedback. Similar to the search results, specific feedback can be seen in detail by clicking the title of the feedback. The feedback detail page, shown in Fig. 15.26, provides users an option to vote on particular feedback. Accumulated votes will then be considered to be prioritized more highly.

15.10.2 Tips

- Voting on existing feedback will help motivate developers to prioritize future work.
- A clear description, including the data used, when encountering problems, is essential to be able to reenact the problem and help the debugging process.

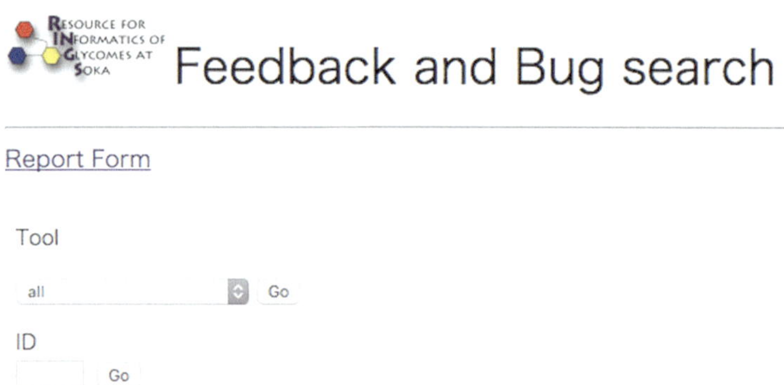

Fig. 15.25 The top of the Feedback Form where users can report new feedback using the "Report Form" or view previous feedback

Fig. 15.26 The detailed page of a particular feedback. Users can vote for certain feedback if they agree and would like it to be prioritized

15.11 Summary

RINGS was developed to provide analytical tools for glycomics analysis, with an easy-to-use interface. However, without user feedback, the interfaces currently provided cannot be improved. Note that we are also under the process of rebuilding the RINGS interface using the latest technologies. Moreover, as mentioned in the Introduction, in the future, RINGS tools will be designed to allow any format to be used as input. We are also considering allowing GlyTouCan IDs to be used as well. Therefore, RINGS will continue to be improved as user demands are met.

Acknowledgements This work was supported in part by Grant-In-Aid for Young Scientists (A); KAKENHI (20016025); the Japan Society for the Promotion of Science (JSPS); the Ministry of Education, Culture, Sports, Science and Technology (MEXT); and the Soka University.

Reference

Akune Y, Hosoda M, Kaiya S, Shinmachi D, Aoki-Kinoshita K (2010) The RINGS resource for glycome informatics analysis and data mining on the web. OMICS 14(4):475–486

Aoki K, Yamaguchi A, Ueda N, Akutsu T, Mamitsuka H, Goto S, Kanehisa M (2004) KCaM (KEGG carbohydrate matcher): a software tool for analyzing the structures of carbohydrate sugar chains. Nucleic Acids Res 32(Web Server issue):W267–W272

Aoki-Kinoshita K, Ueda N, Mamitsuka H, Kanehisa M (2006) Profilepstmm: capturing tree-structure motifs in carbohydrate sugar chains. Bioinformatics 22(14):e25–e34

Aoki-Kinoshita KF, Kanehisa M (2015) Glycomic analysis using KEGG glycan. Methods Mol Biol 1273:97–107

Hashimoto K, Takigawa I, Shiga M, Kanehisa M, Mamitsuka H (2008) Mining significant tree patterns in carbohydrate sugar chains. Bioinformatics 24(16):i167–i173

Jiang H, Aoki-Kinoshita KF, Ching WK (2011) Extracting glycan motifs using a biochemically-weighted kernel. Bioinformation 7(8):405–412

Krambeck FJ, Bennun SV, Narang S, Choi S, Yarema KJ, Betenbaugh MJ (2009) A mathematical model to derive n-glycan structures and cellular enzyme activities from mass spectrometric data. Glycobiology 19(11):1163–1175

Ranzinger R, Herget S, von der Lieth CW, Frank M (2011) Glycomedb – a unified database for carbohydrate structures. Nucleic Acids Res 39(Database issue):D373–D376

Ueda N, Aoki KF, Mamitsuka H (2004) A general probabilistic framework for mining labeled ordered trees. In: SIAM international conference on data mining. SIAM, Philadelphia

Chapter 16
Glycan Data Retrieval and Analysis Using GLYCOSCIENCES.de Applications

Thomas Lütteke

Abstract The GLYCOSCIENCES.de web portal (www.glycosciences.de) combines various databases and applications related to glycobiology and glycomics. This chapter demonstrates the use of these resources to find a variety of information on the Lewis^X (Le^X) epitope such as literature references, NMR data, or corresponding PDB entries. Several query options are presented that enable the users to find the information of interest from different point of views. The main focus of GLYCOSCIENCES.de is put on 3D structural data. Therefore, methods to create 3D structure models with the Sweet-II tool and to analyze conformational properties of Le^X by usage of PDB data and conformational maps from GlycoMapsDB are also introduced in this chapter.

Keywords Glycoinformatics • Database • Substructure search • Glycan 3D structure • NMR • Protein Data Bank • Molecular modeling

16.1 Introduction

The field of glycoinformatics has considerably grown in recent years, with a variety of resources available by now (Berteau and Stenutz 2004; Aoki-Kinoshita and Kanehisa 2006; Frank and Schloissnig 2010; Lütteke 2012; Aoki-Kinoshita 2013; Campbell et al. 2014). The GLYCOSCIENCES.de web portal (Lütteke et al. 2006) combines several tools and databases related to glycobiology and glycomics. The main focus of these applications is on 3D structures of carbohydrates, glycoproteins, and protein-carbohydrate complexes, but other topics such as mass spectrometry (Lohmann and von der Lieth 2004), nuclear magnetic resonance (NMR; Loss et al. 2006), or glycan sequence notation are also covered. The GLYCOSCIENCES.de database, formerly called SweetDB (Loss et al. 2002), originally aimed to make CarbBank (Doubet et al. 1989; Doubet and Albersheim 1992) data available online and, wherever possible, add 3D structure models of the glycans to the

T. Lütteke (✉)
Justus-Liebig-University Giessen, Institute of Veterinary Physiology and Biochemistry, Frankfurter Str. 100, 35392 Giessen, Germany
e-mail: thomas.luetteke@vetmed.uni-giessen.de

K.F. Aoki-Kinoshita (ed.), *A Practical Guide to Using Glycomics Databases*, DOI 10.1007/978-4-431-56454-6_16

database entries. These models are created using the Sweet-II tool (Bohne et al. 1999). Later on, NMR data were imported from SugaBase (van Kuik et al. 1992) or manually extracted from the scientific literature. The Protein Data Bank (PDB, Berman et al. 2000) is another major data source of GLYCOSCIENCES.de. The development of an algorithm to detect carbohydrates in 3D structure data such as PDB entries and the implementation of this algorithm in the pdb2linucs tool (Lütteke et al. 2004) enabled a mostly automatic extraction of glycan data from 3D structures deposited in the PDB and storage of this information in the GLYCOSCIENCES.de database. Further properties such as torsion angles or information on amino acid composition in the spatial vicinity of the detected glycans are extracted from the PDB entries as well and stored in separate datasets. Each individual entry presents only limited information, because it can be considered as a single snapshot that is not necessarily representative. The collection of all these data, however, provides valuable records that can be used to better understand glycan properties and to validate data by identifying potentially erroneous structures via identification of "unusual" data. GLYCOSCIENCES.de offers various tools to statistically analyze these PDB-derived data (Lütteke et al. 2005).

Primary structures ("sequences") of glycans are stored in GLYCOSCIENCES.de using the LINUCS format (Bohne-Lang et al. 2001), which offers a linear representation of branched glycans with a unique sorting of branches. This format, however, is hard to read by humans. Therefore, the CarbBank 2D notation (Doubet and Albersheim 1992), which is close to IUPAC extended notation (McNaught 1997), is used to display glycan structure information to the user and also serves as an input format of some tools or queries. Recently, graphical representations of glycans have become popular for input and display (Campbell et al. 2014). These are currently not yet implemented in GLYCOSCIENCES.de but will be added in a future release.

The following sections will describe some use cases involving several database query options and data analysis tools. Individual pages are referred to via menu links from the start page (www.glycosciences.de). Menu items to be clicked are presented in the following text in the form $\rightarrow A \rightarrow B \rightarrow C$, which means that you should click item A in the main menu, then B in the submenu, and C in the next menu level. B and C can also refer to links that are found outside the menu on the page that is retrieved after clicking A or B, respectively.

16.2 Use Cases

In the following sections, the usage of GLYCOSCIENCES.de is introduced with the example of the Lewis^X (Le^X) epitope β-D-Galp-(1–4)[α-L-Fucp-(1–3)]-β-D-GlcpNAc (Fig. 16.1). This trisaccharide is, for example, found on various cancer cells and is used as a diagnostic marker to discriminate Hodgkin's lymphoma from other lymphocytic cancers (Powlesland et al. 2011). It is also relevant, e.g., for bacterial infections with *Helicobacter pylori* (Moran 2009). GLYCOSCIENCES.de provides a variety of data on the Le^X trisaccharide and on larger glycans that feature

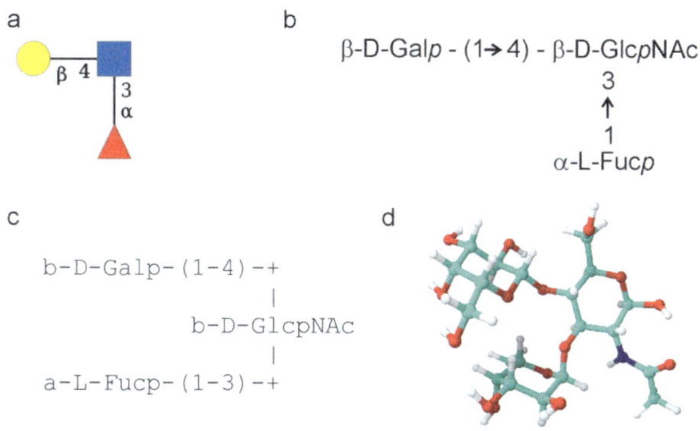

Fig. 16.1 Lewis^X trisaccharide. (**a**) IUPAC condensed notation, (**b**) CFG/Essentials style graphical representation, (**c**) CarbBank-style 2D graph, (**d**) 3D structure

this epitope, including literature references, NMR data, 3D structural information from the PDB, and molecular modeling results.

16.2.1 Use Case: Identification of Lewis^X Data in GLYCOSCIENCES.de Database

16.2.1.1 Exact Structure Search and Description of Database Entries

There are various ways to locate entries that contain Le^X information in the GLYCOSCIENCES.de database. We will start by searching for the Le^X trisaccharide using the exact structure search (→*Databases*→ *Structure*→ *Exact Structure Search*). The search form (Fig. 16.2a) offers a text area where you can enter (by manual typing or copy/paste) a query structure in CarbBank format. LINUCS format is also accepted here. (*Note*: Residue names are not case sensitive in GLYCOSCIENCES.de applications, i.e., upper-/lowercase spelling is not relevant.) Some examples are provided on the page to illustrate the input format. To enter the Le^X trisaccharide into this form, you can simply click on the *Lewis^X* button in the "examples" section below the search field. Click *Search now* to start the query. If the entered structure is present in the database, the corresponding entry (LinucsID 50) is directly displayed now (Fig. 16.3). (*Note*: If you know the LinucsID of an entry, you can directly enter it on the → *Databases* start page to quickly access the entry.)

Data in individual database entries are classified into categories such as composition, literature references, references to PDB entries, or cross-links to corresponding entries in other glycan databases. Each category can be expanded or collapsed separately, so that you can get an overview of the information that you are interested in without having to scroll through data that are currently not important for you. Some categories can contain long lists of data. In such cases only a limited number of items are displayed, and a link to further elements is given. Long lists of literature

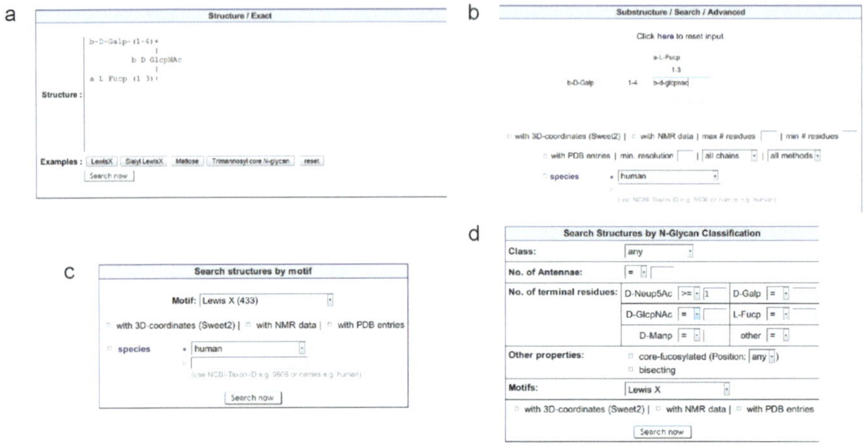

Fig. 16.2 Examples of GLYCOSCIENCES.de query options. (a) Exact structure search, (b) substructure search (advanced mode), (c) motif search, (d) search by N-glycan classification

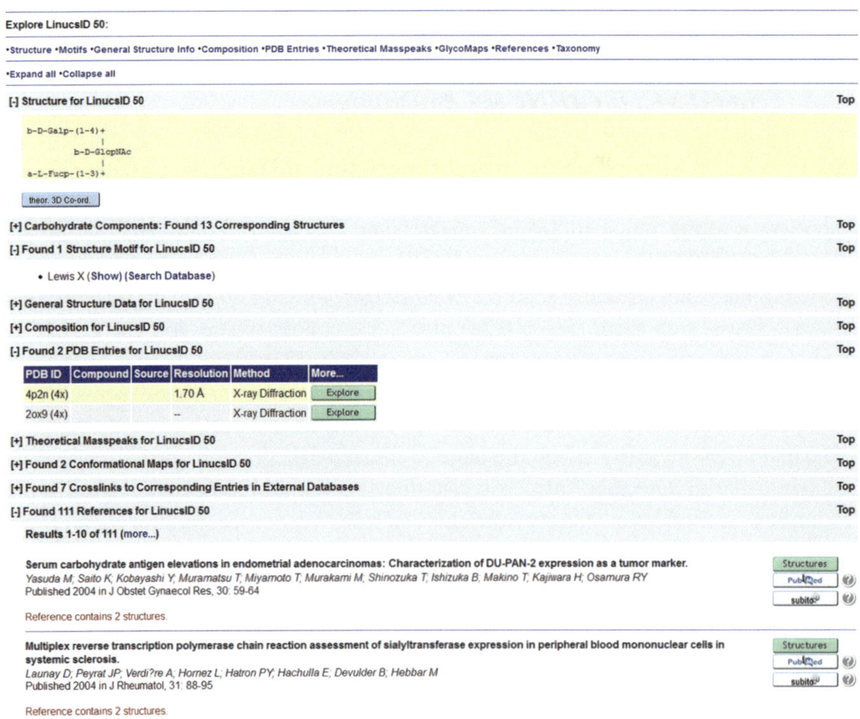

Fig. 16.3 GLYCOSCIENCES.de database entry of Le^X

references can also be searched individually by clicking the *search within references for LinucsID ...* link below the reference list. Click on this link and enter, e.g., *2001–2004* in the *Year* field of the displayed form to limit the presented references to those that are published between 2001 and 2004.

Each entry in the GLYCOSCIENCES.de database is based on a specific glycan structure, which can include an aglycone part. The aglycone can be important if you are, e.g., interested in corresponding 3D structures in the PDB, or if you search for NMR data, on which the exact aglycone can have an influence. In many cases, however, the aglycone part is of limited interest to the user. Therefore, the *Carbohydrate Components* section of a GLYCOSCIENCES.de database entry holds a list of entries that contain the identical glycan part as the current entry but differ in the aglycone part (see Fig. 16.3, click on the bar below the 2D structure notation of the glycan to expand this section). This way the user can get an overview of corresponding entries and quickly find further information without having to perform separate queries for the individual entries. Some structures in the database feature underdetermined information such as undefined linkage positions. Such entries are also linked to the corresponding fully defined structures via the *Carbohydrate Components* section.

GLYCOSCIENCES.de was the first database that systematically assigned glycan motif information to its entries. Glycan motifs are substructures that are found in many natural glycan structures and are often linked to specific functions such as recognition epitopes. Detected motifs are listed in the *Structure Motifs* section of a database entry (Fig. 16.3). For each motif that is listed for a structure, two links are presented. A click on *Show* highlights the occurrence(s) of that motif in the structure, and the *Search Database* link leads to a list of all entries that feature this motif. If applicable, a cross-link to the GlycoCD database (see below) is also given here, which provides further information on the function and expression of the motif.

16.2.1.2 Substructure Search

As a functional epitope, Le^X is recognized by glycan-binding proteins. In this role it is often part of a larger glycan structure. Therefore, facts about the isolated Le^X trisaccharide might not be of major importance to a researcher. Instead, glycans that feature this epitope can be much more valuable in some cases. Entries that contain Le^X as a part of a larger glycan can be detected using the substructure search. There are currently two different input forms available (→*Databases*→ *Structure*→ *Substructure search* (*beginner*) and→ *Databases*→ *Structure*→ *Substructure search* (*advanced*)), which differ in the way the individual residues are entered. While the beginner form features drop-down menus, from which residues can be selected, the advanced form (Fig. 16.2b) accepts free text input of residues. The drop-down menus are limited to frequent residues, while the free text form enables searches for less frequent residues and also accepts wildcards "?" to match any single character or "*" to match any series of characters. However, users need to know the correct residue names they want to search for.

Below the residue input fields, there are several further fields and checkboxes. These can be used to restrict the result lists to entries that contain specific kinds of data. For example, only a limited number of entries contain NMR data or references to the PDB. If you are searching for such entries, you need to tick the checkboxes in front of *with NMR data* or *with PDB entries*, respectively. PDB entries can be further restricted by providing a minimum resolution, an experimental method (X-ray, NMR) that has been used for obtaining the 3D structure, or a carbohydrate chain type (covalently linked N- or O-glycans, non-covalently linked ligands). Results can also be limited to glycans that are reported to be present in a specific species. These options to limit query outputs are available in most query interfaces of the GLYCOSCIENCES.de database. Please note that species information is available only for a limited number of entries and may be incomplete. Therefore, restricting results to entries that are associated with a certain species does not necessarily produce all glycans that are present in that species but only those for which this information is available in the database.

Structures containing Le^X can be searched via both interfaces. After entering or selecting the residue names and linkages as shown in Fig. 16.2b and clicking the *Search now* button, a list of entries that match the restraints you selected in the query form is presented (Fig. 16.4). Residues that match the ones you entered in your query are highlighted in the results. The individual entries are accessed by clicking the *Explore* button below the structures. Some structures feature further buttons such as *theor. 3D Co-ord.*, *NMR*, or *PDB entries*, which indicate that a 3D

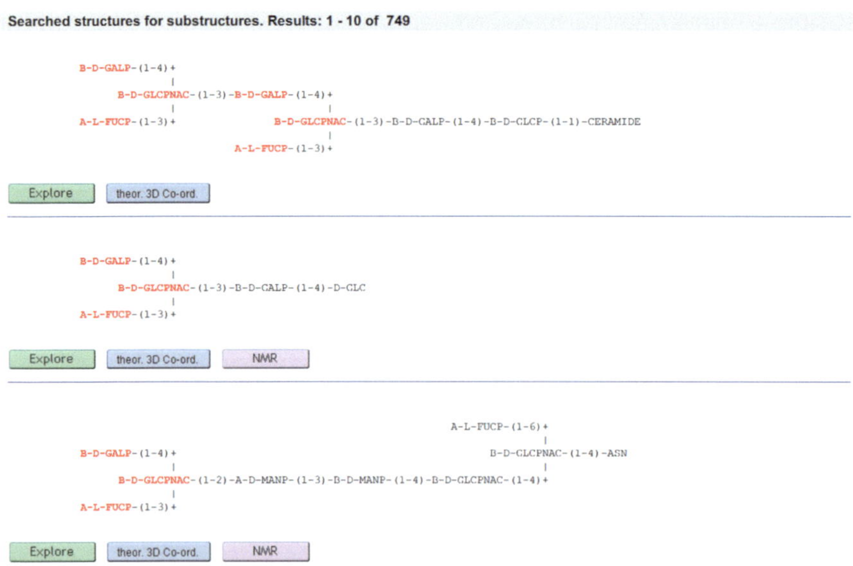

Fig. 16.4 Substructure search results of the query depicted in Fig. 16.2b

structure model, NMR data, or references to PDB entries, respectively, are available in that entry. A click on one of these buttons directly leads to the corresponding section in the entry (see description of database entries above). NMR data can also be queried specifically (→*Databases*→*NMR*). These queries have recently been described in detail elsewhere (Loss and Lütteke 2015).

16.2.1.3 Search by Motifs or N-Glycan Classification

When browsing through the results of the substructure search for the Le^X trisaccharide, it becomes obvious that most results feature the requested substructure at a terminal position. This is not surprising taking into account that Le^X is recognized by specific proteins and thus is presented to them in an accessible position within the glycan structures. In some cases, however, the residues are not at a terminal position but are extended by further residues. For example, in several structures α-D-Neup5Ac is linked to the β-D-Galp part of Le^X, resulting in a sialyl-$Lewis^X$ (Sia-Le^X) motif. Sia-Le^X has different biological functions than Le^X (Gadhoum and Sackstein 2008). A user who wants to employ a substructure search to find structures that feature Le^X as a terminal epitope needs to filter out these entries manually from the results.

This limitation can be bypassed by using the motif search (→*Databases*→*Structure*→*Motifs*). In this interface (Fig. 16.2c), you can select a motif from a predefined list. Motif information is stored together with the entries in the database and thus can be quickly retrieved, so that the list already indicates the number of hits for each motif. This number refers to the total number of entries that feature this motif. The actual number of hits can be smaller if you select further restrictions such as NMR data or PDB entries, which can be used as described in the substructure search section above. Select *Lewis X* from the list of motifs to find all database entries that contain a terminal Le^X epitope. The results look similar to those obtained by the substructure search, but only structures that contain Le^X trisaccharides at terminal positions are included here. Thus, the result list is shorter than in the substructure search and does not require the user to filter out Sia-Le^X, or other extensions of this motif, from the results. A further advantage of the motif search over a substructure search can be found in the fact that the user does not need to know the exact definition of the motif. Instead, knowledge of the motif name, which is more frequently used in the scientific literature than the exact structural definition, is sufficient to perform a query.

A search option that is specific to N-glycans is provided by the N-glycan classification search of GLYCOSCIENCES.de database (→*Databases*→*Structure*→*N-glycan classification*). Here you can search N-glycan structures via properties, such as N-glycan class (high mannose, complex, hybrid), or the number of antennae or of specific terminal residues (Fig. 16.2d). It can also be combined with a motif search. The list of available motifs is much smaller here than in the motif search described above because not all motifs are found in N-glycans. Select *Lewis X* from the motif list to find N-glycans that feature the Le^X epitope. You can add further restrictions here; e.g., select *D-Neup5Ac>= 1* in the *No. of terminal residues* section of this form to find N-glycans that contain at least one sialic acid in addition to the Le^X motif.

16.2.1.4 Search Using GlycoCD

Glycan antigens on cell surfaces can be conveniently detected in the lab using anti-carbohydrate monoclonal antibodies (Schwartz-Albiez 2009). Such antigens are classified by the clusters of differentiation (CD) nomenclature as part of the International Union of Immunological Societies nomenclature (Zola et al. 2003). Several CD antigens are carbohydrates or glycan-binding proteins. The GlycoCD database (Kumar et al. 2012) collects information on both carbohydrate CDs and carbohydrate recognition CDs. GlycoCD is part of the GLYCOSCIENCES.de portal and cross-linked with the main GLYCOSCIENCES.de database. It can be used to access the database from an immunological point of view. The entries are listed on the browse page (→*Databases*→*GlycoCD*→*Browse clusters of differentiation*) and are categorized as *carbohydrate recognition CD* and *Glycan CD*. Below these lists, the carbohydrate recognition CDs are also listed by glycan-binding specificity. Le^X is present within the list of Glycan CDs as "CD15 (Lewis^X)." Click on that link to access information on molecular structure, function, expression, or application of this CD antigen. In the section with links to glycan databases, there is a link *Find glycan structures with CD15 epitope*. This is an alternative to the motif search for finding entries featuring the Le^X motif in the GLYCOSCIENCES.de database.

16.2.2 Use Case: Exploring Conformational Properties of Lewis^X

The "sequence" (or "2D structure") of Le^X is sufficient for finding database entries or literature references of this epitope as described above. If you want to study its actual role in carbohydrate-protein recognition events on a molecular basis, however, you often need to know the 3D structure of the molecule and the conformational space that it can adopt. GLYCOSCIENCES.de provides several ways to access glycan 3D structures. It offers cross-links to PDB entries that feature specific carbohydrates and provides tools for 3D structure modeling of glycans and tools and databases to analyze the adopted conformations.

16.2.2.1 Carbohydrate-Containing PDB Structures

To search the PDB for entries that contain specific carbohydrates, simply use the search queries described above, and tick the *with PDB entries* option to limit the results to structures that feature PDB cross-links. When using carbohydrate data in the PDB, you should always keep in mind that a rather high ratio of carbohydrate-containing PDB entries has errors within the carbohydrate moieties (Lütteke and von der Lieth 2004; Crispin et al. 2007; Nakahara et al. 2008; Lütteke 2009). Therefore, GLYCOSCIENCES.de offers two tools for 3D structure validation: pdb-care (PDB CArbohydrate REsidue check, →*Tools*→*pdb-care*) and CARP (CArbohydrate Ramachandran Plot; →*Tools*→ *CARP*). Both tools accept a PDB ID or an uploaded structure file in PDB file format as input. The purpose of pdb-care (Lütteke and von der Lieth 2004) is detection and naming of carbohydrates in the 3D structure, comparison of detected residue names with names that are used in the PDB file format, and a check of conformity of N-glycan core structures with well-known biosynthesis pathways. Potential problems arising from erroneous atom connections, as well as missing and superfluous atoms (which mainly occur in PDB structures within glycosidic linkages) are also reported. CARP (Lütteke et al. 2005) is meant for validation of glycan conformation by comparison of glycosidic torsions present in a supplied 3D structure with those that are usually observed for a specific glycosidic linkage, similar to the Ramachandran Plot (Ramachandran et al. 1963) for the validation of proteins (Hooft et al. 1997; Read et al. 2011). The usage of pdb-care and CARP has recently been described in detail elsewhere (Emsley et al. 2015).

16.2.2.2 Building 3D Structure Models

Carbohydrate 3D structures can also be created computationally. Such 3D structure models of glycans are in general more reliable than protein 3D structure models because carbohydrates are smaller molecules and thus easier to predict, and they are more flexible than proteins, which means that there usually is not one single "correct" conformation, but multiple conformations can be considered as reliable models of a single glycan. Such 3D structure models of carbohydrate chains can be created using the Sweet-II tool (Bohne et al. 1999). Sweet-II (→*Modeling*→ *Sweet-II*) currently offers three different input modes, which are based on a textual input of residues. A graphical input using the GlycanBuilder tool (Ceroni et al. 2007) will be added in the near future. The *beginner version* of Sweet-II is limited to short linear glycans, and the residues are selected from drop-down menus. Therefore, only a very limited number of glycans can be entered via this mode. The *expert version* is much more flexible but requires more knowledge on how to correctly enter the glycans to be modeled. Each input mode features an *example page* that demonstrates how to enter a glycan. To get an idea of how to use the *expert version* of Sweet-II, you can click on the corresponding *example page* link. If you want to enter your own structure, use the *work page* link. The input form of the *expert version* consists of a table of fields of different size. The residue names are to be entered into the larger fields, whereas the smaller fields are reserved to enter the linkages between the residues. On top of the input form, there is a link *Allowed Templates*, which opens a list of residues that can be entered. This list shows the basic residues without any substituents such as NAc for n-acetyl, SO3 for sulfate, or OMe for o-methyl groups. Substituents can be added (preceded by their position) to the template names. For example, 2-sulfate-β-D-galacturonic acid is entered as "b-D-GalpA2SO3". Deoxy

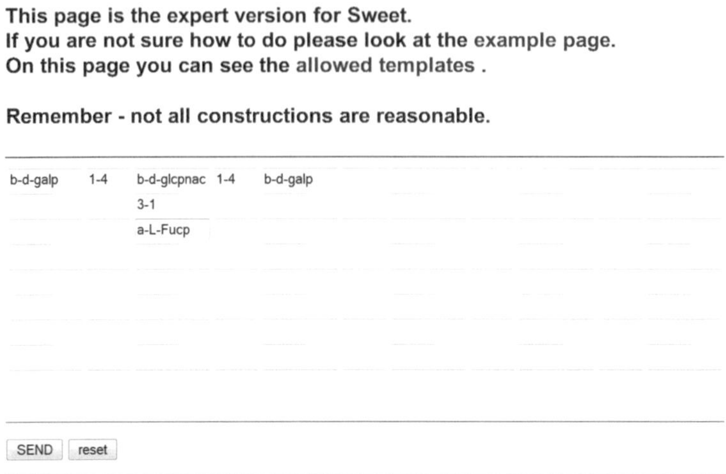

Fig. 16.5 Input of Le^X (1–4)-linked to β-D-Galp in the *expert version* interface of *Sweet-II*. Large fields denote residues, and the small fields in between indicate linkages. Upper-/lowercase spelling is not relevant. In *vertical* linkages the first number is assigned to the residue *above*, and the second number to the residue *below* the linkage field. Therefore, the linkage between b-D-GlcpNAc and a-L-Fucp has to be entered as "3–1" (and not "1–3") in this example to indicate that C1 of a-L-Fucp has to be linked to O3 of b-D-GlcpNAc

modifications can also be added. If you use such modifications, make sure that the basic template name is not changed. For example, following IUPAC nomenclature recommendations (McNaught 1997), a deoxy modification at position 4 of a b-D-Glcp residue would result in a b-D-4-deoxy-xylHexp residue. Such a residue name is currently not accepted by Sweet-II. Instead, you need to enter "4-deoxy-b-D-Glcp" (or "4-deoxy-b-D-Galp", which chemically denotes the same residue) to add this residue to your model structure. The residue names to be used are the same also for the *direct input* form. In this mode you have one large input field similar to the *exact structure search* of the GLYCOSCIENCES.de database. This field also accepts CarbBank-style 2D representations of carbohydrates. If you need to enter a glycan structure manually residue by residue, the *expert version* of Sweet-II is probably easier to use, but if you already have a CarbBank-style representation available, the *direct input* form can be the easier way to enter your structure.

Fill the *expert version* form of Sweet-II as depicted in Fig. 16.5 to build a 3D structure model of Le^X (1–4)-linked to a β-D-Galp residue, and click the *SEND* button to start the calculation of the model. If the generated model contains atomic clashes (i.e., overlaps of atoms that are not linked to each other), a "distance warning" is given at this step. In this case a minimization of the initial model with the Tinker software (Ponder and Richards 1987) is used to solve this issue. You need to click any of the *Optimize – fast* or *Optimize 0.1* buttons to start the minimization. These options differ in the cutoff value to terminate the optimization.

The Sweet-II result page offers access to several Java applets for online view of the modeled 3D structure to quickly get an idea of how the glycan chain actually looks like. The applets will be replaced by JSmol, the HTML5 version of Jmol (www.jmol.org), in the next release of the Sweet-II interface, to circumvent restrictions of Java applets in current web browsers. On the result page, you can also access the structure in PDB format and download it to your computer. This way you can display the structure in any 3D structure viewer of your choice, or you can, e.g., use the coordinates as a starting structure for molecular modeling.

16.2.2.3 Analysis of the Conformational Space Adopted by a Glycan Chain

Modeling approaches can be used to support interpretation of experimental results such as NMR data but can also be used to thoroughly study the conformational properties of a glycan chain. For example, molecular dynamics (MD) simulations of carbohydrates are used to calculate conformational maps of glycosidic linkages (Frank and Schloissnig 2010). Such maps are provided by GlycoMapsDB (Frank et al. 2007). We will use this resource to analyze the conformational space that can be adopted by Le^X. Go to → *Modeling* → *GlycoMapsDB* and click *Search Database*. Here you need to indicate the glycosidic linkage that you want to analyze by specifying the monosaccharides that are involved in the linkage and the linkage position. You can select values from the drop-down menus or directly type them into the search fields below (Fig. 16.6a). Select or type "b-D-Galp", "1–4", and "b-D-GlcpNAc" (again, the query is not case sensitive), make sure that the *show map previews* option is ticked, and set the *results per page* value to at least 50 to get an overview of many results on one page. Click *submit* to start the query. The result page lists the maps that match your input data, the disaccharide fragment whose ϕ/ψ-torsions are plotted in the map, and the complete glycan structure that was used in the MD simulation from which the map was determined. Basic simulation parameters (method, force field, and software used) are shown in the *Methods* column (Fig. 16.6b).

When scrolling through the results, you can find the Le^X trisaccharide in the *Complete Structure* column of MapID 8432. A click on the MapID or the preview map opens the complete entry of this map (Fig. 16.6c). Each GlycoMapsDB entry contains the conformational map itself and information on the underlying glycan structure and the method that was used to obtain the map. A link to the corresponding entry in the main GLYCOSCIENCES.de database is also provided (*Explore* button). Conversely, the GLYCOSCIENCES.de database entry also contains links to the maps in GlycoMapsDB.

Conformational maps indicate the energy landscape of a glycosidic linkage. The lower the energy of a specific conformation (identified by the ϕ/ψ torsions), the more frequently this conformation was adopted over time in the simulation, from which the map was obtained. White parts of the map have not been adopted at all during the simulation. The map with ID 8432 indicates that the β-D-Galp-(1–4)-β-

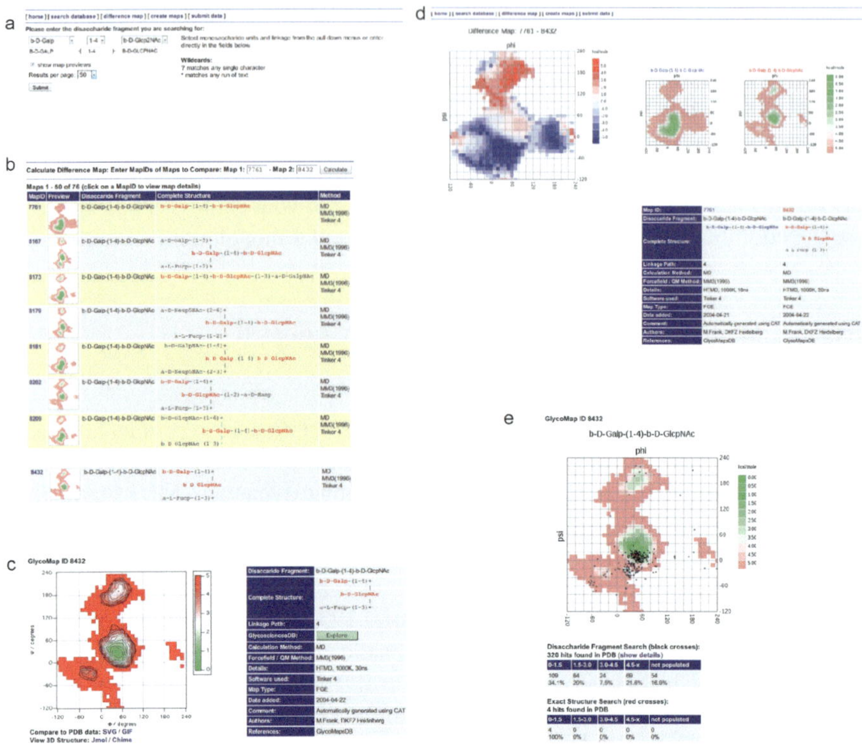

Fig. 16.6 (**a**) GlycoMapsDB query for conformational maps of the β-D-Galp-(1–4)-β-D-GlcpNAc glycosidic linkage. (**b**) Query results include a preview of the map, information on the complete glycan structure from which the map was extracted, and details on the methods that were used to generate the map. An option to create a difference map for comparison of two maps is also available. (**c**) GlycoMapsDB entry 8432. (**d**) A difference map can be used to, e.g., identify the impact of additional residues close to the analyzed glycosidic linkage on the occupied conformational space. (**e**) Comparison of conformational map and torsions derived from the PDB

D-GlcpNAc disaccharide fragment of Le^X has one preferred conformation (global minimum of the energy landscape) with ϕ and ψ angles between 0 and 60° both. Two local minima can also be seen that are occupied less frequently.

Go back to the results of the above query. From the map previews, you can already see that the pure disaccharide (i.e., the entry where the *complete structure* is identical with the *disaccharide fragment*, MapID 7761) occupies the largest conformational space, which means that it is the most flexible molecule of the ones that are found by your query. This is expected, because further residues that are added to the disaccharide often result in limitations of the possible conformations of an individual glycosidic linkage. A convenient way to quantify these restrictions is the calculation of difference maps, where the energy difference of two individual maps is plotted. You can generate such a difference map in GlycoMapsDB by adding the MapIDs of the concerned maps into the form on top of the search results (if you already know the MapIDs, you can also use the *difference maps* link in the GlycoMapsDB menu section). For example, enter "7761" and "8432" into the form (Fig. 16.6b) to quantify the restrictions to the β-D-Galp-(1–4)-β-D-GlcpNAc linkage that are applied by the (1–3)-linked fucose in direct neighborhood of the (1–4)-linked galactose of Le^X. It becomes obvious that, compared to the disaccharide, the global minimum of this linkage is slightly shifted toward higher ψ values in Le^X and the side minimum around $\phi = 60°/\psi = 180°$ is more frequently occupied (Fig. 16.6d).

The maps are computed data, which leads to the question of how reliable they are. One possibility to verify the reliability is comparison of the computed torsions with those observed in experimentally resolved 3D structures from the PDB. Carbohydrate torsions from the PDB are also stored on GLYCOSCIENCES.de and can be accessed with the GlyTorsion tool (→*Tools*→*GlyTorsion*). For a detailed description of how to use GlyTorsion, please refer to (Lütteke and von der Lieth 2009). You can conveniently compare computed conformational maps and corresponding torsions extracted from the PDB within GlycoMapsDB. Go back to MapID 8432 of the GlycoMapsDB search. There you can find a feature *Compare to PDB data* (Fig. 16.6c). A graphical comparison can be provided in SVG or GIF format here. Click on one of these formats to start the comparison. PDB-extracted torsions of β-D-Galp-(1–4)-β-D-GlcpNAc linkages obtained from the GlyTorsion database are now plotted onto the conformational map of GlycoMapsDB (Fig. 16.6e). You can find two kinds of hits in the GlyTorsion dataset: those where the linkage is present within Le^X trisaccharides in the PDB as well (*Exact structure* search, red crosses) and those where the linkage is present in other kinds of oligosaccharides (*Disaccharide fragment* search, black crosses). All of the few red crosses are located in the low-energy area of the map, whereas many black crosses are found outside the preferred areas or even in areas that are not populated in the GlycoMapsDB map at all. Numerical data below the plot supports this impression: all four exact hits are in the low-energy areas of 0–1.5 kcal/mol, whereas almost 40 % of the disaccharide substructure hits are in high-energy areas of more than 1.5 kcal/mol or areas that are not populated at all during the MD simulation, from which the map was calculated. However, this does not necessarily mean that the map or the PDB data are incorrect. Remember that the fucose at position 3 of the GlcNAc residue causes a shift of the preferred conformation of the β-D-Galp-(1–4)-β-D-GlcpNAc linkages compared to the disaccharide. Most torsions that are found in the PDB occupy the area that is preferred by the linkage in the disaccharide and also in other oligosaccharides where additional residues are located farther away from the linkage and thus induce fewer restrictions to this linkage. Therefore, *exact structure* hits are much better suited for validation of the maps than *disaccharide fragment* hits.

Note Torsions outside the preferred low-energy areas are not necessarily erroneous in case of *exact structure* hits as well. The maps are determined for free oligosaccharides, whereas the glycans in PDB entries are usually present as parts of glycoproteins or as protein-carbohydrate complexes. A glycan that is linked to a protein, however, can exhibit a conformation that is different from that in solution (Pederson et al. 2014).

16.3 Conclusions

The GLYCOSCIENCES.de portal combines a series of databases and tools in a common environment, which are linked to each other. This collection of various resources on one server facilitates analyses that make use of multiple tools or datasets, as exemplified for the comparison of conformational data from computationally generated maps and experimentally resolved structures. Many more cross-linkages between individual GLYCOSCIENCES.de applications are used in the background (Rojas-Macias et al. 2014). Nevertheless, in spite of initial attempts to allow cross-database queries (Toukach et al. 2007), the benefits that arise from this combined usage of resources are still mostly limited to individual servers. The GlycoRDF project, a recent initiative to cross-link data from various servers, also including proteomics or genomics databases, will allow more flexible queries that will be able to answer more complex questions that cannot be answered by a single resource (Aoki-Kinoshita et al. 2013; Ranzinger et al. 2014).

References

Aoki-Kinoshita KF (2013) Using databases and web resources for glycomics research. Mol Cell Proteomics 12(4):1036–1045. doi:10.1074/mcp.R112.026252

Aoki-Kinoshita KF, Kanehisa M (2006) Bioinformatics approaches in glycomics and drug discovery. Curr Opin Mol Ther 8(6):514–520

Aoki-Kinoshita KF, Bolleman J, Campbell MP, Kawano S, Kim JD, Lutteke T, Matsubara M, Okuda S, Ranzinger R, Sawaki H, Shikanai T, Shinmachi D, Suzuki Y, Toukach P, Yamada I, Packer NH, Narimatsu H (2013) Introducing glycomics data into the Semantic Web. J Biomed Semant 4(1):39. doi:10.1186/2041-1480-4-39

Berman HM, Westbrook J, Feng Z, Gilliland G, Bhat TN, Weissig H, Shindyalov IN, Bourne PE (2000) The Protein Data Bank. Nucleic Acids Res 28(1):235–242

Berteau O, Stenutz R (2004) Web resources for the carbohydrate chemist. Carbohydr Res 339(5):929–936

Bohne A, Lang E, von der Lieth CW (1999) SWEET – WWW-based rapid 3D construction of oligo- and polysaccharides. Bioinformatics 15(9):767–768

Bohne-Lang A, Lang E, Forster T, von der Lieth CW (2001) LINUCS: linear notation for unique description of carbohydrate sequences. Carbohydr Res 336(1):1–11

Campbell MP, Ranzinger R, Lütteke T, Mariethoz J, Hayes CA, Zhang J, Akune Y, Aoki-Kinoshita KF, Damerell D, Carta G, York WS, Haslam SM, Narimatsu H, Rudd PM, Karlsson NG, Packer NH, Lisacek F (2014) Toolboxes for a standardised and systematic study of glycans. BMC Bioinf 15(Suppl 1):S9. doi:10.1186/1471-2105-15-S1-S9

Ceroni A, Dell A, Haslam SM (2007) The GlycanBuilder: a fast, intuitive and flexible software tool for building and displaying glycan structures. Source Code Biol Med 2:3

Crispin M, Stuart DI, Jones EY (2007) Building meaningful models of glycoproteins. Nat Struct Mol Biol 14(5):354

Doubet S, Albersheim P (1992) CarbBank. Glycobiology 2(6):505

Doubet S, Bock K, Smith D, Darvill A, Albersheim P (1989) The complex carbohydrate structure database. Trends Biochem Sci 14(12):475–477

Emsley P, Brunger AT, Lütteke T (2015) Tools to assist determination and validation of carbohydrate 3D structure data. Methods Mol Biol 1273:229. doi:10.1007/978-1-4939-2343-4_17

Frank M, Schloissnig S (2010) Bioinformatics and molecular modeling in glycobiology. Cell Mol Life Sci 67(16):2749–2772. doi:10.1007/s00018-010-0352-4

Frank M, Lütteke T, von der Lieth CW (2007) GlycoMapsDB: a database of the accessible conformational space of glycosidic linkages. Nucleic Acids Res 35(Database issue):287–290. doi:10.1093/nar/gkl907

Gadhoum SZ, Sackstein R (2008) CD15 expression in human myeloid cell differentiation is regulated by sialidase activity. Nat Chem Biol 4(12):751–757. doi:10.1038/nchembio.116

Hooft RW, Sander C, Vriend G (1997) Objectively judging the quality of a protein structure from a Ramachandran plot. Comput Appl Biosci 13(4):425–430

Kumar S, Lütteke T, Schwartz-Albiez R (2012) GlycoCD: a repository for carbohydrate-related CD antigens. Bioinformatics 28(19):2553–2555. doi:10.1093/bioinformatics/bts481

Lohmann KK, von der Lieth CW (2004) GlycoFragment and GlycoSearchMS: web tools to support the interpretation of mass spectra of complex carbohydrates. Nucleic Acids Res 32:W261–W266

Loss A, Lütteke T (2015) Using NMR data on GLYCOSCIENCES.de. Methods Mol Biol 1273:87. doi:10.1007/978-1-4939-2343-4_6

Loss A, Bunsmann P, Bohne A, Schwarzer E, Lang E, von der Lieth CW (2002) SWEET-DB: an attempt to create annotated data collections for carbohydrates. Nucleic Acids Res 30(1):405–408

Loss A, Stenutz R, Schwarzer E, von der Lieth CW (2006) GlyNest and CASPER: two independent approaches to estimate 1H and 13C NMR shifts of glycans available through a common web-interface. Nucleic Acids Res 34(Web Server issue):W733–W737

Lütteke T (2009) Analysis and validation of carbohydrate three-dimensional structures. Acta Crystallogr D Biol Crystallogr 65(2):156–168

Lütteke T (2012) The use of glycoinformatics in glycochemistry. Beilstein J Org Chem 8:915–929. doi:10.3762/bjoc.8.104

Lütteke T, von der Lieth C-W (2004) pdb-care (PDB carbohydrate residue check): a program to support annotation of complex carbohydrate structures in PDB files. BMC Bioinf 5:69

Lütteke T, von der Lieth CW (2009) Data mining the PDB for glyco-related data. Methods Mol Biol 534:293–310. doi:10.1007/978-1-59745-022-5_21

Lütteke T, Frank M, von der Lieth CW (2004) Data mining the protein data bank: automatic detection and assignment of carbohydrate structures. Carbohydr Res 339(5):1015–1020. doi:10.1016/j.carres.2003.09.038

Lütteke T, Frank M, von der Lieth CW (2005) Carbohydrate Structure Suite (CSS): analysis of carbohydrate 3D structures derived from the Protein Data Bank. Nucleic Acids Res 33(Database Issue):D242–D246

Lütteke T, Bohne-Lang A, Loss A, Goetz T, Frank M, von der Lieth CW (2006) GLYCOSCIENCES.de: an internet portal to support glycomics and glycobiology research. Glycobiology 16(5):71R–81R

McNaught AD (1997) Nomenclature of carbohydrates (recommendations 1996). Adv Carbohydr Chem Biochem 52:43–177

Moran AP (2009) Molecular mimicry of host glycosylated structures by bacteria. In: Moran AP, Holst O, Brennan PJ, von Itzstein M (eds) Microbial glycobiology. Academic, London, pp 847–870

Nakahara T, Hashimoto R, Nakagawa H, Monde K, Miura N, Nishimura S-I (2008) Glycoconjugate Data Bank: structures – an annotated glycan structure database and N-glycan primary structure verification service. Nucleic Acids Res 36(Database issue):D368–D371

Pederson K, Mitchell DA, Prestegard JH (2014) Structural characterization of the DC-SIGN-Lewis(X) complex. Biochemistry 53(35):5700–5709. doi:10.1021/bi5005014

Ponder JW, Richards FM (1987) An efficient newton-like method for molecular mechanics energy minimization of large molecules. J Comput Chem 8:1016–1024

Powlesland AS, Barrio MM, Mordoh J, Hitchen PG, Dell A, Drickamer K, Taylor ME (2011) Glycoproteomic characterization of carriers of the CD15/Lewisx epitope on Hodgkin's Reed-Sternberg cells. BMC Biochem 12:13. doi:10.1186/1471-2091-12-13

Ramachandran GN, Ramakrishnan C, Sasisekharan V (1963) Stereochemistry of polypeptide chain configurations. J Mol Biol 7:95–99

Ranzinger R, Aoki-Kinoshita KF, Campbell MP, Kawano S, Lutteke T, Okuda S, Shinmachi D, Shikanai T, Sawaki H, Toukach P, Matsubara M, Yamada I, Narimatsu H (2014) GlycoRDF: an ontology to standardize glycomics data in RDF. Bioinformatics, in press. doi:10.1093/bioinformatics/btu732

Read RJ, Adams PD, Arendall WB 3rd, Brunger AT, Emsley P, Joosten RP, Kleywegt GJ, Krissinel EB, Lütteke T, Otwinowski Z, Perrakis A, Richardson JS, Sheffler WH, Smith JL, Tickle IJ, Vriend G, Zwart PH (2011) A new generation of crystallographic validation tools for the protein data bank. Structure 19(10):1395–1412. doi:10.1016/j.str.2011.08.006

Rojas-Macias MA, Loss A, Bohne-Lang A, Frank M, Lütteke T (2014) Databases and tools of GLYCOSCIENCES.de web server. In: Taniguchi N, Endo T, Hart GW, Seeberger P, Wong CH (eds) Glycoscience: biology and medicine. Springer, Tokyo, pp 233–239

Schwartz-Albiez R (2009) Inflammation and glycoscience. In: Gabius HJ (ed) The sugar code. Wiley-VCH, Weinheim, pp 447–467

Toukach P, Joshi HJ, Ranzinger R, Knirel Y, von der Lieth CW (2007) Sharing of worldwide distributed carbohydrate-related digital resources: online connection of the bacterial carbohydrate structure DataBase and GLYCOSCIENCES.de. Nucleic Acids Res 35(Database issue):D280–D286

van Kuik JA, Hard K, Vliegenthart JFG (1992) A ^{1}H NMR database computer program for the analysis of the primary structure of complex carbohydrates. Carbohydr Res 235:53–68

Zola H, Swart B, Boumsell L, Mason DY (2003) Human leucocyte differentiation antigen nomenclature: update on CD nomenclature. Report of IUIS/WHO subcommittee. J Immunol Methods 275(1–2):1–8

Chapter 17
Glycobiology Meets the Semantic Web

Shin Kawano

Abstract The improvement and diversification of experimental technologies have caused a flood of data. In order to share and integrate such huge and diverse data, it is important to describe the relationship between data using Semantic Web technology. A goal of the Semantic Web is that computers can automatically process data by linking meaningful data and by forming a web of data. The Semantic Web consists of key technologies such as Resource Description Framework (RDF), ontologies, triple stores (database for RDF), and SPARQL Protocol and RDF Query Language (SPARQL), which is a query language for triple stores. Although the Semantic Web has been used by some specific domains such as government and media, recently it is also applied to the life sciences. In this chapter, I will describe about the Semantic Web and its application to life science including glycobiology. Finally, I will introduce TogoTable, which is a web application using the Semantic Web, used for collecting annotations from distributed databases.

Keywords Annotation • Data integration • Interoperability • LOD • RDF • Semantic Web • SPARQL • TogoTable

17.1 Introduction

The improvement and diversification of research techniques have been producing huge and various kinds of data. In order to deal with such huge and various data in a unified way, it is important to describe the meaning of the data and their relationships as well as representing data in a standardized format. The Semantic Web technology has attracted attention as a means for integrating diverse data.

The Semantic Web is a concept propounded by Tim Berners-Lee (Berners-Lee et al. 2001), who invented the World Wide Web (WWW). A goal of the Semantic Web is that computers can automatically process data by linking meaningful data and by forming a web of data. It has already been utilized in several domains,

S. Kawano (✉)
Database Center for Life Science, Research Organization of Information and Systems, 178-4-4 Wakashiba, 277-0871 Kashiwa, Chiba, Japan
e-mail: kawano@dbcls.rois.ac.jp

K.F. Aoki-Kinoshita (ed.), *A Practical Guide to Using Glycomics Databases*, DOI 10.1007/978-4-431-56454-6_17

e.g., government (Shadbolt et al. 2012, https://data.gov.uk/) and media (Kobilarov et al. 2009, http://www.bbc.com/). In the life science field, activities for which the Semantic Web is utilized have also started. For example, many databases in the European Bioinformatics Institute (EBI) have been published using Semantic Web technologies (Jupp et al. 2014, https://www.ebi.ac.uk/rdf/platform), and developer meetings for the Semantic Web have been held in Japan every year (Katayama et al. 2010, 2011, 2013, 2014, http://www.biohackathon.org/). Even in glycomics, developers of representative glycan databases have agreed to publish their data on the Semantic Web (Aoki-Kinoshita et al. 2013b), and as a result, they have achieved promising results (Aoki-Kinoshita et al. 2013a; Ranzinger et al. 2015). In this way, the Semantic Web technology is utilized in various fields including the glycosciences in order to integrate various kinds of data.

In this chapter, I give a brief introduction of the Semantic Web and explain about the Semantic Web in the life sciences including glycobiology. As an application of the Semantic Web, I will introduce TogoTable (Kawano et al. 2014, http://togotable.dbcls.jp/), which is a web tool to collect annotations from biological databases. TogoTable is applicable not only to glycan databases but also to any databases if the data is provided as the Semantic Web.

17.2 Semantic Web Technology

The WWW has spread widely with the rapid development of the Internet, and it has become indispensable to research activities. WWW consists of documents linked by hyperlinks, and one can find a document by keyword search or get desired information by reading a document and following a link. Therefore, the WWW can be said to be a "web of documents." However, because the documents are written under the premise of being read by humans, computers cannot interpret their contents. For example, it is impossible for computers to interpret what types of relationships exist between documents that are connected by a hyperlink (a HyperText Markup Language (HTML) shows only that there is relation, but it is unclear what kind of relation it is) and distinguish between different concepts written using the same words (e.g., computers cannot discriminate "mouse" the animal from "mouse" a computer device). These interpretations are easy for humans from the surrounding context. On the other hand, computers are good at dealing with structured data rather than unstructured documents. The Semantic Web is an attempt to achieve a "web of data" by connecting data in a computer-interpretable form. This will lead to the development of applications such as advanced search and mash-up services.

The Semantic Web consists of several key components. First, the Semantic Web has a common data model called the Resource Description Framework (RDF), which is standardized by the World Wide Web Consortium (W3C, https://www.w3.org/RDF/). It represents any resource using combinations of subject-predicate-object, called triples (Fig. 17.1). Subject is a description of a resource

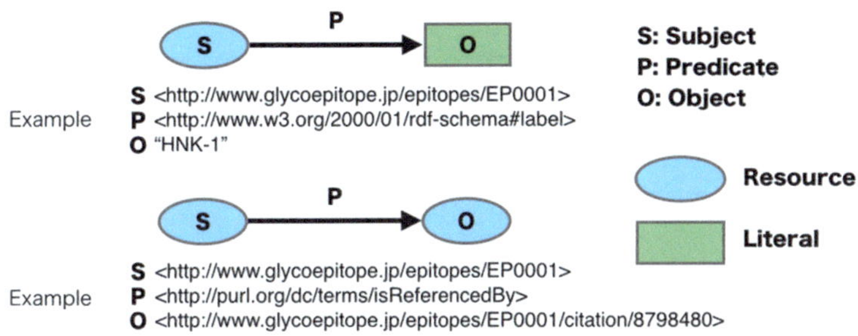

Fig. 17.1 Examples of triples. A triple consists of a subject-predicate-object ternary. Whereas subject and predicate must be URIs, object can be a literal or a URI

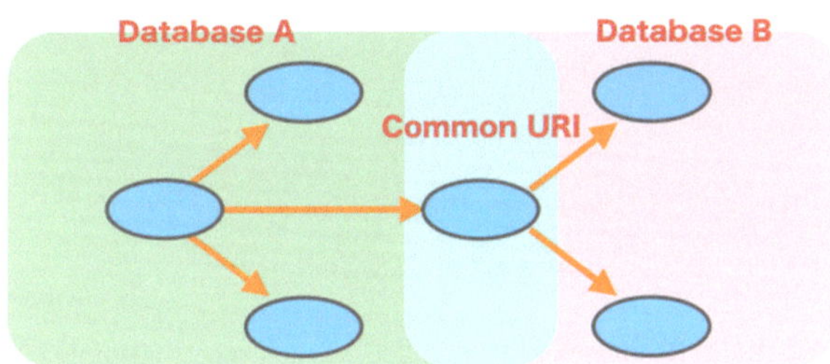

Fig. 17.2 A common URI connects resources in the different databases

target, and object is a value (literal) of the subject or another resource which is related to the subject. Predicate represents the relation between subject and object.

RDF data are stored in dedicated databases called triple stores such as Virtuoso (Erling and Mikhailov 2009, http://virtuoso.openlinksw.com/), Ontotext GraphDB (http://ontotext.com/products/graphdb/), and Stardog (http://stardog.com/). There are advantages and disadvantages of each triple store (Wu et al. 2014). Because the triple stores are still developing, there are some weaknesses in terms of performance, scalability, and stability. However, because vendors have actively continued the development of triple stores, these problems are expected to be solved in the near future.

Second, all resources are represented as Uniform Resource Identifiers (URIs). In particular, the uniform resource locator (URL) is used commonly on the Internet. By using unique identifiers for representing resources, a computer can determine whether or not resources in different triples are identical. Although the content page of a URL need not necessarily exist in practice, if there some content is made available such as a description of a resource, a human would be able to know about the resource by accessing the URL. Another benefit of using a URI (URL) is being able to connect triples through common URIs (Fig. 17.2). A connection of triples builds a network of resources called the Linked Open Data (LOD) cloud (Fig. 17.3). Since the LOD cloud is a graph structure, any resource can be searched from a variety of perspectives by graph pattern matching algorithms.

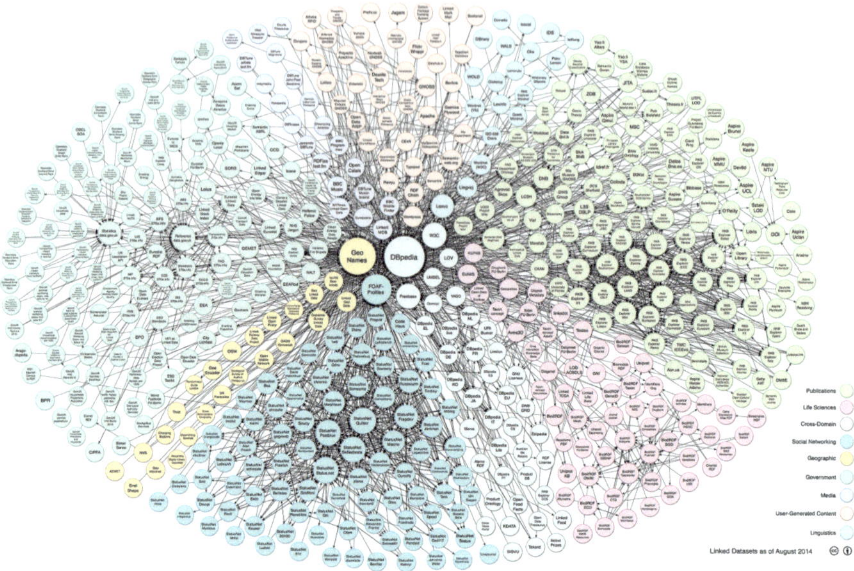

Fig. 17.3 The LOD cloud. Resources are connected to each other, forming a web of resources. This figure is a reprint from "Linking Open Data cloud diagram 2014" (http://lod-cloud.net/), by Max Schmachtenberg, Christian Bizer, Anja Jentzsch, and Richard Cyganiak, and it is provided under the Creative Commons Attribution license

Third, an ontology is used to represent "classes" of resources and "properties" of predicates on the Semantic Web. An ontology defines not only standardized terminologies (controlled vocabularies) but also hierarchical relationships among them. By describing hierarchical dependencies using the ontology, reasoning by a computer becomes possible. For example, "amylase activity" is defined as a subordinate concept of "hydrolase activity" in the Gene Ontology (The Gene Ontology Consortium et al. 2000, http://geneontology.org/). When there is a gene annotated as having "amylase activity," computers can recognize that the gene also has "hydrolase activity" by reasoning using the ontology.

Finally, RDF has a query language called SPARQL (SPARQL Protocol And RDF Query Language), which is also standardized by the W3C (https://www.w3.org/TR/sparql11-overview/). SPARQL finds subgraphs from the LOD network by graph pattern matching. All triple stores have a search interface called a SPARQL endpoint. One of the most characteristic features of SPARQL is federated search. It is possible to search across databases which are located physically at different locations. This means that data from various databases can be integrated by cross database search using SPARQL, without having to transfer data across networks.

In the life science field, several major databases have adopted the Semantic Web and formed parts of the LOD. For example, UniProt, which provides protein sequences and their functional annotations, is an early adopter of RDF (The UniProt

Consortium 2014, http://www.uniprot.org/), and the Worldwide Protein Data Bank (wwPDB), which is a repository of protein tertiary structures, also provides their data in the RDF data model (Kinjo et al. 2012, http://www.wwpdb.org/). Recently, many databases provided by EBI have completed RDFization (Jupp et al. 2014, https://www.ebi.ac.uk/rdf/platform), and PubChem (Fu et al. 2015, https://pubchem.ncbi.nlm.nih.gov/), which is a database of chemical compounds provided by the National Center for Biotechnology Information (NCBI), employs the RDF data model (https://pubchem.ncbi.nlm.nih.gov/rdf/). In addition, the Bio2RDF project has provided RDF data as a third-party developer (Belleau et al. 2008, http://bio2rdf.org/). In this way, an infrastructure to handle different kinds of data in an integrated manner is being developed rapidly. Representative databases providing data using the RDF data model are listed in Table 17.1.

Table 17.1 Representative databases providing data in the RDF data model

Database	URL	Provider	References
UniProt	http://sparql.uniprot.org/	UniProt Consortium	The UniProt Consortium (2014)
wwPDB	http://rdf.wwpdb.org/	Worldwide Protein Data Bank	Kinjo et al. (2012)
EBI RDF platform	https://www.ebi.ac.uk/rdf/platform	European Bioinformatics Institute	Jupp et al. (2014)
Ensembl	https://www.ebi.ac.uk/rdf/services/ensembl/sparql	European Bioinformatics Institute	
BioModels	https://www.ebi.ac.uk/rdf/services/biomodels/sparql	European Bioinformatics Institute	Wimalaratne et al. (2014)
BioSamples	https://www.ebi.ac.uk/rdf/services/biosamples/sparql	European Bioinformatics Institute	Faulconbridge et al. (2014)
ChEMBL	https://www.ebi.ac.uk/rdf/services/chembl/sparql	European Bioinformatics Institute	Willighagen et al. (2013)
Expression Atlas	https://www.ebi.ac.uk/rdf/services/atlas/sparql	European Bioinformatics Institute	
Reactome	https://www.ebi.ac.uk/rdf/services/reactome/sparql	European Bioinformatics Institute	
PubChem	https://pubchem.ncbi.nlm.nih.gov/rdf/	National Center for Biotechnology Information	Fu et al. (2015)

(continued)

Table 17.1 (continued)

Database	URL	Provider	References
Medical Subject Headings (MeSH)	https://id.nlm.nih.gov/mesh/	National Center for Biotechnology Information	Bushman et al. (2015)
NBDC RDF Portal	http://integbio.jp/rdf/	National Bioscience Database Center	
LinkDB	http://www.genome.jp/linkdb/linkdb_rdf.html	Kyoto University	
Microbial Genome Database (MBGD) for comparative analysis	http://mbgd.genome.ad.jp/rdf/wiki/index.php/Main_Page	National Institutes of Basic Biology	Chiba et al. (2015)
Bio2RDF	http://bio2rdf.org/	Université Laval and Stanford University	Belleau et al. (2008)
Identifiers.org	http://identifiers.org/services/sparql	European Bioinformatics Institute	Wimalaratne et al. (2015)

Note that these URLs point to RDF data or SPARQL endpoints rather than user-readable web pages of the database

17.3 The Semantic Web in the Glycobiology

Databases in glycobiology have been developed from the 1990s. First, CarbBank (Doubet and Albersheim 1992) was established as a glycan structure database, and its derivatives were developed including KEGG GLYCAN (Hashimoto et al. 2006, http://www.genome.jp/kegg/glycan/), GlycomeDB (Ranzinger et al. 2011, http://www.glycome-db.org/), and UniCarbKB (Campbell et al. 2014, http://www.unicarbkb.org/). The Carbohydrate Structure Database (CSDB, http://csdb.glycoscience.ru/) has collected glycan structures from bacteria, plants, and fungi (Egorova and Toukach 2014). There are not only glycan structure databases but also other characteristic databases such as MonosaccharideDB (http://www.monosaccharidedb.org/), which collects monosaccharides as building blocks of glycans; GlycoEpitope (Okuda et al. 2014, http://www.glycoepitope.jp/), which provides a collection of carbohydrate antigens and antibodies; and GlycoProtDB (Kaji et al. 2012, http://jcggdb.jp/rcmg/gpdb/), which is a database of N-glycosylation sites on amino acid sequences. However, because each database has been developed independently, it was difficult to integrate data between glycan databases.

In 2012, developers of major glycan databases gathered to discuss ways to integrate glycan-related databases, and they agreed to use the RDF data model

(Aoki-Kinoshita et al. 2013b; Katayama et al. 2014). In order to adopt glycan data to the RDF data model, they have developed several components.

The Web3 Unique Representation of Carbohydrate Structures (WURCS, http://www.wurcs-wg.org/) is a new representation of glycan structures using a linear string (Tanaka et al. 2014). Although linear notations of glycan structures have been developed before such as LinearCode (Banin et al. 2002), LINUCS (Bohne-Lang et al. 2001), and IUPAC (IUPAC-IUB JCBN 1982), WURCS have requisite features for the Semantic Web. WURCS can represent glycans which have ambiguous structures such as unknown structural isomers of monomers and unknown linkages/linkage positions. In addition, because WURCS can represent any glycan structures by unique notations, it can be used as a URI for the Semantic Web.

The minimum information required for a glycomics experiment (MIRAGE) specifies the minimum required metadata items (reporting guidelines) in order to report glycan experiments (York et al. 2014). It includes contact address of the user generating the data, experimental conditions such as glycan separation methods and analytical devices, and data processing methods such as analytical tools and their parameters. It is possible to collect a standardized and structured metadata of glycan-related data using the MIRAGE, and standardized structural data facilitates RDFization of glycan data.

A glycan ontology (GlycoRDF ontology) has also been developed (Ranzinger et al. 2015, http://purl.jp/bio/12/glyco/glycan), which includes the following classes:

The "compound" class, which represents biological molecules and has subclasses such as the "saccharide" class and the "N-glycan" and "O-glycan" classes

The "citation" class, which represents reference documents

The "source" class, which represents the biological or chemical origin of a glycan molecule and has subclasses such as the "source_natural" class (a sample was extracted from a biological organism) and the "source_synthetic" class (a sample was synthesized chemically)

The "evidence" class, which represents experimental techniques used to determine glycan structures and has subclasses such as the "evidence_nmr" class, the "evidence_lc" class, and the "evidence_ms" class

The "referenced_compound" class, which connects the above classes

For example, an N-glycan synthesized chemically and confirmed using NMR would use a combination of the "N-glycan," "source_synthetic," and "evidence_nmr" along with its citations when published.

Predicates have also been defined in the glycan ontology. For example, "has_compound," "published_in," "is_from_source," and "has_evidence" associate the "referenced_compound" class with the "compound" class, the "citation" class, the "source" class, and the "evidence" class, respectively. Using the GlycoRDF ontology, some major glycan databases including CSDB, GlycomeDB, MonosaccharideDB, UniCarbKB, GlycoEpitope, and GlycoProtDB have provided their data as the RDF data model (https://github.com/ReneRanzinger/GlycoRDF/wiki).

These components allow the RDFization of glycan databases, and the glycan data are provided as a part of the LOD network. Providing RDF of glycan data enables anyone to integrate data not only between glycan databases but also between glycan databases and other databases in the life sciences. The resources of glycoinformatics, especially the Semantic Web, are available at http://glycoinfo.org/.

17.4 Data Collection Using TogoTable

As described above, the Semantic Web will be a key technology of the life sciences. As one of the applications of the Semantic Web, I have developed a web tool called TogoTable. It adds annotations collected from the LOD network into a tabular form. Because the data used by TogoTable are connected with each other, it is possible to retrieve annotations which are stored in different databases. For example, annotations in the UniProt database can be retrieved from PDB IDs. Since some glycan databases provide their data using the RDF data model, TogoTable can collect glycan annotations. In this section, I introduce how to use TogoTable.

17.4.1 How to Use TogoTable

Here, GlycomeDB IDs and KCF (KEGG Chemical Function, Aoki-Kinoshita 2009) representations of glycans in the GlycomeDB RDF are collected using KEGG Glycan IDs, as an example. In this example, the file contains three columns: GlycomeDB IDs, KEGG Glycan IDs, and JCGGDB IDs (the example file can be downloaded from http://togotable.dbcls.jp/help/glycobook_example.txt):

1. Prepare a tab-separated value (TSV) file.
   First, a TSV file is prepared from tabular data, which contains IDs from a biological database. Major spreadsheet software such as Microsoft Excel, Google Spreadsheet, and Apache OpenOffice Calc can save files as TSV files. Figures 17.4 and 17.5 show an example using Google Spreadsheet.
2. Upload the TSV file to TogoTable.
   The TSV file can be uploaded on the TogoTable website (http://togotable.dbcls.jp/, Figs. 17.6, 17.7, and 17.8). When uploading it, you need to choose whether the first line in the file is the header line or not. The uploaded data is stored in the TogoTable inner database.

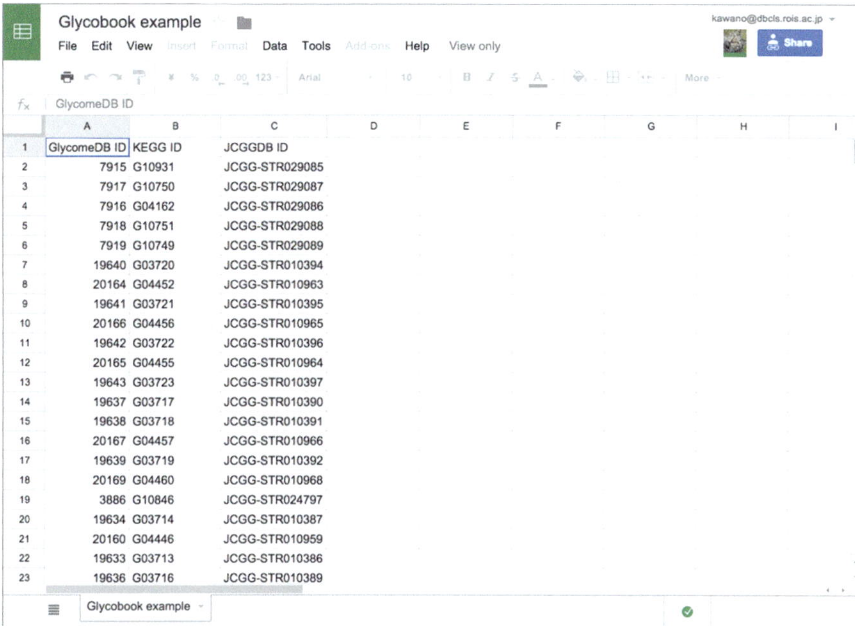

Fig. 17.4 Open the example data (glycobook_example.txt) using Google Spreadsheet

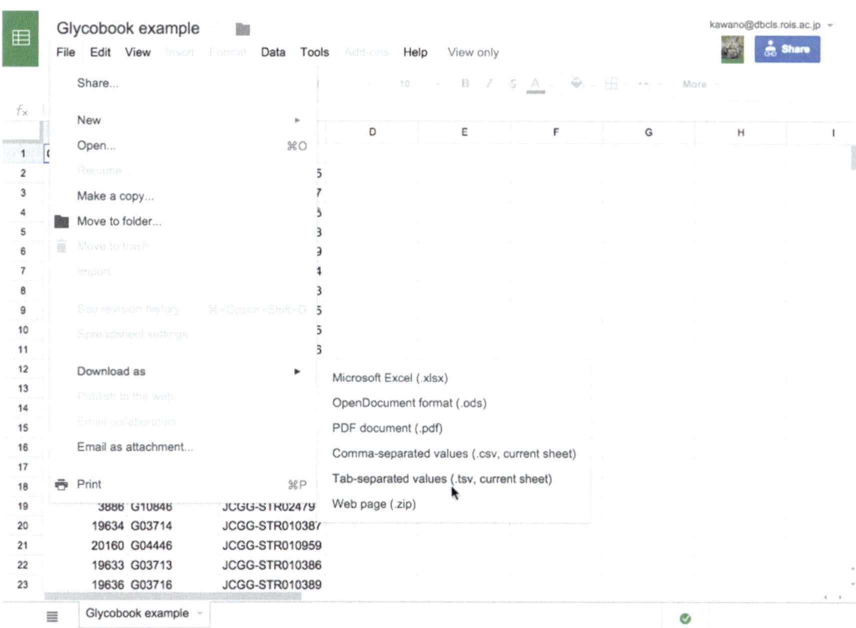

Fig. 17.5 Choose “Tab-separated values (.tsv, current sheet)” in “File” menu to save the data as a TSV file

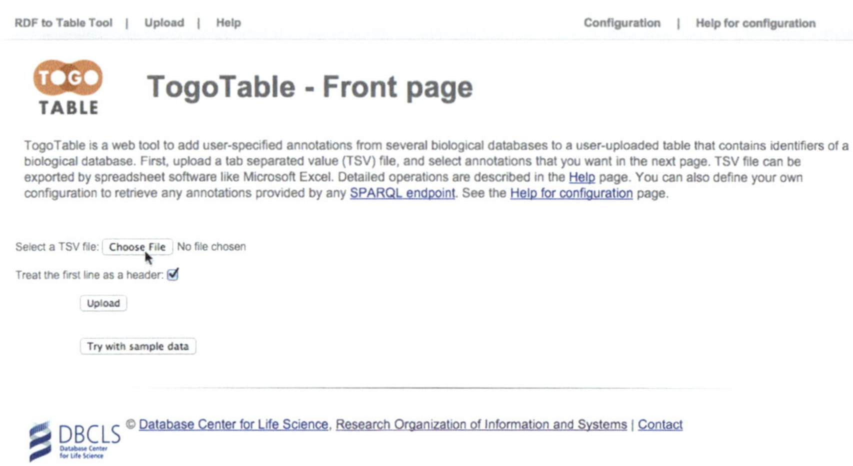

Fig. 17.6 Click "Choose File" on the front page of TogoTable

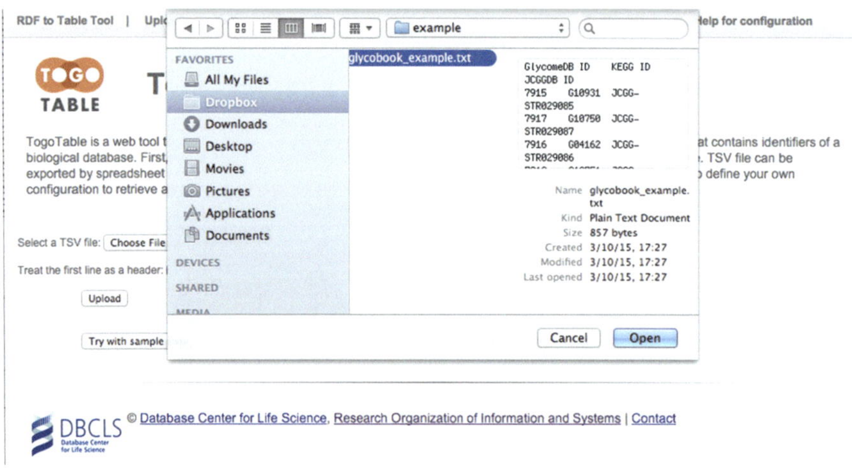

Fig. 17.7 Select a TSV file from your computer

3. Select annotations that you want.
   If upload is successful, a tabular table is displayed (Fig. 17.9). When you click any cell with an ID, a pop-up window will appear (Fig. 17.10). In the window, you can specify annotations, which you want. First, select a database to which the selected ID belongs (Fig. 17.11). Next, select a database from which you want to

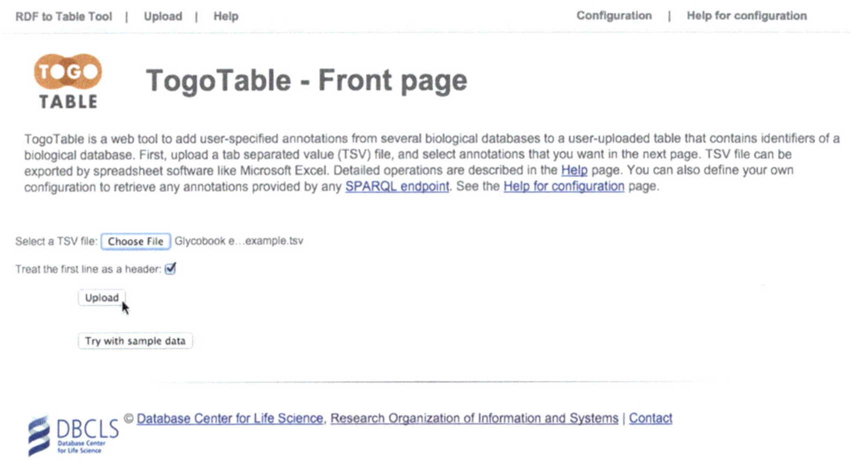

Fig. 17.8 Upload the selected file. If the first line in the file is a header line, check "Treat the first line as a header" before uploading. If not, clear the check box. In this example, the first line is a header line (GlycomeDB ID, KEGG ID, JCGGDB ID, see Fig. 17.4); thus, the check box is selected

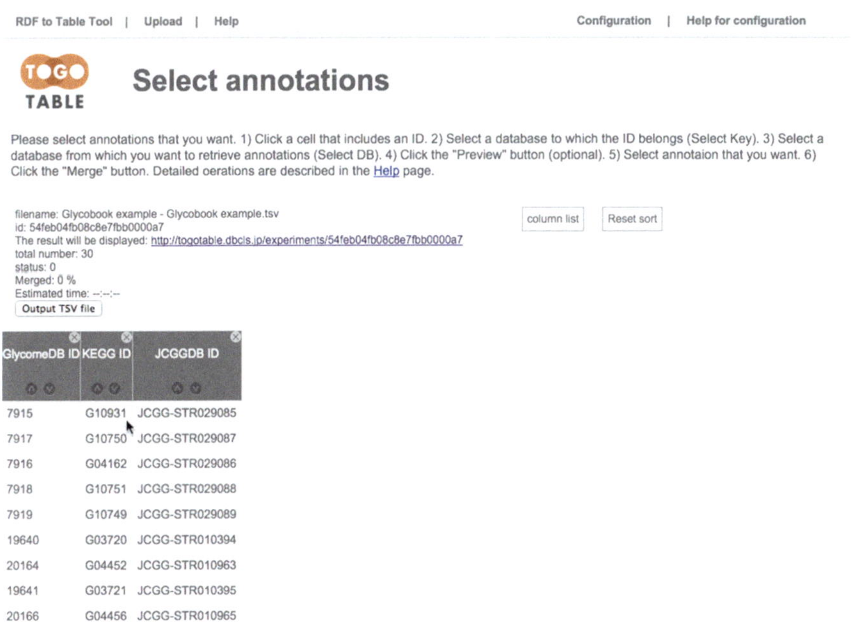

Fig. 17.9 Click a cell with an ID. Here, G10931 in the KEGG ID column is clicked

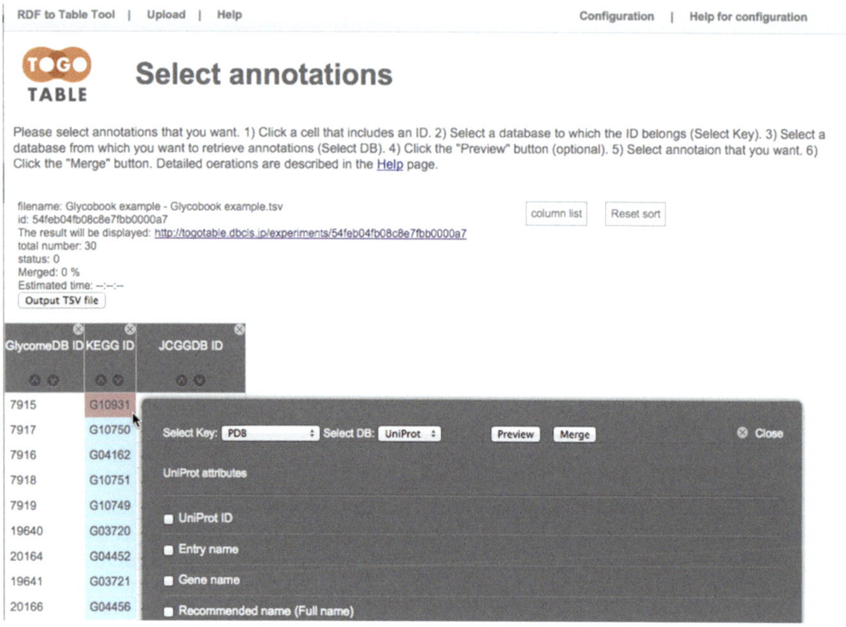

Fig. 17.10 Pop-up window for selecting annotations will appear

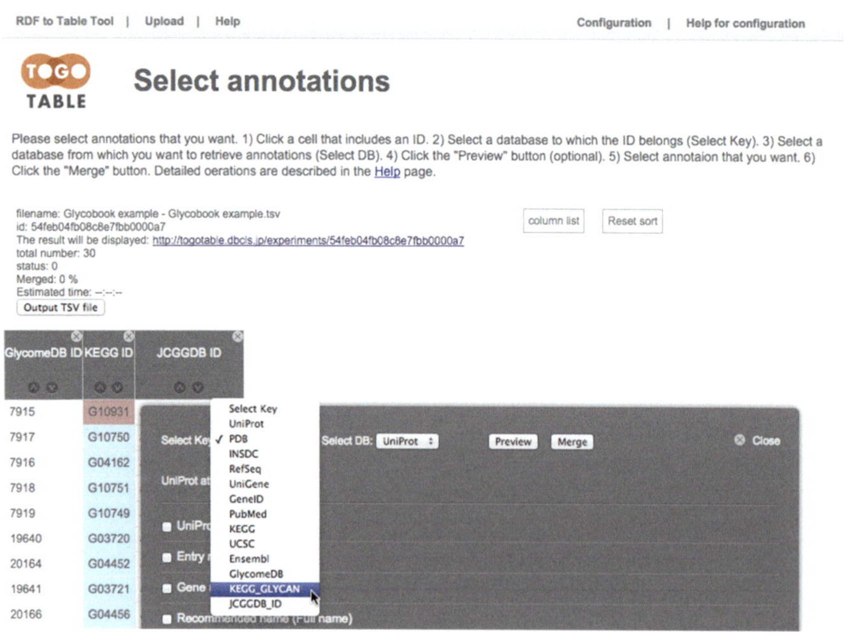

Fig. 17.11 Select a database to which the selected ID belongs, from the menu labeled "Select Key." G10931 is an ID in the KEGG GLYCAN database; thus, select "KEGG_GLYCAN" here

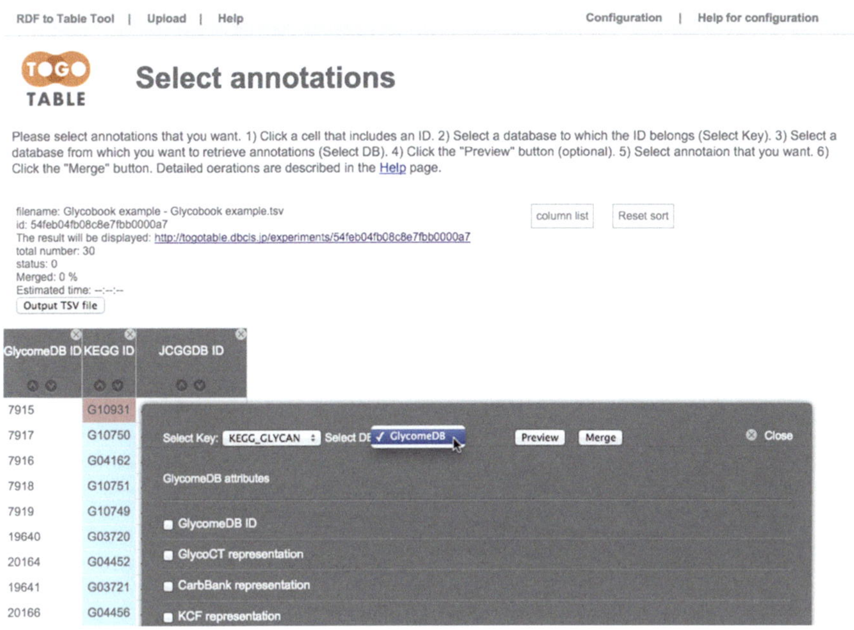

Fig. 17.12 Select a database from which you want to retrieve annotations, from the menu labeled "Select DB." Since KEGG GLYCAN database is connected only to GlycomeDB in the current TogoTable, select "GlycomeDB" here

retrieve annotations (Fig. 17.12). Here, database(s) connected from the selected ID will be listed. After choosing a database, annotations that can be retrieved will appear and you can select desired annotation(s) (Fig. 17.13). You can show actual annotations of the selected ID when you click the "Preview" button (Fig. 17.14).

4. Click the "Merge" button.
   After checking the annotations to import, push the "Merge" button. The selected annotations of all rows in the table are retrieved from the appropriate triple store (Fig. 17.15). If there are no annotations for the ID, the corresponding cells will be blank.

5. Shuffle, sort, and hide the columns.
   The layout of the resulting table can be changed. The order of columns can be replaced by dragging the header cell (Fig. 17.16), and the data can be sorted by

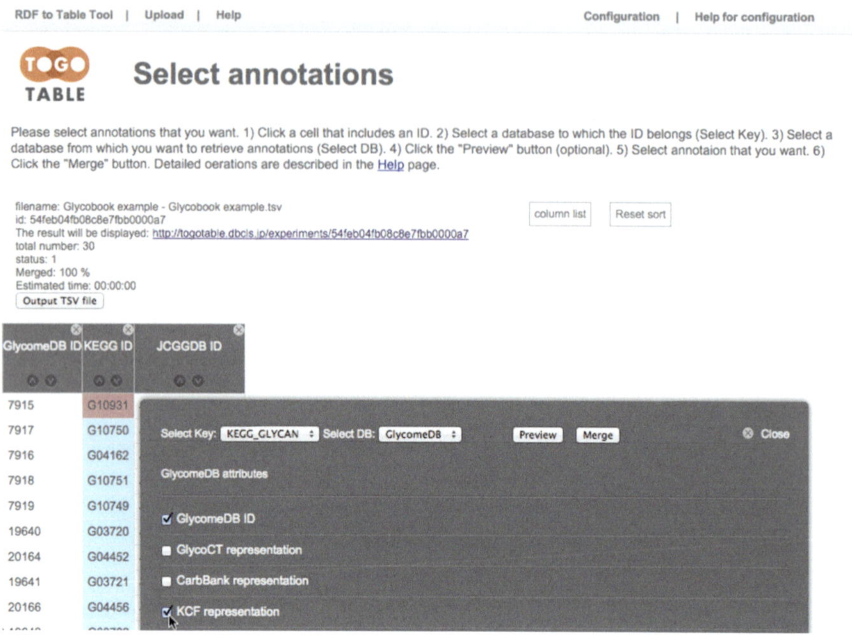

Fig. 17.13 Select annotation(s)

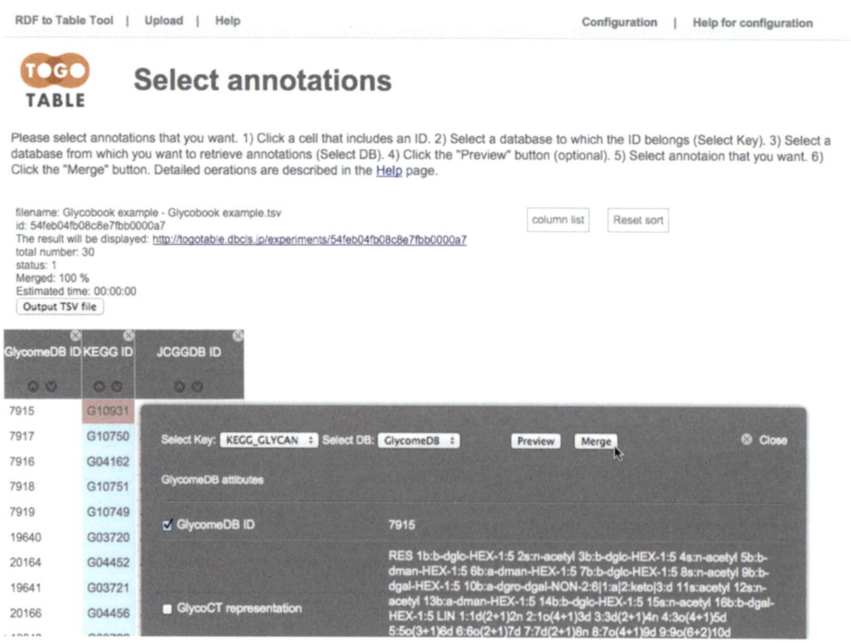

Fig. 17.14 When you click the "Preview" button, the actual annotations of the selected ID will be displayed. Finally, click the "Merge" button to retrieve annotations

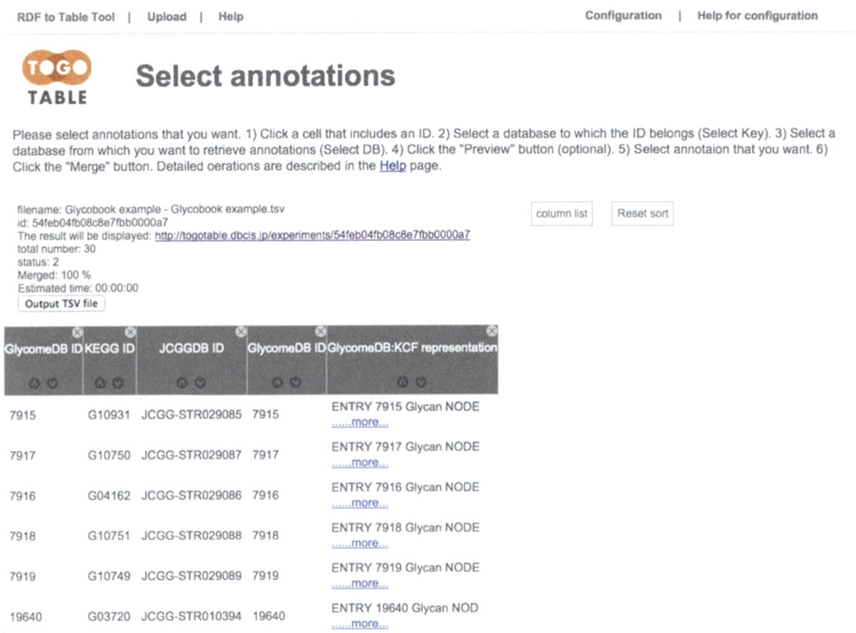

Fig. 17.15 After clicking the "Merge" button, the selected annotations of all rows are retrieved. Since annotations of GlycomeDB ID and KCF representation from GlycomeDB have been selected using KEGG ID as a key ID in this example, their columns are added to the right side of the table. Note that retrieved GlycomeDB IDs are identical to the original GlycomeDB IDs (the leftmost column). Because the KCF representations are too large to display in a cell, they are shortened

clicking the up arrow or down arrow on a header cell. It is also possible to hide a column by clicking the "x" button on the upper right of a header cell, and it can be redisplayed from the pop-up column list, which appears when clicking the "column list" button (Fig. 17.17).

6. Download the result.
   Finally, the annotation-added table data can be downloaded by clicking the "Output TSV file" button (Fig. 17.18). Because the downloaded file is a TSV file, it can be reopened using any spreadsheet software (Fig. 17.19).

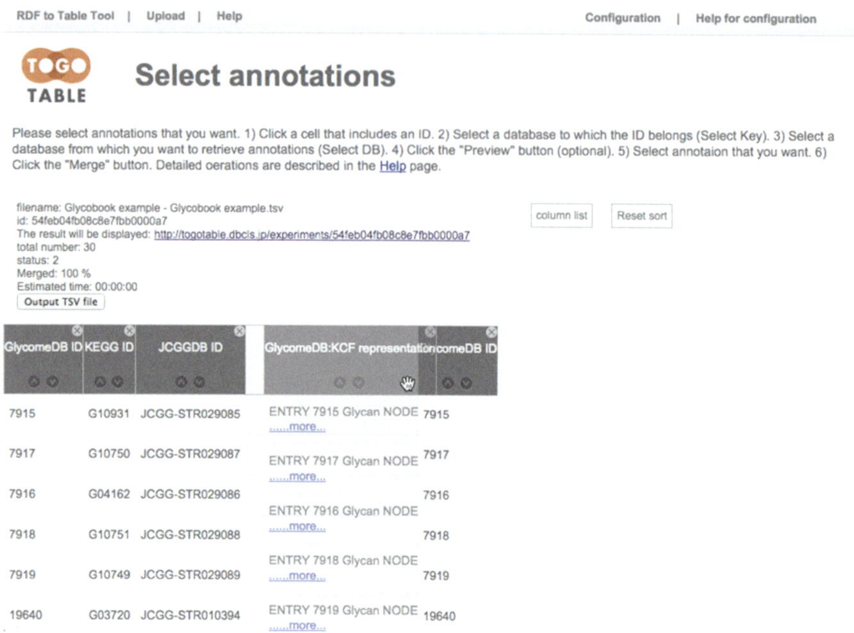

Fig. 17.16 The order of columns can be changed by drag-and-drop of a header cell

17.4.2 Configuring to Retrieve User-Specified Annotations in TogoTable

In the TogoTable, users can retrieve not only from preset annotations but also from any annotations that are provided using RDF format. You can go to the user configuration page from the "Configuration" link located on the upper right of the TogoTable front page. Here, I omit the details because SPARQL knowledge is needed and exceeds the scope of this chapter. Briefly, (1) create your own configuration based on the system configuration, (2) configure the database (SPARQL endpoint) for searching, and (3) configure attributes (annotations) for retrieving using SPARQL. You can retrieve any annotations using your own configuration. Since the configuration file is stored in the web storage of your browser, you can reuse the configuration as long as you use the same browser. For more information, refer to "Help for configuration."

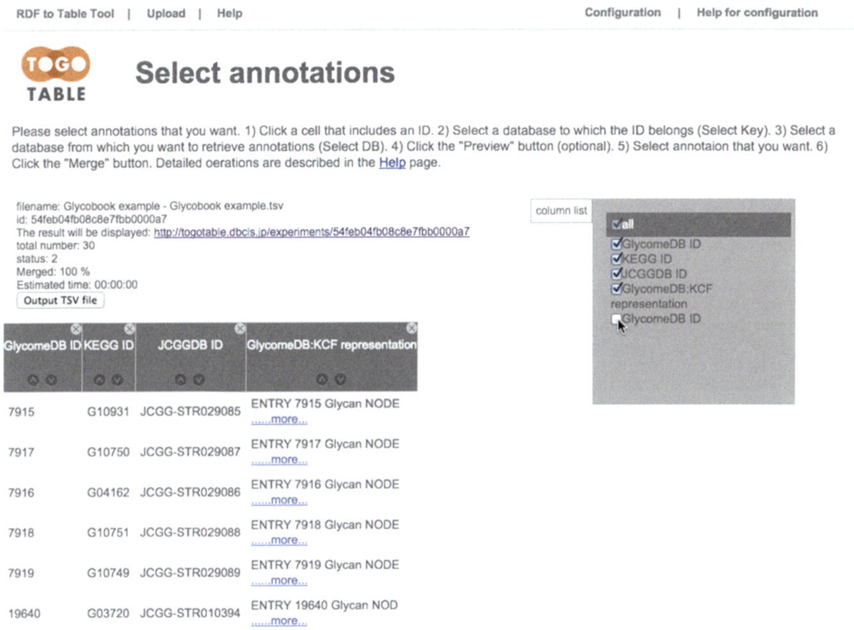

Fig. 17.17 It is possible to hide a column by clicking the "x" on the upper right of a header column. Here, the GlycomeDB ID column, which was originally located in the fourth column, is removed. It can be redisplayed by checking the box in the pop-up of the "column list"

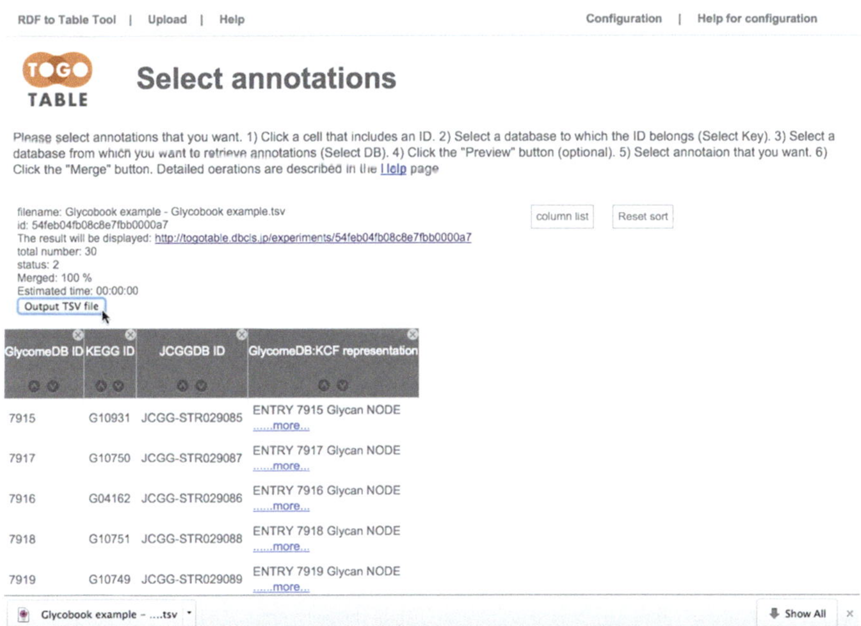

Fig. 17.18 It is possible to download the final tabular data as a TSV file by clicking the "Output TSV file" button

	GlycomeDB ID	KEGG ID	JCGGDB ID	GlycomeDB:KCF representation
2	7915	G10931	JCGG-STR02	ENTRY 7915 Glycan NODE 10 1 GlcNAc 0.0 0.0 2
3	7917	G10750	JCGG-STR02	ENTRY 7917 Glycan NODE 11 1 GlcNAc 0.0 0.0 2
4	7916	G04162	JCGG-STR02	ENTRY 7916 Glycan NODE 10 1 GlcNAc 0.0 0.0 2
5	7918	G10751	JCGG-STR02	ENTRY 7918 Glycan NODE 11 1 GlcNAc 0.0 0.0 2
6	7919	G10749	JCGG-STR02	ENTRY 7919 Glycan NODE 11 1 GlcNAc 0.0 0.0 2
7	19640	G03720	JCGG-STR01	ENTRY 19640 Glycan NODE 17 1 GlcNAc 0.0 0.0 2
8	20164	G04452	JCGG-STR01	ENTRY 20164 Glycan NODE 11 1 GlcNAc 0.0 0.0 2
9	19641	G03721	JCGG-STR01	ENTRY 19641 Glycan NODE 17 1 GlcNAc 0.0 0.0 2
10	20166	G04456	JCGG-STR01	ENTRY 20166 Glycan NODE 9 1 GlcNAc 0.0 0.0 2
11	19642	G03722	JCGG-STR01	ENTRY 19642 Glycan NODE 17 1 GlcNAc 0.0 0.0 2
12	20165	G04455	JCGG-STR01	ENTRY 20165 Glycan NODE 8 1 GlcNAc 0.0 0.0 2
13	19643	G03723	JCGG-STR01	ENTRY 19643 Glycan NODE 17 1 GlcNAc 0.0 0.0 2
14	19637	G03717	JCGG-STR01	ENTRY 19637 Glycan NODE 17 1 GlcNAc 0.0 0.0 2
15	19638	G03718	JCGG-STR01	ENTRY 19638 Glycan NODE 17 1 GlcNAc 0.0 0.0 2
16	20167	G04457	JCGG-STR01	ENTRY 20167 Glycan NODE 11 1 GlcNAc 0.0 0.0 2
17	19639	G03719	JCGG-STR01	ENTRY 19639 Glycan NODE 17 1 GlcNAc 0.0 0.0 2
18	20169	G04460	JCGG-STR01	ENTRY 20169 Glycan NODE 11 1 GlcNAc 0.0 0.0 2
19	3886	G10846	JCGG-STR02	ENTRY 3886 Glycan NODE 6 1 GalNAc 0.0 0.0 2
20	19634	G03714	JCGG-STR01	ENTRY 19634 Glycan NODE 17 1 GlcNAc 0.0 0.0 2
21	20160	G04446	JCGG-STR01	ENTRY 20160 Glycan NODE 9 1 LKdo 0.0 0.0 2
22	19633	G03713	JCGG-STR01	ENTRY 19633 Glycan NODE 17 1 GlcNAc 0.0 0.0 2
23	19636	G03716	JCGG-STR01	ENTRY 19636 Glycan NODE 17 1 GlcNAc 0.0 0.0 2
24	19635	G03715	JCGG-STR01	ENTRY 19635 Glycan NODE 17 1 GlcNAc 0.0 0.0 2
25	51	G00126	JCGG-STR02	ENTRY 51 Glycan NODE 6 1 Glc 0.0 0.0 2 Gal
26	52	G04349	JCGG-STR02	ENTRY 52 Glycan NODE 6 1 Glc 0.0 0.0 2 Gal
27	7906	G11960	JCGG-STR02	ENTRY 7906 Glycan NODE 11 1 GlcNAc 0.0 0.0 2
28	20154	G04438	JCGG-STR01	ENTRY 20154 Glycan NODE 15 1 GlcNAc 0.0 0.0 2
29	20153	G04437	JCGG-STR01	ENTRY 20153 Glycan NODE 15 1 GlcNAc 0.0 0.0 2

Glycobook example - Glycobook e

Fig. 17.19 The downloaded file can be reopened by spreadsheet software

17.5 Conclusion

In this chapter, I provided a brief description of Semantic Web technology and an application using glycan RDF databases. The attempt to integrate data by giving computer-interpretable meaning to data has just begun. In the future, many data and datasets not only from glycobiology but also from general life science are expected to connect each other and form a part of LOD. When this happens, we will be able to execute more advanced searches throughout the life sciences and beyond.

References

Aoki-Kinoshita KF (2009) KCF format. In: Glycome informatics: methods and applications. Chapman and Hall/CRC, Boca Raton, pp 31–32

Aoki-Kinoshita KF, Bolleman J, Campbell MP et al (2013a) Introducing glycomics data into the Semantic Web. J Biomed Semant 4:39

Aoki-Kinoshita KF, Sawaki H, An HJ et al (2013b) The third ACGG-DB meeting report: towards an international collaborative infrastructure for glycobioinformatics. Glycobiology 23:144–146

Banin E, Neuberger Y, Altshuler Y et al (2002) A novel linear code® nomenclature for complex carbohydrates. Trends Glycosci Glycotechnol 14:127–137

Belleau F, Nolin MA, Tourigny N et al (2008) Bio2RDF: towards a mashup to build bioinformatics knowledge systems. J Biomed Inform 41:706–716

Berners-Lee T, Hendler J, Lassila O (2001) The Semantic Web. Sci Am 284:28–37
Bohne-Lang A, Lang E, Förster T et al (2001) LINUCS: linear notation for unique description of carbohydrate sequences. Carbohydr Res 336:1–11
Bushman B, Anderson D, Fu G (2015) Transforming the medical subject headings into linked data: creating the authorized version of MeSH in RDF. J Libr Metadata 15:157–176
Campbell MP, Peterson R, Mariethoz J et al (2014) UniCarbKB: building a knowledge platform for glycoproteomics. Nucleic Acids Res 42:D215–D221
Chiba H, Nishide H, Uchiyama I (2015) Construction of an ortholog database using the semantic web technology for integrative analysis of genomic data. PLoS One 10:e0122802
Doubet S, Albersheim P (1992) CarbBank. Glycobiology 2:505
Egorova KS, Toukach PV (2014) Expansion of coverage of Carbohydrate Structure Database (CSDB). Carbohydr Res 389:112–114
Erling O, Mikhailov I (2009) RDF support in the virtuoso DBMS. In: Tassilo Pellegrini T, Auer S, Tochtermann K et al (eds) Networked knowledge – networked media. Springer, Berlin/Heidelberg, pp 7–24
Faulconbridge A, Burdett T, Brandizi M et al (2014) Updates to BioSamples database at European bioinformatics institute. Nucleic Acids Res 42:D50–D52
Fu G, Batchelor C, Dumontier M et al (2015) PubChemRDF: towards the semantic annotation of PubChem compound and substance databases. J Cheminform 7:34
Hashimoto K, Goto S, Kawano S et al (2006) KEGG as a glycome informatics resource. Glycobiology 16:63R–70R
IUPAC-IUB Joint Commission on Biochemical Nomenclature (JCBN) (1982) Abbreviated terminology of oligosaccharide chains. Recommendations 1980. J Bio Chem 257:3347–3351
Jupp S, Malone J, Bolleman J et al (2014) The EBI RDF platform: linked open data for the life sciences. Bioinformatics 30:1338–1339
Kaji H, Shikanai T, Sasaki-Sawa A et al (2012) Large-scale identification of N-glycosylated proteins of mouse tissues and construction of a glycoprotein database, GlycoProtDB. J Proteome Res 11:4553–4566
Katayama T, Arakawa K, Nakao M et al (2010) The DBCLS BioHackathon: standardization and interoperability for bioinformatics web services and workflows. J Biomed Semant 1:8
Katayama T, Wilkinson MD, Vos R et al (2011) The 2nd DBCLS BioHackathon: interoperable bioinformatics web services for integrated applications. J Biomed Semant 2:4
Katayama T, Wilkinson MD, Micklem G et al (2013) The 3rd DBCLS BioHackathon: improving life science data integration with Semantic Web technologies. J Biomed Semant 4:6
Katayama T, Wilkinson MD, Aoki Kinoshita KF et al (2014) BioHackathon series in 2011 and 2012: penetration of ontology and linked data in life science domains. J Biomed Semant 5:5
Kawano S, Watanabe T, Mizuguchi S et al (2014) TogoTable: cross-database annotation system using the Resource Description Framework (RDF) data model. Nucleic Acids Res 42:W442–W448
Kinjo AR, Suzuki H, Yamashita R et al (2012) Protein Data Bank Japan (PDBj): maintaining a structural data archive and resource description framework format. Nucleic Acids Res 40:D453–D460
Kobilarov G, Scott T, Raimond Y et al (2009) Media meets semantic web–how the bbc uses dbpedia and linked data to make connections. In: Aroyo L, Traverso P, Ciravegna F et al (eds) The semantic web: research and applications. Springer, Berlin/Heidelberg, pp 723–737
Okuda S, Nakao H, Kawasaki T (2014) GlycoEpitope: database for carbohydrate antigen and antibody. In: Taniguchi N, Endo T, Hart GW et al (eds) Glycoscience: biology and medicine. Springer, Tokyo, pp 267–273
Ranzinger R, Herget S, von der Lieth CW et al (2011) GlycomeDB-a unified database for carbohydrate structures. Nucleic Acids Res 39:D373–D376
Ranzinger R, Aoki-Kinoshita KF, Campbell MP et al (2015) GlycoRDF: an ontology to standardize glycomics data in RDF. Bioinfomatics 31:919–925
Shadbolt N, O'Hara K, Berners-Lee T et al (2012) Linked open government data: lessons from data.gov.uk. IEEE Intell Syst 27:16–24

Tanaka K, Aoki-Kinoshita KF, Kotera M et al (2014) WURCS: the Web3 unique representation of carbohydrate structures. J Chem Inf Model 54:1558–1566
The Gene Ontology Consortium, Ashburner M, Ball CA et al (2000) Gene ontology: tool for the unification of biology. Nat Genet 25:25–29
The UniProt Consortium (2014) Activities at the Universal Protein Resource (UniProt). Nucleic Acids Res 42:D191–D198
Willighagen EL, Waagmeester A, Spjuth O et al (2013) The ChEMBL database as linked open data. J Cheminform 5:23
Wimalaratne SM, Grenon P, Hermjakob H et al (2014) BioModels linked dataset. BMC Syst Biol 8:91
Wimalaratne SM, Bolleman J, Juty N et al (2015) SPARQL-enabled identifier conversion with Identifiers.org. Bioinformatics 31:1875–1877
Wu H, Fujiwara T, Yamamoto Y et al (2014) BioBenchmark Toyama 2012: an evaluation of the performance of triple stores on biological data. J Biomed Semant 5:32
York WS, Agravat S, Aoki-Kinoshita KF et al (2014) MIRAGE: the minimum information required for a glycomics experiment. Glycobiology 24:402–406

MIX
Papier aus verantwortungsvollen Quellen
Paper from responsible sources
FSC® C105338

If you have any concerns about our products,
you can contact us on
ProductSafety@springernature.com

In case Publisher is established outside the EU,
the EU authorized representative is:
Springer Nature Customer Service Center GmbH
Europaplatz 3, 69115 Heidelberg, Germany

Printed by Libri Plureos GmbH
in Hamburg, Germany